BAYESIAN PROGRAMMING

Chapman & Hall/CRC
Machine Learning & Pattern Recognition Series

SERIES EDITORS

AIMS AND SCOPE

This series reflects the latest advances and applications in machine learning and pattern recognition through the publication of a broad range of reference works, textbooks, and handbooks. The inclusion of concrete examples, applications, and methods is highly encouraged. The scope of the series includes, but is not limited to, titles in the areas of machine learning, pattern recognition, computational intelligence, robotics, computational/statistical learning theory, natural language processing, computer vision, game AI, game theory, neural networks, computational neuroscience, and other relevant topics, such as machine learning applied to bioinformatics or cognitive science, which might be proposed by potential contributors.

PUBLISHED TITLES

MACHINE LEARNING: An Algorithmic Perspective
Stephen Marsland

HANDBOOK OF NATURAL LANGUAGE PROCESSING,
Second Edition
Nitin Indurkhya and Fred J. Damerau

UTILITY-BASED LEARNING FROM DATA
Craig Friedman and Sven Sandow

A FIRST COURSE IN MACHINE LEARNING
Simon Rogers and Mark Girolami

COST-SENSITIVE MACHINE LEARNING
Balaji Krishnapuram, Shipeng Yu, and Bharat Rao

ENSEMBLE METHODS: FOUNDATIONS AND ALGORITHMS
Zhi-Hua Zhou

MULTI-LABEL DIMENSIONALITY REDUCTION
Liang Sun, Shuiwang Ji, and Jieping Ye

BAYESIAN PROGRAMMING
Pierre Bessière, Emmanuel Mazer, Juan-Manuel Ahuactzin, and Kamel Mekhnacha

Chapman & Hall/CRC
Machine Learning & Pattern Recognition Series

BAYESIAN PROGRAMMING

PIERRE BESSIÈRE
CNRS, PARIS, FRANCE

EMMANUEL MAZER
CNRS, GRENOBLE, FRANCE

JUAN-MANUEL AHUACTZIN
PROBAYES, PUEBLA, MEXICO

KAMEL MEKHNACHA
PROBAYES, GRENOBLE, FRANCE

CRC Press is an imprint of the
Taylor & Francis Group, an **informa** business
A CHAPMAN & HALL BOOK

CRC Press
Taylor & Francis Group
6000 Broken Sound Parkway NW, Suite 300
Boca Raton, FL 33487-2742

First issued in paperback 2022

ISBN 13: 978-1-03-247740-4 (pbk)
ISBN 13: 978-1-4398-8032-6 (hbk)

DOI: 10.1201/b16111

Publisher's Note
The publisher has gone to great lengths to ensure the quality of this reprint but points out that some imperfections in the original copies may be apparent.

Library of Congress Cataloging-in-Publication Data

Bessiére, Pierre.
Bayesian programming / authors, Pierre Bessiére, Emmanuel Mazer, Juan-Manuel Ahuactzin, Kamel Mekhnacha.
pages cm -- (Chapman & Hall/CRC machine learning & pattern recognition series)
Includes bibliographical references and index.
ISBN 978-1-4398-8032-6 (hardcover : alk. paper)
1. Computer programming. 2. Bayesian statistical decision theory--Data processing. 3. Computer simulation. 4. Mathematical models. I. Title.

QA76.6.B4764 2014
519.5'42--dc23 2013039509

Visit the Taylor & Francis Web site at
http://www.taylorandfrancis.com

and the CRC Press Web site at
http://www.crcpress.com

To the late Edwin Thompson Jaynes

for his doubts about certitudes

and for his certitudes about probabilities

Contents

Foreword

Modern artificial intelligence (AI) has embraced the Bayesian paradigm, in which inferences are made with respect to a given probability model, on the basis of observed evidence, according to the laws of probability. The development of Bayesian networks and other related formalisms for expressing probability models has dramatically broadened the class of problems that can be addressed by Bayesian methods. Yet, practitioners have found that such models by themselves are too limited in their expressive power. For real-world applications, additional machinery is needed to construct more complex models, especially those with repetitive substructures.

Bayesian Programming comprises a methodology, a programming language, and a set of tools for developing and applying these kinds of complex models. The programming language contains iteration constructs to facilitate the inclusion of repetitive submodels as well as a useful extension to standard Bayesian network modeling: the ability to specify conditional distributions for multiple variables jointly, rather than just for individual variables.

The approach is described in great detail, with many worked examples backed up by an online code repository. Unlike other books that tend to focus almost entirely on mathematics, this one gives equal time to conceptual and methodological guidance for the model-builder. It grapples with the knotty problems that arise in practice, some of which do not yet have clear solutions.

Stuart Russell
University of California, Berkeley

Preface

The Bayesian Programming project first began in the late 1980s, when we came across Jaynes' ideas and writings on the subject of "Probability as Logic."

We were convinced that dealing with uncertainty and incompleteness was one of the main challenges for robotics and a *sine qua non* condition to the advancement toward autonomous robots, but we had no idea of how to tackle these questions.

Edwin T. Jaynes' proposition of probability as an alternative and an extension of logic to deal with both incompleteness and uncertainty was a revelation for us. The theoretical solution was described in his writings, especially in the early version of his book *Probability Theory: The Logic of Science* [Jaynes, 2003] that he was distributing, already aware that illness may prevent him from finishing it.

He described the principles of what he called "the robot," which was not a physical device, but an inference engine to automate probabilistic reasoning — a kind of Prolog for probability instead of logic. We decided to try to implement such an inference engine and to apply it to robotics.

The main lines of the Bayesian programming formalism were designed, notably, by Olivier Lebeltel. The first prototype versions of the inference engine were developed in Lisp and several applications to robotic programming were designed.

In the late 1990s we realized, first, for research applications that probabilistic modeling could be successfully applied to any sensory-motor systems, whether artificial or living, and, second, that in industry, Bayesian programming has numerous potential applications beyond robotics.

To investigate the research applications along with our international partners, we set up two successive European projects: BIBA (Bayesian Inspired Brain and Artifacts) and BACS (Bayesian Approach to Cognitive Systems). In these projects we made progress on three different scientific questions:

- How can we develop better artifacts using Bayesian approaches?
- Is Bayesian reasoning biologically plausible at the behavioral level?
- Is Bayesian reasoning biologically plausible at a neuronal scale?

To advance with industrial applications, we started the company ProbaYes, which applies probabilistic reasoning to help in decision making, whenever the information at hand is incomplete and uncertain. ProbaYes accomplished a considerable task by completely redesigning and implementing the inference engine (called ProBT) in C++ and provides a free distribution of this software for the academic community.

We think that the Bayesian Programming methodology and tools are reaching maturity. The goal of this book is to present them so that anyone is able to use them.

We will, of course, continue to improve tools and develop new models. However, pursuing the idea that probability is an alternative to Boolean logic, we now have a new important research objective, which is to design specific hardware, inspired from biology, to build a Bayesian computer.

Along with the coauthors of this book, the main developers of ProBT were David Raulo, Ronan Le Hy, and Christopher Tay. We would like to emphasize their hard work and their invaluable contributions, which have led to an effective programming tool. Professor Philippe Lerray and Linda Smail also contributed to the definition and implementation of key algorithms found in ProBT, namely the structural identification and symbolic simplification algorithms.

None of this would have been possible without the work of all the PhD students and postdocs who have used and improved Bayesian programming along the years either within our group at Grenoble University (France) or elsewhere in Europe (in approximate chronological order): Eric Dedieu, Olivier Lebeltel, Christophe Coué, Ruben Senen García Ramírez, Julien Diard, Cédric Pradalier, Jihene Serkhane, Guy Ramel, Adriana Tapus, Carla Maria Chagas e Cavalcante Koike, Miriam Amavizca, Jean Laurens, Francis Colas, Ronan Le Hy, Pierre-Charles Dangauthier, Shrihari Vasudevan, Joerg Rett, Estelle Gilet, Xavier Perrin, João Filipe Ferreira, Clement Moulin-Frier, Gabriel Synnaeve, and Raphael Laurent. Many of these former students have been mentored or coadvised by Julien Diard, Jorge Dias, Thierry Fraichard, Christian Laugier, or Roland Siegwart, whom we also thank for the countless discussions we had together.

A very special thanks to Jacques Droulez and Jean-Luc Schwartz for their inspiring and essential collaborations.

We would also like to thank the team who assisted us during the actual making of the book: Mr. Peter Bjornsen for correcting the many errors found in the initial version, Mr. Shashi Kumar for tuning the configuration file needed by LaTeX, Ms. Linda Leggio and Mr. David Grubbs for the organization.

Finally thanks to CNRS (Centre National de la Recherche Scientifique), INRIA, University of Grenoble, and the European Commission, who actively supported this project.

Pierre Bessière
Emmanuel Mazer
Juan-Manuel Ahuactzin
Kamel Mekhnacha

Chapter 1

Introduction

> The most incomprehensible thing about the world is that it is comprehensible.
>
> Albert Einstein or Immanuel Kant[1]

1.1 Probability an alternative to logic

Computers have brought a new dimension to modeling. A model, once translated into a program and run on a computer, may be used to understand, measure, simulate, mimic, optimize, predict, and control. During the last 50 years science, industry, finance, medicine, entertainment, transport, and communication have been completely transformed by this revolution.

However, models and programs suffer from a fundamental flaw: *incompleteness*. Any model of a real phenomenon is incomplete. Hidden variables, not taken into account in the model, influence the phenomenon. The effect of the hidden variables is that the model and the phenomenon never have the exact same behaviors. *Uncertainty* is the direct and unavoidable consequence

[1] This is how Antonina Vallentin quotes Einstein in the biography she wrote [Vallentin, 1954]. However, it seems that the exact quotation should rather be "the eternal mystery of the world is its comprehensibility" [Einstein, 1936]. Furthermore, this sentence appears itself between quotes in Einstein's original text in a context where it could be attributed to Immanuel Kant.

of incompleteness. A model may neither exactly foresee the future observations of a phenomenon nor predict the consequences of its decisions as both its observations and actions are biased by the hidden variables.

Computing a cost price to decide on a sell price may seem a purely arithmetic operation consisting of adding elementary costs. However, often these elementary costs may not be known exactly. For instance, a part's cost may be biased by exchange rates, production cost may be biased by the number of orders, and transportation costs may be biased by the time of year. Exchange rates, the number of orders, and the time of year when unknown are hidden variables, which induce uncertainty in the computation of the cost price.

Analyzing the content of an e-mail to filter spam is a difficult task, because no word or combination of words can give you an absolute certitude about the nature of the e-mail. At most, the presence of certain words is a strong clue that an e-mail is spam. It may never be a conclusive proof, because the context may completely change its meaning. For instance, if one of your friends is forwarding you a spam for discussion about the spam phenomenon, its whole content is suddenly not spam any longer. A linguistic model of spam is irremediably incomplete because of this boundless contextual information. Filtering spam is not hopeless and some very efficient solutions exist, but the perfect result is a chimera.

Machine control and dysfunction diagnosis is very important to industry. However, the dream of building a complete model of a machine and all its possible failures is an illusion. One should recall the first "bug" of the computer era: the moth located in relay 70 panel F of the Harvard Mark II computer. Once again, it does not mean that control and diagnosis are hopeless, it only means that models of these machines should take into account their own incompleteness and the resulting uncertainty.

In 1781, Sir William Herschel discovered Uranus, the seventh planet of the solar system. In 1846, Johann Galle observed for the first time, Neptune, the eighth planet. In the meantime, both Urbain Leverrier, a French astronomer, and John Adams, an English one, became interested in the "uncertain" trajectory of Uranus. The planet was not following the exact trajectory that Newton's theory of gravitation predicted. They both came to the conclusion that these irregularities could be the result of a hidden variable not taken into account by the model: the existence of an eighth planet. They even went much further, finding the most probable position of this eighth planet. The Berlin observatory received Leverrier's prediction on September 23, 1846, and Galle observed Neptune the very same day!

Logic is both the mathematical foundation of rational reasoning and the fundamental principle of present day computing. However, logic, by essence, is restricted to problems where information is both *complete* and *certain*. An *alternative mathematical framework* and an *alternative computing framework* are both needed to deal with incompleteness and uncertainty.

Probability theory is this alternative mathematical framework. It is a model of rational reasoning in the presence of incompleteness and uncertainty. It is an extension of logic where both certain and uncertain information have their places.

James C. Maxwell stated this point synthetically:

> The actual science of logic is conversant at present only with things either certain, impossible, or entirely doubtful, none of which (fortunately) we have to reason on. Therefore the true logic for this world is the calculus of Probabilities, which takes account of the magnitude of the probability which is, or ought to be, in a reasonable man's mind.
>
> James C. Maxwell
> quoted in *Probability Theory — The Logic of Science*
> by Edwin T. Jaynes [2003]

Considering probability as *a model of reasoning* is called the *subjectivist* or *Bayesian* approach. It is opposed to the *objectivist* approach, which considers probability as *a model of the world.* This opposition is not only an epistemological controversy; it has many fundamental and practical consequences.[2]

To model reasoning, you must take into account the preliminary knowledge of the subject who is doing the reasoning. This preliminary knowledge plays the same role as the axioms in logic. Starting from different preliminary knowledge may lead to different conclusions. Starting from wrong preliminary knowledge will lead to wrong conclusions even with perfectly correct reasoning. Reaching wrong conclusions following correct reasoning proves that the preliminary knowledge was wrong, offers the opportunity to correct it, and eventually leads you to learning.

Incompleteness is simply the irreducible[3] gap between the preliminary knowledge and the phenomenon and uncertainty is a direct and measurable consequence of this imperfection.

In contrast, modeling the world by denying the existence of a "subject" and consequently rejecting preliminary knowledge leads to complicated situations and apparent paradoxes. This rejection implies that if the conclusions are wrong, either the reasoning could be wrong or the data could be aberrant, leaving no room for improvement or learning. Incompleteness does not mean anything without preliminary knowledge, and uncertainty and noise must be mysterious properties of the physical world.[4]

The objectivist school has been dominant during the 20th century, but the subjectivist approach has a history as long as probability itself. It can be traced back to Jakob Bernoulli in 1713:

---

[2]See FAQ-FAM: Objectivism vs. subjectivism controversy and the "mind projection fallacy," Section 16.13.

[3]See FAQ-FAM: Incompleteness irreducibility, Section 16.10.

[4]See FAQ-FAM: Noise or ignorance, Section 16.12.

> Uncertainty is not in things but in our head: uncertainty is a lack of knowledge.
>
> *Ars Conjectandi*
> Jakob Bernouilli [1713]

to the Marquis Simon de Laplace, one century later, in 1814:

> One sees, from this Essay, that the theory of probabilities is basically just common sense reduced to calculus; it makes one appreciate with exactness that which accurate minds feel with a sort of instinct, often without being able to account for it.[5]
>
> *Essai philosophique sur les probabilités*
> Marquis Simon de Laplace [1814]

to the already quoted James C. Maxwell in 1850 and to the visionary Henri Poincaré in 1902:

> Randomness is just the measure of our ignorance.
>
> To undertake any probability calculation, and even for this calculation to have a meaning, we have to admit, as a starting point, an hypothesis or a convention, that always comprises a certain amount of arbitrariness. In the choice of this convention, we can be guided only by the principle of sufficient reason.
>
> From this point of view, every science would just be unconscious applications of the calculus of probabilities. Condemning this calculus would be condemning the whole science.
>
> *La science et l'hypothèse*
> Henri Poincaré [1902]

and finally, by Edwin T. Jaynes in his book *Probability theory: The logic of science* where he brilliantly presents the subjectivist alternative and sets clearly and simply the basis of the approach:

> By inference we mean simply: deductive reasoning whenever enough information is at hand to permit it; inductive or probabilistic reasoning when — as is almost invariably the case in real problems — all the necessary information is not available. Thus the topic of "Probability as Logic" is the optimal processing of *uncertain* and *incomplete* knowledge.
>
> *Probability Theory: The Logic of Science*
> Edwin T. Jaynes [2003]

[5] On voit, par cet Essai, que la théorie des probabilités n'est, au fond, que le bon sens réduit au calcul; elle fait apprécier avec exactitude ce que les esprits justes sentent par une sorte d'instinct, sans qu'ils puissent souvent s'en rendre compte.

1.2 A need for a new computing paradigm

Bayesian probability theory is clearly the sought mathematical alternative to logic.[6]

However, we want working solutions to incomplete and uncertain problems. Consequently, we require an alternative computing framework based on Bayesian probabilities.

To create such a complete computing Bayesian framework, we require a new *modeling methodology* to build probabilistic models, we require new *inference algorithms* to automate probabilistic calculus, we require new *programming languages* to implement these models on computers, and finally, we will eventually require new *hardware* to run these Bayesian programs efficiently.

The ultimate goal is a Bayesian computer. The purpose of this book is to describe the current first steps in this direction.

1.3 A need for a new modeling methodology

The existence of a systematic and generic method to build models is a *sine qua non* requirement for the success of a modeling and computing paradigm. This is why algorithms are taught in the basic course of computer science giving students the elementary and necessary methods to develop classical programs.

We propose *Bayesian Programming* as this generic methodology to build subjective probabilistic models. It is very simple even if it is atypical and a bit worrisome at the beginning.

The purpose of Chapters 2 to 11 is to present this new modeling methodology.

The presentation is intended for the general public and does not suppose any prerequisites other than a basic foundation in mathematics.

Its purpose is to introduce the fundamental concepts, to present the novelty and interest of the approach, and to initiate the reader to the subtle art of Bayesian modeling. Numerous simple examples of applications are presented in different fields.

It is divided in two parts, Chapters 2 to 6 which present the principles of Bayesian programming and Chapters 7 to 11 which offer a cookbook for the good practice of probabilistic modeling.

[6] See FAQ-FAM: Cox theorem, Section 16.8.

Chapter 2 — Basic Concepts: The purpose of this chapter is to gently introduce the basic concepts of Bayesian Programming.

We start with a simple example of Bayesian spam filtering, which helps to eliminate junk e-mails. Commercially available software is based on a similar approach.

The problem is very easy to formulate. We want to classify texts (e-mail) into one of two categories, either nonspam or spam. The only information we can use to classify the e-mails is their content: a set of words.

The classifier should furthermore be able to adapt to its user and to learn from experience. Starting from an initial standard setting, the classifier should modify its internal parameters when the user disagrees with its own decision. It will hence adapt to the user's criteria to categorize nonspam and spam. It will improve its results as it encounters increasingly classified e-mails.

The classifier uses an N words dictionary. Each e-mail will be classified according to the presence or absence of each of the words.

Chapter 3 — Incompleteness and Uncertainty: The goal of this chapter is twofold: (i) to present the concept of incompleteness and (ii) to demonstrate how incompleteness is a source of uncertainty.

Chapter 4 — Description = Specification + Identification: In this chapter, we come back to the fundamental notion of description. A description is a probabilistic model of a given phenomenon. It is obtained after two phases of development:

1. A Specification phase where the programmer expresses in probabilistic terms his own knowledge about the modeled phenomenon.

2. An Identification phase where this starting probabilistic canvas is refined by learning from data.

Descriptions are the basic elements that are used, combined, composed, manipulated, computed, compiled, and questioned in different ways to build Bayesian programs.

Chapter 5 — The Importance of Conditional Independence: The goal of this chapter is both to explain the notion of Conditional Independence and to demonstrate its importance in actually solving and computing complex problems.

Chapter 6 — Bayesian Program = Description + Question: In the two previous chapters, as an example, we built a description (Bayesian model) of a water treatment center. In this chapter, we use this description to solve different problems: prediction of the output, choice of the best control strategy, and diagnosis of failures. This shows that multiple questions may be asked with the same description to solve very different problems. This clear separation between the model and its use is a very important feature of Bayesian Programming.

Chapters 2 to 6 present the concept of the Bayesian program. Chapters 7 to 11 are used to show how to combine elementary Bayesian programs to build more complex ones. Some analogies are stressed between this probabilistic mechanism and the corresponding algorithmic ones, for instance the use of subroutines or conditional and case operators.

Chapter 7 — Information Fusion: The most common application of Bayesian technics is to merge sensor information to estimate the state of a given phenomenon.

The situation is always the same: you want information about a given phenomenon; this phenomenon influences sensors that you can read and from these readings you try to estimate the phenomenon.

Usually the readings are neither completely informative about the phenomenon, nor completely consistent with one another. Consequently, you are compelled to a probabilistic approach and the question you want to address is what is the state knowing the readings.

A very common difficulty is the profusion of sensors which leads to a very high dimensionality state space for the joint distribution. A very common solution to break this curse of dimensionality is to make the very strong assumption that, knowing the phenomenon, the sensors may be considered to provide independent readings. Knowing the common cause, the different consequences are considered independent.

However, this hypothesis is often caricatured. In this chapter we present this basic approach but, also, different ways to relax this "naive" hypothesis.

Chapter 8 — Bayesian Programming with Coherence Variables: What does "equality" mean for Bayesian variables?

Two different calculi may lead to the same result. It is the case if you try to compute the same thing with two different methods in a "consistent" or "coherent" calculus system. You can impose it as a constraint of your model by specifying that a given equation should be respected. Solving the equation then consists in finding the conditions on the two terms of the equation in order to make them "equal." It can finally be used as a programming notion when you "assign" the result of a calculus to a given variable in order to use it in a subsequent calculus. However, for all these fundamental notions of logic, mathematics and computing the results of the calculus are always values either Boolean, numeric, or symbolic.

In probabilistic computing, the basic objects that are manipulated are not values but rather probability distributions on variables. In this context, the "equality" has a different meaning as it should say that two variables have the same probability distribution. To realize this, we introduce in this chapter the notion of a "coherence variable" linking two variables of any nature.

A coherence variable is a Boolean variable. If the coherence variable is equal to 1 (or "true") it imposes that the two variables are "coherent" which means that they should share the same probability distribution knowing the same premisses.

Chapter 9 — Bayesian Programming Subroutines: The purpose of this chapter is to exhibit a first mean to combine descriptions with one another in order to incrementally build more and more sophisticated probabilistic models. This is obtained by including in the decomposition, calls to Bayesian subroutines. We show that, as in standard programming, it is possible to use existing probabilistic models to build more complex ones and to further structure the definition of complex descriptions, as some reusability of a previously defined model is possible.

Chapter 10 — Bayesian Programming Conditional Statement: The purpose of this chapter is to introduce probabilistic branching statements. We will start by describing the probabilistic "if-then-else" statement which, as in standard programming, can naturally be extended to a probabilistic "case" statement. From an inference point of view, the probabilistic if-then-else statement is simply the integration over the probability distribution on a binary variable representing the truth value of the condition used in the classical "if" statement. The main difference with the classical approach is that a Bayesian program will explore both branches when the truth value of the condition is given by a probability distribution. This allows us to mix behaviors and to recognize models.

Chapter 11 — Bayesian Programming Iteration: In this chapter we propose to define a description with distributions indexed by integers. By setting the range of indexes, we define the number of distributions used in the description. This way we define generic descriptions only depending on the range of the indexes, just as we fixed the number of iterations in a "for" loop.

In pursuing this idea, we can revisit the notion of the filter, where each new evidence is incorporated into the result of a previous inference. If the index represents successive time intervals, we can then use these techniques to study time sequences and use the Markov assumption to simplify the description. The approach is useful for implementing dynamic Bayesian networks with the Bayesian programming formalism.

1.4 A need for new inference algorithms

A modeling methodology is not sufficient to run Bayesian programs. We also require an efficient Bayesian inference engine to automate the probabilistic calculus. This assumes we have a collection of inference algorithms adapted and tuned to more or less specific models and a software architecture to combine them in a coherent and unique tool.

Numerous such Bayesian inference algorithms have been proposed in the literature. The purpose of this book is not to present these different computing techniques and their associated models once more. Instead, we offer a synthesis of this work and a number of bibliographic references for those who would like more detail on these subjects.

Chapters 12 to 15 are dedicated to that.

Chapter 12 — Bayesian Programming Formalism: The purpose of this chapter is to present Bayesian Programming formally and to demonstrate that it is very simple and very clear but, nevertheless, very powerful and very subtle. Probability is an extension of logic, as mathematically sane and simple as logic, but with more expressive power than logic.

It may seem unusual to present the formalism at the end of the book. We have done this to help comprehension and to assist intuition without sacrificing rigor. After reading this chapter, anyone can check that all the examples and programs presented earlier comply with the formalism.

Chapter 13 — Bayesian Models Revisited: The goal of this chapter is to review the main probabilistic models currently used.

We systematically use the Bayesian Programming formalism to present these models, because it is precise and concise, and it simplifies their comparison. We mainly concentrate on the definition of these models. Discussions about inference and computation are postponed to Chapter 14 and discussions about learning and identification are postponed to Chapter 15.

We chose to divide the different probabilistic models into three categories: the general purpose probabilistic models, the engineering oriented probabilistic models, and the cognitive oriented probabilistic models.

In the first category, the modeling choices are made independently of any specific knowledge about the modeled phenomenon. Most of the time, these choices are essentially made to keep the inference tractable. However, the technical simplifications of these models may be compatible with large classes of problems and consequently may have numerous applications.

In the second category, on the contrary, the modeling choices and simplifications are decided according to some specific knowledge about the modeled phenomenon. These choices could eventually lead to very poor models from a computational viewpoint. However, most of the time, problem-dependent knowledge, such as conditional independence between variables, leads to very significant and effective simplifications and computational improvements.

Finally, in the cognitive-oriented probabilistic models category, different models are presented according to a cognitive classification where common cognitive problems are linked to common Bayesian solutions.

Several of these models were already presented with more detail in the previous chapters. Certain models will appear several times in different categories but presented with a different point of view for each presentation. We think that these repetitions are useful as our goal in this chapter is to give a synthetic overview of all these models.

Chapter 14 — Bayesian Inference Algorithms Revisited: This chapter surveys the main available general purpose algorithms for Bayesian inference.

It is well known that general Bayesian inference is a very difficult problem, which may be practically intractable. Exact inference has been proved to be NP-hard [Cooper, 1990], as has the general problem of approximate inference [Dagum and Luby, 1993].

Numerous heuristics and restrictions to the generality of possible inferences have been proposed to achieve admissible computation time. The purpose of this chapter is to make a short review of these heuristics and techniques.

Before starting to crunch numbers, it is usually possible (and wise) to make some symbolic computations to reduce the amount of numerical computation required. The first section of this chapter presents the different possibilities. We will see that these symbolic computations can be either exact or approximate.

Once simplified, the expression obtained must be numerically evaluated. In a few cases exact (exhaustive) computation may be possible thanks to the previous symbolic simplification, but most of the time, even with the simplifications, only approximate calculations are possible. The second section of this chapter describes the principles of the main algorithms.

Chapter 15 — Bayesian Learning Revisited: In Chapter 4 we have seen how data are used to transform a "specification" into a "description": the free parameters of the distributions are instantiated with the data making the joint distribution computable for any value of the variables. This identification process may be considered as a learning mechanism allowing the data to shape the description before any inferences could be made. In this chapter, we consider learning problems in more detail and show how some of them may be expressed as special instances of Bayesian programs.

1.5 A need for a new programming language and new hardware

A modeling methodology and new inference algorithms are not sufficient to make these models operational. We also require new programming languages to implement them on classical computers and, eventually, new specialized hardware architectures to run these programs efficiently.

However captivating these topics may be, we chose not to deal with them in this book.

It is premature to discuss new hardware dedicated to probabilistic inference, and the book is already too long to make room for one more topic!

However, we would like to stress that 25 years ago, no one dared to dream about graphical computers. Today, no one dares to sell a computer without a graphical display with millions of pixels able to present real-time 3D animations or to play high quality movies thanks to specific hardware that makes such marvels feasible.

We are convinced that 25 years from now, the ability to treat incomplete and uncertain data will be as inescapable for computers as graphical abilities are today. We hope that you will also be convinced of this at the end of this book. Consequently, we will require specific hardware to face the huge computing burden that some Bayesian inference problems may generate.

Many possible directions of research may be envisioned to develop such new hardware. Some, especially promising, are inspired by biology. Indeed, some researchers are currently exploring the hypothesis that the central nervous system (CNS) could be a probabilistic machine either at the level of individual neurons or assemblies of neurons. Feedback from these studies could provide inspiration for this necessary hardware.

1.6 A place for numerous controversies

We believe that Bayesian modeling is an elegant matter that can be presented simply, intuitively, and with mathematical rigor. We hope that we succeed in doing so in this book. However, the subjectivist approach to probability has been and still is a subject of countless controversies.

Some questions must be asked, discussed, and answered, such as for instance: the comparison between Bayesian Programming and possibility theories; the computational complexity of Bayesian inference; the irreducibility of incompleteness; and, last but not least, the subjectivist versus objectivist epistemological conceptions of probability itself.

To make the main exposition as clear and simple as possible, none of these controversies, historical notes, epistemological debates, and tricky technical questions are discussed in the body of the book. We have made the didactic choice to develop all these questions into a special chapter (Chapter 16) titled "FAQ and FAM" (Frequently Asked Questions and Frequently Argued Matters).

This chapter is organized as a collection of "record cards," at most one page long, presented in alphabetical order. Cross references to these subjects are included in the main text for readers interested in going further than a simple presentation of the principles of Bayesian modeling. You already encountered a few of them earlier in this introduction.

Finally, Chapter 17 is a short summary of the book, where the central concepts are recalled in an extensive glossary.

1.7 Running real programs as exercises

One way to read this book and learn Bayesian programming is to run and modify the Python programs given as examples.

Each example will be presented under the following format:

The program in file "chapter1/dice.py" emulates throwing a dice twice with the pypl package.

```
from pyplpath import *

# import all
from pypl import *

# define a probabilistic variable
dice= plSymbol("Dice", plIntegerType(1,6))
#define a way to adress the values
dice_value = plValues(dice)

# define a  uniform probability distribution on the variable
P_dice = plUniform(dice)

# print it
print 'P_dice = ', P_dice

# perform two random draws with the distribution
# and print the result
for i in range(2):
  P_dice.draw(dice_value)
  print i+1,'th trow', dice_value
```

This may require some computer science proficiency, which is not required from the readers of this book. Running these programs is a plus but is not necessary in the comprehension of this book. To run these programs on a computer, a Python package called pypl is needed. The source code of the examples as well as the Python package can be downloaded free of charge from "http:/www.probayes.com/Bayesian-Programming-Book/." The Python package is based on ProBT, a C++ multiplatform professional library used to automate probabilistic calculus.

Additional exercises and programs are available on this Web site.

Part I

Bayesian Programming Principles

Chapter 2

Basic Concepts

> Life, as many people have spotted, is, of course, terribly unfair. For instance, the first time the Heart of Gold ever crossed the galaxy the massive improbability field it generated caused two-hundred-and-thirty-nine thousand lightly-fried eggs to materialize in a large, wobbly heap on the famine-struck land of Poghril in the Pansel system. The whole Poghril tribe had just died out from famine, except for one man who died of cholesterol-poisoning some weeks later.
>
> *The Hitchhiker's Guide to the Galaxy*
> Douglas Adams [1995]

The purpose of this chapter is to gently introduce the basic concepts of Bayesian Programming.

These concepts will be extensively used and developed in Chapters 4 to 11 and they will be revisited, summarized, and formally defined in Chapter 12.

We start with a simple example of Bayesian spam filtering, which helps to eliminate junk e-mails. Commercially available software is based on a similar approach.

The problem is very easy to formulate. We want to classify texts (e-mail) into one of two categories either nonspam or spam. The only information we can use to classify the e-mails is their content: a set of words.

The classifier should furthermore be able to adapt to its user and to learn from experience. Starting from an initial standard setting, the classifier should modify its internal parameters when the user disagrees with its own decision. It will hence adapt to the user's criteria to differentiate between nonspam and spam. It will improve its results as it encounters increasingly classified e-mails.

The classifier uses an N word dictionary. Each e-mail will be classified according to the presence or absence of each of the words.

2.1 Variable

The variables necessary to write this program are as follows:

1. *Spam*[1]: a binary variable, false if the e-mail is not spam and true otherwise.

2. $W_0, W_1, ..., W_{N-1}$: N binary variables. W_n is true if the nth word of the dictionary is present in the text.

These $N+1$ binary variables sum up all the information we have about an e-mail.

2.2 Probability

A variable can have one and only one value at a given time, so the value of Spam is either true[2] or false, as the e-mail may either be spam or not.

However, this value may be unknown. Unknown does not mean that you do not have any information concerning *Spam*. For instance, you may know that the average rate of nonspam e-mail is 25%.

This information may be formalized, writing:

- $P([Spam = \text{false}]) = 0.25$ which stands for "the probability that an e-mail is not spam is 25%"

[1]Variables will be denoted by their names in italics with initial capital.

[2]Variable values will be denoted by their names in Roman, in lowercase.

- $P([Spam = \text{true}]) = 0.75$

2.3 The normalization postulate

According to our hypothesis, an e-mail is either interesting to read or spam. It means that it cannot be both but it is necessarily one of them. This implies that:

$$P([Spam = \text{true}]) + P([Spam = \text{false}]) = 1.0 \tag{2.1}$$

This property is true for any discrete variable (not only for binary ones) and consequently the probability distribution on a given variable X should necessarily be normalized:

$$\sum_{\forall x \in X} P([X = \text{x}]) = 1.0 \tag{2.2}$$

For the sake of simplicity, we will use the following notation:

$$\sum_{X} P(X) = 1.0 \tag{2.3}$$

2.4 Conditional probability

We may be interested in the probability that a given variable assumes a value based on some information. This is called a conditional probability. For instance, we may be interested in the probability that a given word appears in spam: $P([W_n = \text{true}] \mid [Spam = \text{true}])$. The sign "|" separates the variables into two sets: on the right are the variables with values known with certainty, on the left the probed variables. This notation may be generalized as: $P(W_n \mid [Spam = \text{true}])$ which stands for the probability distribution on W_n knowing that the e-mail is spam. This distribution is defined by two probabilities corresponding to the two possible values of W_n. For instance:

1. $P([W_n = \text{false}] \mid [Spam = \text{true}]) = 0.9996$
2. $P([W_n = \text{true}] \mid [Spam = \text{true}]) = 0.0004$

Analogously to Expression 2.3 for any two variables X and Y we have:

$$\forall \text{y} \in Y \sum_{\forall x \in X} P([X = \text{x}] \mid [Y = \text{y}]) = 1.0 \tag{2.4}$$

Again we will use shorthand to denote the same formal equation:

$$\sum_{X} P(X \mid Y) = 1.0 \tag{2.5}$$

Consequently, $\sum_{w \in \{\text{true,false}\}} P([W_n = \text{w}] \mid [Spam = \text{true}]) = 1.0$

2.5 Variable conjunction

We may also be interested in the probability of the conjunction of two variables: $P(Spam \wedge W_n)$.

$Spam \wedge W_n$, the conjunction of the two variables $Spam$ and W_n, is a new variable which can take four different values:

$$\{(false, false), (false, true), (true, false), (true, true)\} \tag{2.6}$$

This may be generalized as the conjunction of an arbitrary number of variables. For instance, in the sequel, we will be very interested in the joint probability distribution of the conjunction of N + 1 variables:

$$P(Spam \wedge W_0 \wedge ... \wedge W_n ... \wedge W_{\text{N}-1}) \tag{2.7}$$

2.6 The conjunction postulate (Bayes theorem)

The probability of a conjunction of two variables X and Y may be computed according to the Conjunction Rule:

$$\begin{aligned} P(X \wedge Y) &= P(X) P(Y \mid X) \\ &= P(Y) P(X \mid Y) \end{aligned} \tag{2.8}$$

This rule is better known under the form of the so-called Bayes theorem:

$$P(Y \mid X) = \frac{P(Y) P(X \mid Y)}{P(X)} \tag{2.9}$$

However, we prefer the first form, which clearly states that it is a means of computing the probability of a conjunction of variables according to both the probabilities of these variables and their relative conditional probabilities.

For instance, we have:

$$\begin{aligned}
& P([Spam = true] \wedge [W_n = true]) \\
= \; & P([Spam = true])\, P([W_n = true] \mid [Spam = true]) \\
= \; & 0.75 \times 0.0004 \\
= \; & 0.0003 \\
= \; & P([W_n = true])\, P([Spam = true] \mid [W_n = true])
\end{aligned} \tag{2.10}$$

2.7 Syllogisms

It is very important to acquire a clear intuitive feeling of what a conditional probability and the conjunction rule mean. A first step toward this understanding may be to restate the classical logical syllogisms in their probabilistic forms.

Let us first recall the two logical syllogisms:

1. Modus Ponens: $a \wedge [a \Rightarrow b] \rightarrow b$
   if a is true and if a implies b then b is true[3].

2. Modus Tollens: $\neg b \wedge [a \Rightarrow b] \rightarrow \neg a$
   if b is false and if a implies b then a is false.

For instance, if a stands for "x may be divided by 9" and b stands for "x may be divided by 3," we know that $a \Rightarrow b$, and we have:

1. Modus Ponens: If "x may be divided by 9" then "x may be divided by 3".

2. Modus Tollens: If "x may be divided by 3" is false then "x may be divided by 9" is also false.

Using probabilities, we may state:

1. Modus Ponens: $P(b \mid a) = 1$, which means that knowing that a is true then we may be sure that b is true.

2. Modus Tollens: $P(\neg a \mid \neg b) = 1$, which means that knowing that b is false then we may be sure that a is false.

[3] Logical propositions will be denoted by names in italics and lowercase.

$P(\neg a \mid \neg b) = 1$ may be derived from $P(b \mid a) = 1$, using the normalization and conjunction postulates:

$$\begin{aligned} P(\neg a|\neg b) &= 1 - P(a|\neg b) && \text{from (2.5)} \\ &= 1 - \frac{P(\neg b|a)\,P(a)}{P(\neg b)} && \text{from (2.9)} \\ &= 1 - \frac{(1 - P(b|a))\,P(a)}{P(\neg b)} && \text{from (2.5)} \\ &= 1 && \text{because } P(b \mid a) = 1 \end{aligned} \tag{2.11}$$

However, using probabilities we may go further than with logic:

1. From $P(b \mid a) = 1$, using normalization and conjunction postulates we may derive that $P(a \mid b) \geq P(a)$, which means that if we know that b is true, the probability that a is true is higher than it would be if we knew nothing about b.

   Obviously, the probability that "x may be divided by 9" is higher if you do know that "x may be divided by 3" than if you do not.

   This very common reasoning is beyond the scope of pure logic but is very simple in the Bayesian framework.

2. From $P(b \mid a) = 1$, using the normalization and conjunction postulates we may derive that $P(\neg b \mid \neg a) \geq P(\neg b)$, which means that if we know that a is false the probability that b is false is more than it would be if we knew nothing about a.

   The probability that "x may be divided by 3" is less if you know that x may not be divided by 9 than if you do not know anything about x.

2.8 The marginalization rule

A very useful rule, called the marginalization rule, may be derived from the normalization and conjunction postulates. This rule states:

$$\sum_X P(X \wedge Y) = P(Y) \tag{2.12}$$

It may be derived as follows:

$$\begin{aligned}
\sum_X [P(X \wedge Y)] &= \sum_X [P(Y) P(X|Y)] && \text{from (2.8)} \\
&= P(Y) \sum_X [P(X|Y)] && (2.13) \\
&= P(Y) && \text{from (2.5)}
\end{aligned}$$

2.9 Joint distribution and questions

The joint distribution on a set of two variables X and Y is the distribution on their conjunction: $P(X \wedge Y)$. If you know the joint distribution, then you know everything you may want to know about the variables. Indeed, using the conjunction and marginalization rules you have:

1. $P(Y) = \sum_X P(X \wedge Y)$
2. $P(X) = \sum_Y P(X \wedge Y)$
3. $P(Y \mid X) = \dfrac{P(X \wedge Y)}{\sum_Y P(X \wedge Y)}$
4. $P(X \mid Y) = \dfrac{P(X \wedge Y)}{\sum_X P(X \wedge Y)}$
5. $P(X \wedge Y) = P(X \wedge Y)$

There are five and only five interesting possible computations with two variables and these five calculi all come down to sum, product, and division on the joint probability distribution $P(X \wedge Y)$.

This is of course also true for a joint distribution on more than two variables.

For our spam instance, if you know the joint distribution:

$$P(Spam \wedge W_0 \wedge \ldots \wedge W_n \wedge \ldots \wedge W_{N-1}) \tag{2.14}$$

you can compute any of the $3^{N+1} - 2^{N+1}$ possible questions that you can imagine on this set of $N + 1$ variables.

A question is defined by partitioning a set of variables in three subsets: the searched variables (on the left of the conditioning bar), the known variables (on the right of the conditioning bar), and the free variables. The searched variables set must not be empty.

Examples of these questions are:

1. The joint distribution itself:

$$P(Spam \wedge W_0 \wedge \ldots \wedge W_n \wedge \ldots \wedge W_{N-1}) \tag{2.15}$$

2. The a priori probability to be a spam:

$$P(Spam) = \sum_{W_0 \wedge \cdots \wedge W_{N-1}} \left[P(Spam \wedge W_0 \wedge \cdots \wedge W_{N-1}) \right] \tag{2.16}$$

3. The a priori probability for the nth word of the dictionary to appear:

$$P(W_n) = \frac{\sum\limits_{Spam \wedge W_0 \wedge \cdots \wedge W_{n-1} \wedge W_{n+1} \wedge \cdots \wedge W_{N-1}} \left[P(Spam \wedge W_0 \wedge \cdots \wedge W_{N-1}) \right]}{\sum\limits_{Spam \wedge W_0 \wedge \cdots \wedge W_{N-1}} \left[P(Spam \wedge W_0 \wedge \cdots \wedge W_{N-1}) \right]} \tag{2.17}$$

4. The probability for the nth word to appear, knowing that the text is a spam:

$$P(W_n | [Spam = true]) = \frac{\sum\limits_{W_0 \wedge \cdots \wedge W_{n-1} \wedge W_{n+1} \wedge \cdots \wedge W_{N-1}} \left[P([Spam = true] \wedge W_0 \wedge \cdots \wedge W_{N-1}) \right]}{\sum\limits_{W_0 \wedge \cdots \wedge W_{N-1}} \left[P([Spam = true] \wedge W_0 \wedge \cdots \wedge W_{N-1}) \right]} \tag{2.18}$$

5. The probability for the e-mail to be a spam knowing that the nth word appears in the text:

$$P(Spam | [W_n = true]) = \frac{\sum\limits_{W_0 \wedge \cdots W_{n-1} \wedge W_{n+1} \cdots \wedge W_{N-1}} \left[P(Spam \wedge W_0 \wedge \cdots w_n \cdots \wedge W_{N-1}) \right]}{\sum\limits_{Spam \wedge \cdots W_{n-1} \wedge W_{n+1} \cdots \wedge W_{N-1}} \left[P(Spam \wedge W_0 \wedge \cdots w_n \cdots \wedge W_{N-1}) \right]} \tag{2.19}$$

6. Finally, the most interesting one, the probability that the e-mail is a spam knowing for all N words in the dictionary if they are present or not in the text:

$$P(Spam | w_0 \wedge \cdots \wedge w_{N-1}) = \frac{P(Spam \wedge w_0 \wedge \cdots \wedge w_{N-1})}{\sum\limits_{Spam} \left[P(Spam \wedge w_0 \wedge \cdots \wedge w_{N-1}) \right]} \tag{2.20}$$

2.10 Decomposition

The key challenge for a Bayesian programmer is to specify a way to compute the joint distribution that has the three main qualities of being a *good model*, *easy to compute*, and *easy to learn*.

This is done using a *decomposition* that restates the joint distribution as a product of simpler distributions.

Starting from the joint distribution and applying recursively the conjunction rule we obtain:

$$\begin{aligned} & P(Spam \wedge W_0 \wedge \cdots \wedge W_{N-1}) \\ = \; & P(Spam) \times P(W_0|Spam) \times P(W_1|Spam \wedge W_0) \\ & \times \cdots \\ & \times P(W_{N-1}|Spam \wedge W_0 \wedge \cdots \wedge W_{N-2}) \end{aligned} \tag{2.21}$$

This is an exact mathematical expression.

We simplify it drastically by assuming that the probability of appearance of a word knowing the nature of the text (spam or not) is independent of the appearance of the other words.

For instance, we assume that:

$$P(W_1 \mid Spam \wedge W_0) = P(W_1 \mid Spam) \tag{2.22}$$

We finally obtain:

$$P(Spam \wedge W_0 \wedge \ldots \wedge W_{N-1}) = P(Spam) \prod_{n=0}^{N-1} P(W_n \mid Spam) \tag{2.23}$$

Figure 2.1 shows the graphical model of this expression.

Observe that the assumption of independence between words is clearly not completely true. For instance, it completely neglects that the appearance of pairs of words may be more significant than isolated appearances. However, as subjectivists, we assume this hypothesis and may develop the model and the associated inferences to test how reliable it is.

The file "chapter2/spam.py" contains the Bayesian program used to set or to compute the numerical values found in this chapter. For example, the encapsulated Postscript of Figure 2.1 was obtained with the following instruction:

```
model.draw_graph(ProBT_Examples_Dir+"chapter2/data/spam_graph")
```

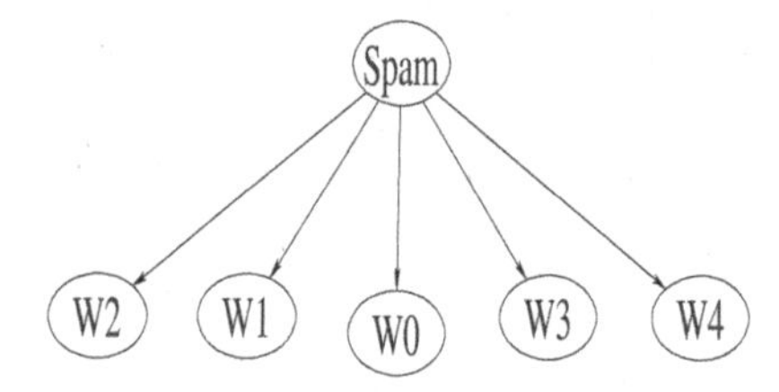

FIGURE 2.1: The graphical model of a small Bayesian spam filter based on five words.

2.11 Parametric forms

To be able to compute the joint distribution, we must now specify the $N+1$ distributions appearing in the decomposition. We already specified $P(Spam)$ in Section 2.2:

- $P(Spam)$:
  - $P([Spam = \text{true}]) = 0.75$
  - $P([Spam = \text{false}]) = 0.25$

In file "chapter2/spam.py": a simple example of a variable declaration and of a probability table based on nf and nt

```
#define a binary type
    binary_type = plIntegerType(0,1)
#define a binary variable
    Spam = plSymbol(''Spam",binary_type)
#define a prior distribution probability on Spam
    P_Spam = plProbTable(Spam,[nf,nt])
```

Each of the N forms $P(W_n \mid Spam)$ must in turn be specified. The first idea is to simply count the number of times the nth word of the dictionary appears in both spam and nonspam. This would naively lead to histograms:

- $P(W_n \mid Spam)$:
  - $P(W_n \mid [Spam = \text{false}]) = \dfrac{a_f^n}{a_f}$
  - $P(W_n \mid [Spam = \text{true}]) = \dfrac{a_t^n}{a_t}$

where a_f^n stands for the number of appearances of the nth word in nonspam e-mails and a_f stands for the total number of nonspam e-mails. Similarly, a_t^n stands for the number of appearances of the nth word in spam e-mails and a_t stands for the total number of spam e-mails.

The drawback of histograms is that when no observation has been made, the probabilities are null. For instance, if the nth word has never been observed in spam then:

$$P([W_n = \text{true}] \mid [Spam = \text{true}]) = 0.0 \tag{2.24}$$

A very strong assumption indeed, which says that what has not yet been observed is impossible! Consequently, we prefer to assume that the parametric forms $P(W_n \mid Spam)$ are Laplace succession laws rather than histograms:

- $P(W_n \mid Spam)$:
  - $P(W_n \mid [Spam = \text{false}]) = \dfrac{1 + a_f^n}{|W_n| + a_f}$
  - $P(W_n \mid [Spam = \text{true}]) = \dfrac{1 + a_t^n}{|W_n| + a_t}$

where $|W_n|$ stands for the number of possible values of variable W_n. Here, $|W_n| = 2$ as W_n is a binary variable.

If the nth word has never been observed in spam then:

$$P([W_n = \text{true}] \mid [Spam = \text{true}]) = \frac{1}{2 + a_t} \tag{2.25}$$

which tends toward zero when a_t tends toward infinity but never equals zero. An event not yet observed is not completely impossible, even if it becomes very improbable if it has never been observed in a long series of experiments.

It is possible to declare an array of probabilistic variables, for example in chapter2/spam.py:

```
#define N binary variable with
W = plArray(''W",binary_type,1,N)
```

The following instruction is used to define a conditional distribution as a table of probability distributions indexed by $Spam = 0$ and $Spam = 1$.

```
#define a conditional distribution of each word i
P_Wi_K_Spam = plDistributionTable(W[i],Spam)
```

The next two instructions are used to define these two distributions (according to the previous definitions) and to store them in the distribution table.

```
#define the two distributions on Wi:
#one for Spam = 0
P_Wi_K_Spam.push(plProbTable(W[i],\
 [ 1-((float(nfi[i])+1)/(2+nf)), \
 (float(nfi[i])+1)/(2+nf)]) ,0)
#the other for Spam = 1
P_Wi_K_Spam.push(plProbTable(W[i],\
 [ 1-((float(nti[i])+1)/(2+nt)),\
 (float(nti[i])+1)/(2+nt)]) ,1)
```

2.12 Identification

The N forms $P(W_n \mid Spam)$ are not yet completely specified because the $2N+2$ parameters $a_f^{n=0,\ldots,N-1}$, $a_t^{n=0,\ldots,N-1}$, a_f, and a_t have no values yet.

The identification of these parameters could be done either by batch processing of a series of classified e-mails or by an incremental updating of the parameters using the user's classifications of the e-mails as they arrive.

Both methods could be combined: the system could start with initial standard values of these parameters issued from a generic database, then some incremental learning customizes the classifier to each individual user.

2.13 Specification = Variables + Decomposition + Parametric forms

We call *specification* the part of the Bayesian program specified by the programmer. This part is always made of the same three subparts:

1. *Variables*: The choice of the relevant variables for the problem.

2. *Decomposition*: The expression of the joint probability distribution as the product of simpler distributions.

3. *Parametric forms*: The choice of mathematical function forms of each of these distributions.

2.14 Description = Specification + Identification

We call *description* the probabilistic model of our problem.

The description is the joint probability distribution on the relevant variables. It is completely specified when the eventual free parameters of the *specification* are given values after an *identification* (learning) phase.

2.15 Question

Once you have a description (a way to compute the joint distribution), it is possible to ask any question, as we saw in Section 2.9.

For instance, after some simplification, the answers to our six questions are:

1.
$$P\left(Spam \wedge W_0 \wedge \ldots \wedge W_{N-1}\right) = P\left(Spam\right) \prod_{n=0}^{N-1} P\left(W_n \mid Spam\right) \tag{2.26}$$

By definition, the joint distribution is equal to the decomposition.

2.
$$P\left(Spam\right) = P\left(Spam\right) \tag{2.27}$$

as $P\left(Spam\right)$ appears as such in the decomposition.

3.
$$P(W_n) = \sum_{Spam} P(Spam)\, P(W_n \mid Spam) \tag{2.28}$$

The a priori probability for the nth word of the dictionary to appear, which gives:

$$P([W_n = \text{true}]) = \left(0.25 \times \frac{1 + a_f^n}{2 + a_f}\right) + \left(0.75 \times \frac{1 + a_t^n}{2 + a_t}\right) \tag{2.29}$$

We see that the denomination "a priori" is here misleading as $P(W_n)$ is completely defined by the description and cannot be fixed in this model.

4.
$$P(W_n \mid [Spam = \text{true}]) = \frac{1 + a_t^n}{2 + a_t} \tag{2.30}$$

as the probability for the nth word to appear knowing that the text is a spam is already specified in the description.

5.
$$P(Spam \mid [W_n = \text{true}]) = \frac{P(Spam)\, P([W_n = \text{true}] \mid Spam)}{\sum_{Spam} P(Spam)\, P([W_n = \text{true}] \mid Spam)} \tag{2.31}$$

as the probability for the e-mail to be spam knowing that the nth word appears in the text.

6.
$$P(Spam | w_0 \wedge \cdots \wedge w_{N-1}) = \frac{P(Spam) \prod_{n=0}^{N-1} [P(w_n | Spam)]}{\sum_{Spam} \left[P(Spam) \prod_{n=0}^{N-1} [P(w_n | Spam)] \right]} \tag{2.32}$$

The denominator appears to be a normalization constant. It is not necessary to compute it to decide if we are dealing with spam. For instance, an easy trick is to compute the ratio:

$$\frac{P([Spam = true] | w_0 \wedge \cdots \wedge w_{N-1})}{P([Spam = false] | w_0 \wedge \cdots \wedge w_{N-1})} = \frac{P([Spam = true])}{P([Spam = false])} \times \prod_{n=0}^{N-1} \left[\frac{P(w_n | [Spam = true])}{P(w_n | [Spam = false])} \right] \tag{2.33}$$

This computation is faster and easier because it requires only $2N$ products.

Generally speaking, any partition of the set of relevant variables in three subsets defines a question. These subsets are:

1. The set of searched variables "*Searched*" which should not be empty,
2. The set of known variables "$Known$" which are often referred to as evidence variables,
3. The complementary set of free variables "$Free$".

These three sets define a valid question. For example, assuming you only know two words of the message, the question $P(Spam \mid W_0 \wedge W_1)$ is defined by the subsets:

1. $Searched = \{Spam\}$
2. $Known = \{W_0, W_1\}$
3. $Free = \{W_2, \ldots, W_{n-1}\}$

Each question defines a set of distributions on the $Searched$ variables, each of them corresponding to a possible value of the $Known$ variables. For example $P(Spam \mid W_0 = false \wedge W_1 = false)$ is one of the four corresponding distributions which is the probability for a message to be spam knowing that the words W_0 and W_1 are not in the message and knowing nothing else about the other words.

A simple but inefficient way to compute these distributions is to use the following equation:

$$\begin{aligned} & P(Searched|known) \\ = & \frac{\sum_{Free}[P(Searched \wedge known \wedge Free)]}{\sum_{Searched \wedge Free}[P(Searched \wedge known \wedge Free)]} \end{aligned} \tag{2.34}$$

In principle it is always possible to compute Equation 2.34 because $P(Searched \wedge Known \wedge Free)$ is completely defined by the description.

2.16 Bayesian program = Description + Question

Finally, a Bayesian program will always have the following simple structure:

$$Program\begin{cases}Description.\begin{cases}Specification(\pi)\begin{cases}Variables\\Decomposition\\Forms\end{cases}\\Identification\ (based\ on\ \delta)\end{cases}\\Question\end{cases}\tag{2.35}$$

The Bayesian spam filter program is completely defined by:

$$Pr\begin{cases}Ds\begin{cases}Sp(\pi)\begin{cases}Va: Spam, W_0, W_1 \ldots W_{N-1}\\Dc:\begin{cases}P\left(Spam \wedge W_0 \wedge \ldots \wedge W_n \wedge \ldots \wedge W_{N-1}\right)\\=P\left(Spam\right)\prod_{n=0}^{N-1}P\left(W_n \mid Spam\right)\end{cases}\\Fo:\begin{cases}P\left(Spam\right):\begin{cases}P\left([Spam=\text{false}]\right)=0.25\\P\left([Spam=\text{true}]\right)=0.75\end{cases}\\P\left(W_n \mid Spam\right):\begin{cases}P\left(W_n \mid [Spam=\text{false}]\right)\\=\dfrac{1+a_f^n}{2+a_f}\\P\left(W_n \mid [Spam=\text{true}]\right)\\=\dfrac{1+a_t^n}{2+a_t}\end{cases}\end{cases}\end{cases}\\Identification\ (based\ on\ \delta)\end{cases}\\Qu: P\left(Spam \mid w_0 \wedge \ldots \wedge w_n \wedge \ldots \wedge w_{N-1}\right)\end{cases}\tag{2.36}$$

2.17 Results

If we consider a spam filter with an N word dictionary, then any given e-mail contains one and only one of the 2^N possible subsets of the dictionary.

Here we restrict our spam filter to a five word dictionary so that we can analyze the $2^5 = 32$ subsets. Assume that a set of 1000 e-mails is used in the identification phase and that the resulting numbers of nonspam are 250 and 750 respectively. Assume also that the resulting counter tables for a_f^n and a_t^n are those shown in Table 2.1 and the corresponding distribution $P(W_n|Spam)$ is given in Table 2.2.

It is now possible to compute the probability for an e-mail to be a spam or not given it contains or not each of the N words. This may be done using equation 2.32.

Table 2.3 shows the obtained results for different subsets of words present in the e-mail.

TABLE 2.1: Counters Resulting from an Analysis of 1000 E-mails. (The values a_f^n and a_t^n denote the number of e-mails that contained the nth word in nonspam and spam e-mails, respectively.)

n	Word n	a_f^n	a_t^n
0	fortune	0	375
1	next	125	0
2	programming	250	0
3	money	0	750
4	you	125	375

TABLE 2.2: The Resulting Distribution $P(W_n|Spam)$ from Table 2.1

n	$P(W_n \mid [Spam = \text{false}])$		$P(W_n \mid [Spam = \text{true}])$	
	$W_n = \text{false}$	$W_n = true$	$W_n = \text{false}$	$W_n = true$
0	0.996032	0.00396825	0.5	0.5
1	0.5	0.5	0.99867	0.00132979
2	0.00396825	0.996032	0.99867	0.00132979
3	0.996032	0.00396825	0.00132979	0.99867
4	0.5	0.5	0.5	0.5

TABLE 2.3: Adding or Subtracting a Single Word from the Subset of Words Present in the E-mail Can Greatly Change the Probability of It Being Spam

Subset number	Words present	$P(Spam \mid w_0 \wedge \ldots \wedge W_4)$	
		$[Spam = \text{false}]$	$[Spam = \text{true}]$
3	{money}	5.24907e-06	0.999995
11	{next,money}	0.00392659	0.996073
12	{next,money you}	0.00392659	0.996073
15	{next,programming,money}	0.998656	0.00134393
27	{fortune,next,money}	1.57052e-05	0.999984

A single word can provide much, little or no information when classifying an e-mail. For instance, an e-mail with {next,money} has a probability of 0.996073 of being a spam, adding the word "you" does not change anything, adding "programming" contradicts completely that this e-mail is a spam (probability of 0.00134393), and adding "fortune" confirms it (probability of 0.999984).

The file "chapter2/spam.py" contains the two basic functions to program a spam filter: *build_spam_question* and *use_spam_question*. The following example shows how to use these two functions.

```
#tests
#number of e-mails considered as nonspam:
nf = 250
#number of e-mails considered as spam: nt
nt = 750
#number of time words i appears in nonspam messages
nfi = [0, 125, 250, 0, 125 ]
#number of time words i appears in spam messages
nti = [375 ,0 ,0 ,750 , 375]
# build the question and print informations about the model
my_question = build_spam_question (nf, nfi, nt ,nti)
#what is the probability distribution for a mail containing
#''next'' ''programming'' and ''you''
use_spam_question(my_question,[0,1,1,0,1])
```

Chapter 3

Incompleteness and Uncertainty

What we know is not much. What we do not know is immense.[1]

Marquis Simon de Laplace

The goal of this chapter is twofold: (i) to present the concept of incompleteness and (ii) to demonstrate how incompleteness is a source of uncertainty.

3.1 Observing a water treatment unit

The uncertainty of a phenomenon has two causes: (i) inaccuracies in the model and (ii) ignored variables. In this section, we demonstrate this fact by a second experiment: the water treatment unit. This experiment consists of two stages. In the first stage, we describe the complete model of the water treatment unit, giving all the variables and functions involved in the model. In the

[1] "Ce que nous connaissons est peu de chose, ce que nous ignorons est immense." Reported as his nearly last words by Joseph Fourier in his "Historical praise for M. le Marquis de Laplace" in front of the French Royal Academy of Science in 1829 [Fourier, 1829]. The complete citation is the following: "Les personnes qui ont assisté à ses derniers instants lui rappelaient les titres de sa gloire, et ses plus éclatantes découvertes. Il répondit: "Ce que nous connaissons est peu de chose, ce que nous ignorons est immense." C'est du moins, autant qu'on l'a pu saisir, le sens de ses dernières paroles à peine articulées. Au reste, nous l'avons entendu souvent exprimer cette pensée, et presque dans les mêmes termes. Il s'éteignit sans douleur."

second stage, we pretend that some of the variables and functions of the model are not available. In other words, we generate a synthetic incompleteness of our model. The goal is to show the consequences of this incompleteness and to present a first step toward Bayesian modeling.

3.1.1 The elementary water treatment unit

We now describe the complete model of the water treatment unit. Figure 3.1 is a schematic representation.

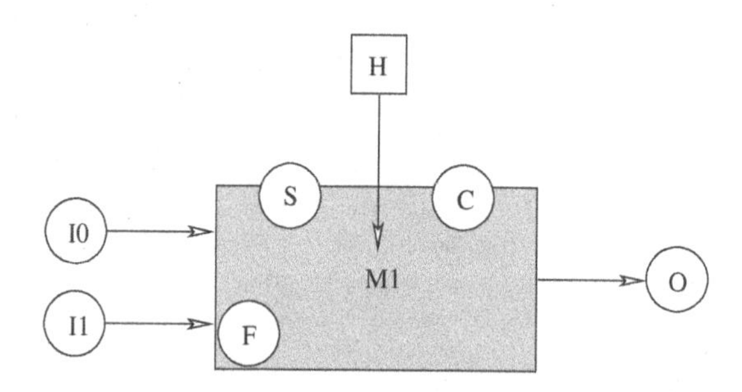

FIGURE 3.1: The treatment unit receives two water streams of quality I_0 and I_1 and generates an output stream of quality O. The resulting quality depends on I_0, I_1, two unknown variables H and F, and a control variable C. An operator regulates C, while the value of F is estimated by a sensor variable S.

The unit takes two water streams as inputs with respective water qualities I_0 and I_1. Two different streams are used because partly purified water is recycled to dilute the more polluted stream, to facilitate its decontamination.

The unit produces an output stream of quality O.

The internal functioning state of the water treatment unit is described by the variable F. This variable F quantifies the efficiency of the unit but is not directly measurable. For instance, as the sandboxes become more loaded with contaminants the purification becomes less and less efficient and the value of F becomes lower and lower.

A sensor S helps to estimate the efficiency F of the unit.

A controller C is used to regulate and optimize O, the quality of the water in the output stream.

Finally, some external factor H may disturb the operation of the unit. For instance, this external factor could be the temperature or humidity of the air.

For didactic purposes, we consider that these seven variables may each take 11 different integer values ranging from 0 to 10. The value 0 is the worst value for I_0, I_1, F, and O, and 10 is the best.

When all variables have their nominal values, the ideal quality Q of the output stream is given by the equation:

$$Q = Int\left(\frac{I_0 + I_1 + F}{3}\right) \tag{3.1}$$

Where $Int\,(x)$ is the integer part of x.

The value of Q never exceeds the value O^*, reached when the unit is in perfect condition, with:

$$O^* = Int\left(\frac{I_0 + I_1 + 10}{3}\right) \tag{3.2}$$

The external factor H may reduce the ideal quality Q and the control C may try to compensate for this disturbance or the bad condition of the treatment unit because of F. Consequently, the output quality O is obtained according to the following equations:

$$\alpha = Int\left(\frac{I_0 + I_1 + F + C - H}{3}\right) \tag{3.3}$$

$$O = \begin{cases} \alpha & \text{if } (0 \leq \alpha \leq O^*) \\ (2O^* - \alpha) & \text{if } (\alpha \geq O^*) \\ 0 & \text{Otherwise} \end{cases} \tag{3.4}$$

We consider the example of a unit directly connected to the sewer: $[I_0 = 2]$, $[I_1 = 8]$.

When $[C = 0]$ (no control) and $[H = 0]$ (no disturbance), Figure 3.2 gives the value of the quality O according to F, $(O^* = 6)$.

When the state of operation is not optimal (F different from 10), it is possible to compensate using C. However, if we over-control, then it may happen that the output deteriorates. For instance, if $[I_0 = 2]$, $[I_1 = 8]$, $[F = 8]$, $[H = 0]$, the outputs obtained for the different values of C are shown in Figure 3.3.

The operation of the unit may be degraded by H. For instance, if $[I_0 = 2]$, $[I_1 = 8]$, $[F = 8]$,$[C = 0]$, the output obtained for the different values of H are shown in Figure 3.4.

Finally, the value of the sensor S depends on I_0 and F as follows:

$$S = Int\left(\frac{I_0 + F}{2}\right) \tag{3.5}$$

The outputs of S in the 121 possible situations for I_0 and F are shown in Figure 3.5. Note that, if we know I_0, I_1, F, H, and C, we know with certainty the values of both S and O. At this stage, our water treatment unit is a completely deterministic process. Consequently, a complete model can be constructed. Now consider what happens if we ignore the exact equations that rule the water treatment unit and, of course, the existence of the external factor H. The starting point for constructing our own model is limited to

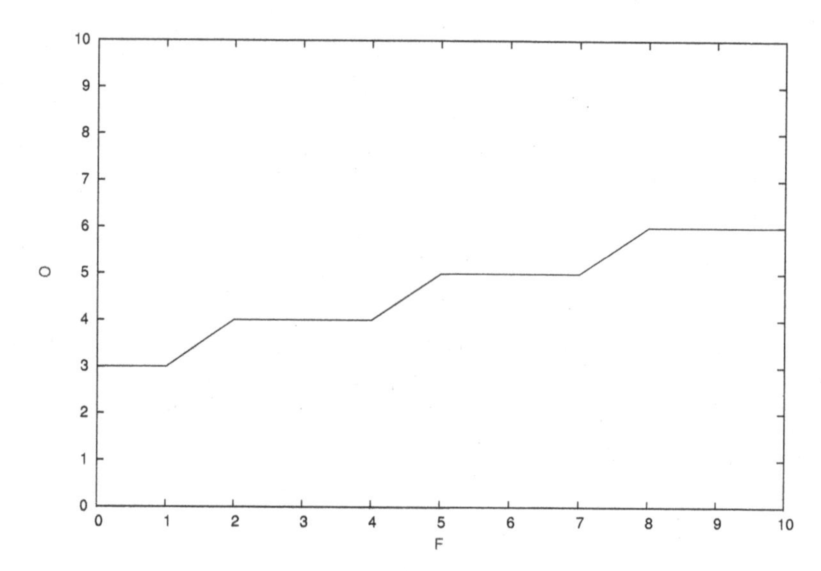

FIGURE 3.2: The output O as a function of the functioning state F with inputs, control, and external factor fixed to: $[I_0 = 2] \wedge [I_1 = 8] \wedge [C = 0] \wedge [H = 0]$.

knowing the existence of the variables I_0, I_1, F, S, C, and O, and that the value of O depends on $I_0 \wedge I_1 \wedge S \wedge C$ and that of S depends on $I_0 \wedge F$.

What do we need to do now? Observe the behavior of the water treatment unit, in particular the quality of the output stream O, for different values of $I_0 \wedge I_1 \wedge S \wedge C$, as well as the sensor value S for different values of I_0 (remember that F cannot be observed). During these observations you will note that there are different situations in which uncertainty appears. The goal of the following section is to discuss this uncertainty.

3.1.2 Experimentation and uncertainty

3.1.2.1 Uncertainty on O because of inaccuracy of the sensor S

A given value of S corresponds to several possible values of I_0 and F. For instance, seven pairs of values of $I_0 \wedge F$ correspond to $[S = 1]$ in Figure 3.5. Worse than this, even knowing I_0 and S, two values of F are possible most of the time (see Figure 3.6). This fact will introduce some "noise" in the prediction of O.

To illustrate this effect let us first experiment with $H = 0$: the operation of the water treatment unit is not disturbed. For $[I_0 = 2]$, $[I_1 = 8]$, $[C = 2]$, we can explore the different possible values of the output O when F varies. However, as F is not directly observable, we can only collect data concerning S and O. These data are presented on Figure 3.7.

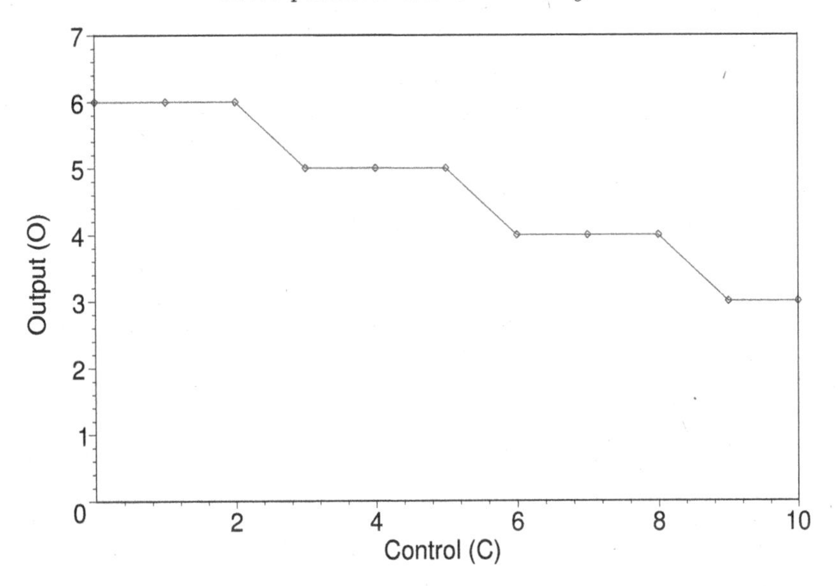

FIGURE 3.3: The output O as a function of control C with inputs, functioning state, and external factor, fixed to: $[I_0 = 2]$, $[I_1 = 8]$, $[F = 8]$, $[H = 0]$.

For some S it is possible to predict exactly the output O:

1. $[S = 1] \Rightarrow [O = 4]$
2. $[S = 3] \Rightarrow [O = 5]$
3. $[S = 4] \Rightarrow [O = 6]$
4. $[S = 6] \Rightarrow [O = 5]$

For some other values of S it is not possible to predict the output O with certainty:

1. If $[S = 2]$, then O may take the value either four or five, with a slightly higher probability for four. Indeed, when $[S = 2]$, then F may be either two or three (see Figure 3.6) and, O will, respectively, be either four or five.

2. If $[S = 5]$, then O may take the value either five or six, with a slightly lower probability for five. When $[S = 5]$, F may be either eight or nine.

3.1.2.2 Uncertainty because of the hidden variable H

Let us now do the same experiment for a disturbed process (value of H drawn at random from the 11 possible values). Of course, we obtain different

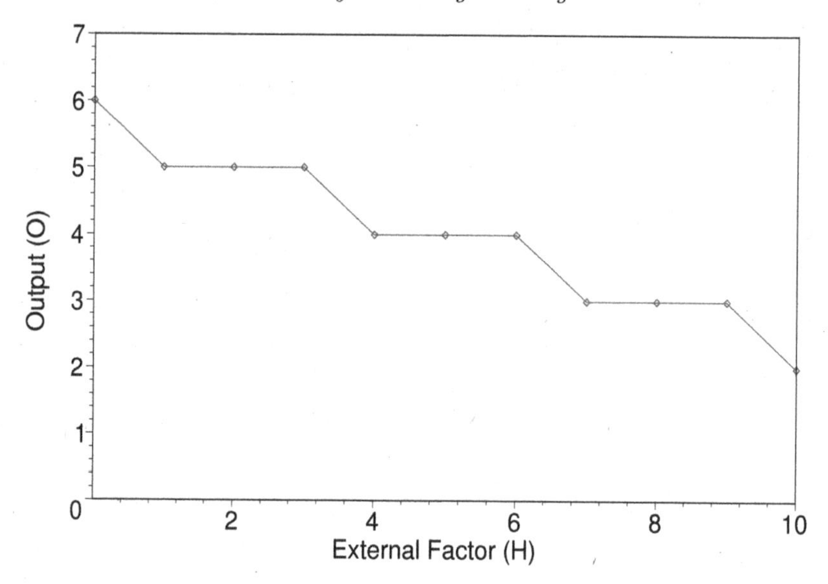

FIGURE 3.4: The output O as a function the external factor H with inputs, functioning state, and control fixed to: $[I_0 = 2]$, $[I_1 = 8]$, $[F = 8]$, $[C = 0]$.

results with more uncertainty due to the effect on the output of the hidden variable H. The obtained data when $[I_0 = 2]$, $[I_1 = 8]$, $[C = 2]$ is presented on Figure 3.8.

In contrast with our previous experiment, this time no value of S is sufficient to infer the value of O exactly.

The dispersion of the observations is the direct translation of the effect of H. Taking into account the effect of hidden variables such as H and even measuring their importance is one of the major challenges that Bayesian Programming must face. This is not an easy task when you are not even aware of the nature and number of these hidden variables!

3.2 Lessons, comments, and notes

3.2.1 The effect of incompleteness

We assume that any model of a "real" (i.e., not formal) phenomenon is incomplete. There are always some hidden variables, not taken into account in the model, that influence the phenomenon. Furthermore, this incompleteness is irreducible: for any physical phenomenon, there is no way to build an exact

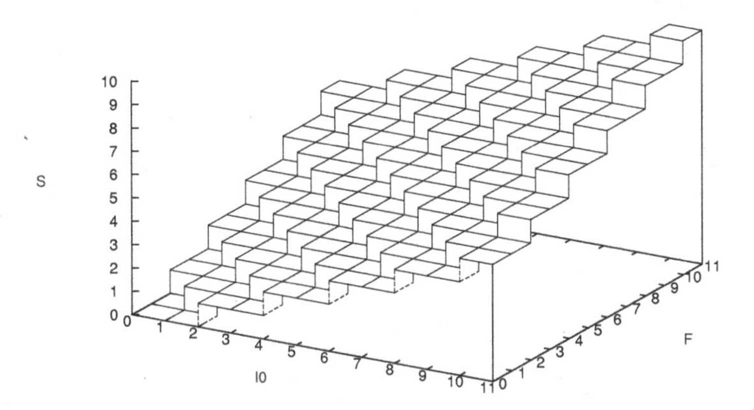

FIGURE 3.5: The sensor S as a function of input I_0 and functioning state F.

model with no hidden variables.[2] The effect of these hidden variables is that the model and the phenomenon never have exactly reproducible behavior. Uncertainty appears as a direct consequence of this incompleteness. Indeed, the model may not completely take into account the data and may not predict exactly the behavior of the phenomenon.[3] For instance, in the above example, the influence of the hidden variable H makes it impossible to predict with certainty the output O given the inputs I_0 and I_1, the reading of the sensor S, and the control C.

3.2.2 The effect of inaccuracy

The above example also demonstrates that there is another source of uncertainty: the *inaccuracy* of the sensors.

By inaccuracy, we mean that a given sensor may read the same value for different underlying situations. Here, the same reading on S may correspond to different values of F.

F is not a hidden variable as it is taken into account by the model. However, F cannot be measured directly and exactly. The values of F can only be inferred indirectly through the sensor S and they cannot be inferred with certainty. It may be seen as a weak version of incompleteness, where a variable

[2]See FAQ/FAM, Section 16.10 "Incompleteness irreducibility" for further discussion of that matter.

[3]See FAQ/FAM, Section 16.12 "Noise or ignorance?" for more information on this subject.

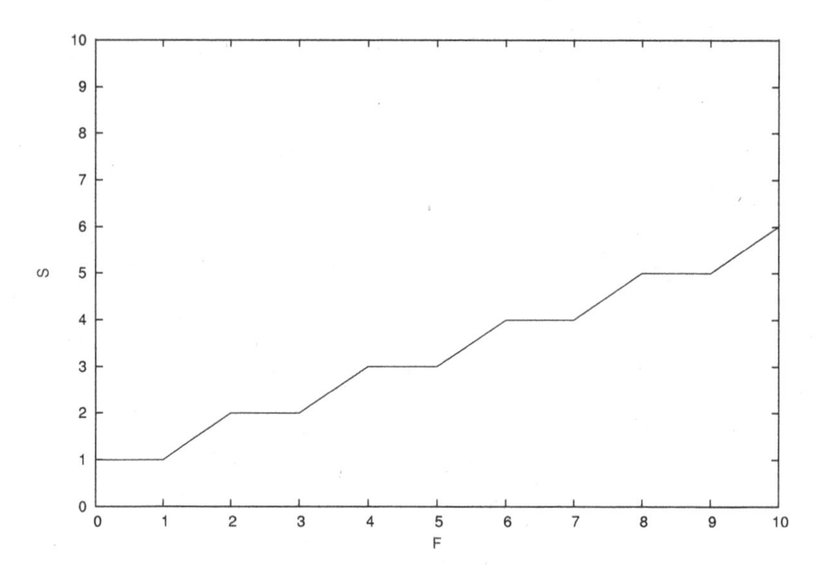

FIGURE 3.6: The sensor reading S as a function of the functioning state F when the input $[I_0 = 2]$.

is not completely hidden but is only partially known and accessible. Even though it is weak, this incompleteness still generates uncertainty.

3.2.3 Not taking into account the effect of ignored variables may lead to wrong decisions

Once the effects of the irreducible incompleteness of models are recognized, a programmer must deal with them either ignoring them and using the incomplete models, or trying to take incompleteness into account using a probabilistic model.

Using a probabilistic model clearly appears to be the better choice as it will always lead to better decisions, based on more information than the nonprobabilistic one.

For instance, a nonprobabilistic model of our production unit, not taking into account the variable H, would be, for instance[4]:

$$\alpha = Int\left(\frac{I_0 + I_1 + F + C}{3}\right) \tag{3.6}$$

$$O = \begin{cases} \alpha & \text{if } (0 \leq \alpha \leq O^*) \\ (2O^* - \alpha) & \text{if } (\alpha \geq O^*) \\ 0 & \text{Otherwise} \end{cases} \tag{3.7}$$

[4]Note the absence of H.

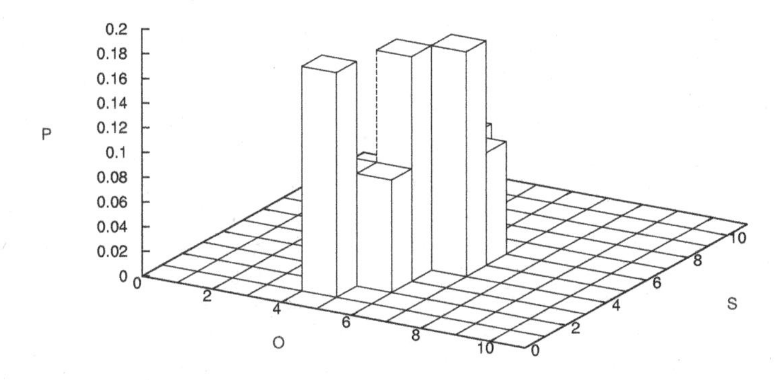

FIGURE 3.7: The histogram of the observed sensor state S and the output O when the inputs, the control, and the external factor are fixed to $[I_0 = 2]$, $[I_1 = 8]$, $[C = 2]$, $[H = 0]$, and the internal function F is generated randomly with a uniform distribution.

$$S = Int\left(\frac{I_0 + F}{2}\right) \tag{3.8}$$

It would lead to false predictions of the output O and, consequently, to wrong control decision on C to optimize this output.

For instance, scanning the 11 different possible values for C when $[I_0 = 2]$, $[I_1 = 8]$, $[F = 8]$ and consequently $[S = 5]$, the above model predicts that indifferently for $[C = 0]$, $[C = 1]$, and $[C = 2]$, O will take its optimal value: six (see Figure 3.3).

The observations depict a somewhat different and more complicated "reality" as shown in Figure 3.9. The choice of C to optimize O is now more complicated but also more informed. The adequate choice of C to produce the optimal output $[O = 6]$ is now, with nearly equivalent probabilities, to select a value of C greater than or equal to two. Indeed, this is a completely different choice from when the "exact" model is used!

3.2.4 From incompleteness to uncertainty

Any program that models a real phenomenon must face a central difficulty: how should it use an incomplete model of the phenomenon to reason, decide, and act efficiently?

The purpose of Bayesian Programming is precisely to tackle this problem

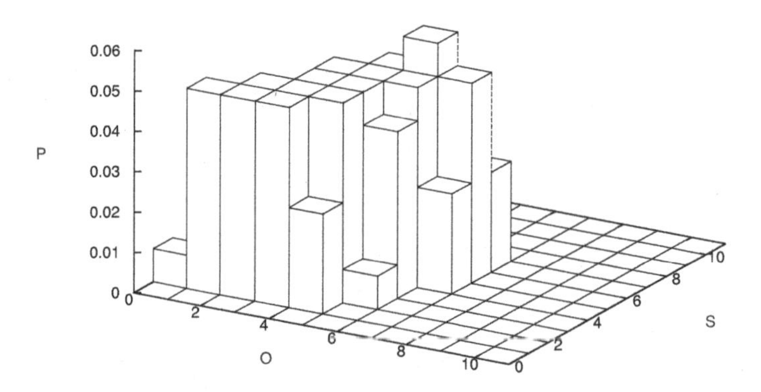

FIGURE 3.8: The histogram of the observed sensor state S and the output O when the inputs and the control are set to $[I_0 = 2]$, $[I_1 = 8]$, $[C = 2]$, and the values of the external factor and the internal functioning $H \wedge F$ are drawn at random.

with a well established formal theory: probability calculus. The sequel of this book will try to explain how to do this.

In the Bayesian Programming approach, the programmer does not propose an exact model but rather expresses a probabilistic canvas in the specification phase. This probabilistic canvas gives some hints about what observations are expected. The specification is not a fixed and rigid model purporting completeness. Rather, it is a framework, with open parameters, waiting to be shaped by the experimental data. Learning is the means of setting these parameters. The resulting probabilistic descriptions come from both: (i) the views of the programmer and (ii) the physical interactions specific of each phenomenon. Even the influence of the hidden variables is taken into account and quantified; the more important their effects, the more noisy the data, and the more uncertain the resulting descriptions.

The theoretical foundations of Bayesian Programming may be summed up by Figure 3.10.

The first step in Figure 3.10 transforms the irreducible incompleteness into uncertainty. Starting from the specification and the experimental data, learning builds probability distributions.

The maximum entropy principle is the theoretical foundation of this first step. Given some specifications and some data, the probability distribution that maximizes the entropy is the distribution that best represents the com-

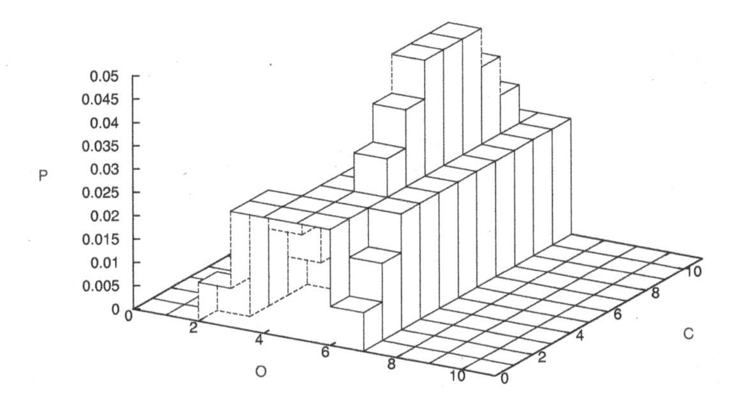

FIGURE 3.9: The histogram of the observed output O and the control C when the inputs are set to $[I_0 = 2]$, $[I_1 = 8]$, and the internal functioning F is set to $[F = 8]$ with H drawn at random.

bined specification and data. Entropy gives a precise, mathematical, and quantifiable meaning to the quality of a distribution.[5]

Two extreme examples may help to understand what occurs:

1. Suppose that we are studying a formal phenomenon. There may not be any hidden variables. A complete model may be proposed. The phenomenon and the model could be identical. For instance, this would be the case if we take the equations of Section 3.1.1 as the model of the phenomenon described in that same section. If we select this model as the specification, any data set will lead to a description made of Diracs. There is no uncertainty; any question may be answered either by true or false. Logic appears as a special case of the Bayesian approach in that particular context (see Cox [1979]).

2. At the opposite extreme, suppose that the specification consists of very poor hypotheses about the modeled phenomenon, for instance, by ignoring H and also the inputs I_0 and I_1 in a model of the above process. Learning will only lead to flat distributions, containing no information. No relevant decisions can be made, only completely random ones.

Specifications allow us to build general models where inaccuracy and hidden

[5]See FAQ/FAM, Section 16.11 "Maximum entropy principle justifications" for justifications for the use of the maximum entropy principle.

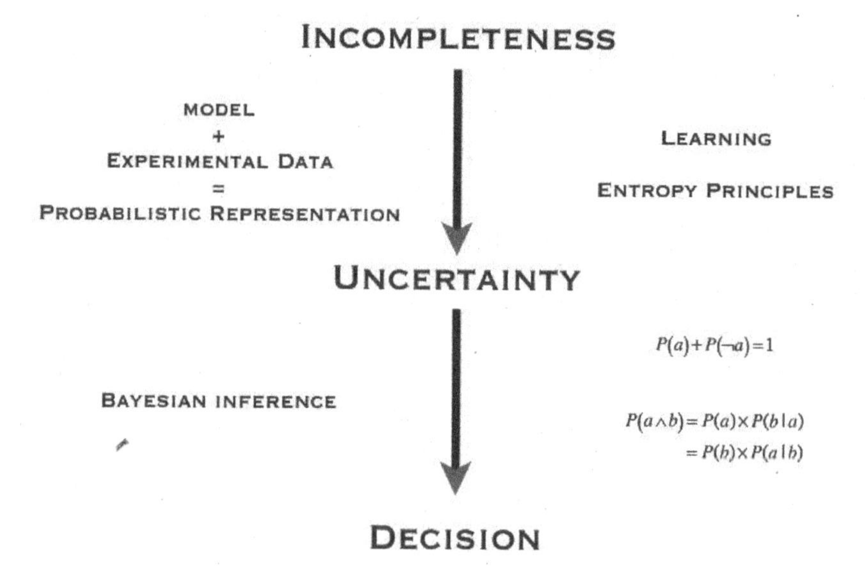

FIGURE 3.10: Theoretical foundation: from Incompleteness to Uncertainty and from Uncertainty to Decision.

variables may be explicitly represented. These models may lead to good prediction and decision. The formalism also allows us to take into account missing variables. In real life, such models are in general poorly informative and may not be useful in practical applications. They give no certitudes, although they provide a means of taking the best possible decision according to the available information. This is the case here when the only hidden variable is H.

The second step in Figure 3.10 consists of reasoning with the probability distributions obtained by the first step. To do so, we only require the two basic rules of Bayesian inference presented in Chapter 2. These two rules are to Bayesian inference what the resolution principle is to logical reasoning (see Robinson [1965], Robinson [1979], Robinson and Silbert [1982a], and Robinson and Silbert [1982b]). These inferences may be as complex and subtle as those usually achieved with logical inference tools, as will be demonstrated in the different examples presented in the sequel of this book.

Chapter 4

Description = Specification + Identification

The scientific methodology forbids us to have an opinion on questions which we do not understand, on questions which we do not know how to put clearly. Before anything else, we must know how to state problems. In science, problems do not appear spontaneously. This "sense of problem" is precisely what characterizes a true scientific mind. For such a mind, any knowledge is an answer to a question. Without a question there cannot be scientific knowledge. Nothing is obvious. Nothing is given. Everything is built.[1]

La Formation de l'Esprit Scientifique
Gaston Bachelard [1938]

[1] "L'esprit scientifique nous interdit d'avoir une opinion sur des questions que nous ne comprenons pas, sur des questions que nous ne savons pas poser clairement. Avant tout, il faut savoir poser les problèmes. Et quoi qu'on dise, dans la vie scientifique, les problèmes ne se posent d'eux-mêmes. C'est précisément ce 'sens du problème' qui donne la marque du véritable esprit scientifique. Pour un esprit scientifique, toute connaissance est une réponse à une question. S'il n'y a pas eu de question, il ne peut y avoir connaissance scientifique. Rien ne va de soi. Rien n'est donné. Tout est construit" [Bachelard, 1938].

In this chapter, we come back to the fundamental notion of description. A description is a probabilistic model of a given phenomenon. It is obtained after two phases of development:

1. A specification phase where the programmer expresses in probabilistic terms his own knowledge about the modeled phenomenon.

2. An identification phase where this starting probabilistic canvas is refined by learning from data.

Descriptions are the basic elements that are used, combined, composed, manipulated, computed, compiled, and questioned in different ways to build Bayesian programs.

4.1 Pushing objects and following contours

To introduce this notion of description we present two very simple robotic experiments where we want a small mobile robot named Khepera either to push objects or to follow their contours.

4.1.1 The Khepera robot

Khepera is a two-wheeled mobile robot, 57 millimeters in diameter and 29 millimeters in height, with a total weight of 80 grams (see Figure 4.1). It was designed at EPFL[2] and is commercialized by K-Team.[3] The robot is equipped with eight light sensors (six in front and two behind), which take values between 0 and 511 in inverse relation to light intensity, stored in variables $L_0, \ldots, L_7$ (see Figure 4.2). These eight sensors can also be used as infrared proximeters, taking values between 0 and 1023 in inverse relation to the distance from the obstacle, stored in variables $Px_0, \ldots, Px_7$ (see Figure 4.2). The robot is controlled by the rotation speeds of its left and right wheels, stored in variables M_l and M_r respectively. From these 18 basic sensory and motor variables, we derive two new sensory variables (*Dir* and *Prox*) and one new motor variable (*Rot*). They are described below:

- *Dir* is a variable that approximately corresponds to the bearing of the closest obstacle (see Figure 4.2). It takes values between -10 (obstacle to the left of the robot) and $+10$ (obstacle to the right of the robot), and is defined as follows:

---

[2]Ecole Polytechnique Fédérale de Lausanne (Switzerland).
[3]http://www.K-team.com/.

FIGURE 4.1: The Khepera mobile robot.

$$Dir = \mathbf{Floor}\left(\frac{90\,(Px_5 - Px_0) + 45\,(Px_4 - Px_1) + 5\,(Px_3 - Px_2)}{9\,(1 + Px_0 + Px_0 + Px_1 + Px_2 + Px_3 + Px_4 + Px_5)}\right) \tag{4.1}$$

- *Prox* is a variable that approximately corresponds to the proximity of the closest obstacle (see Figure 4.2). It takes values between 0 (obstacle a long way from the robot) and 15 (obstacle very close to the robot), and is defined as follows:

$$Prox = \mathbf{Floor}\left(\frac{\mathbf{Max}\,(Px_0, Px_1, Px_2, Px_3, Px_4, Px_5)}{64}\right) \tag{4.2}$$

- The robot is piloted solely by its rotation speed (the translation speed is fixed). It receives motor commands from the *Rot* variable, calculated from the difference between the rotation speeds of the left and right wheels. *Rot* takes on values between −10 (fastest to the left) and +10 (fastest to the right).

4.1.2 Pushing objects

The goal of the first experiment is to teach the robot how to push objects.

First, in a specification phase, the programmer specifies his knowledge about this behavior in probabilistic terms.

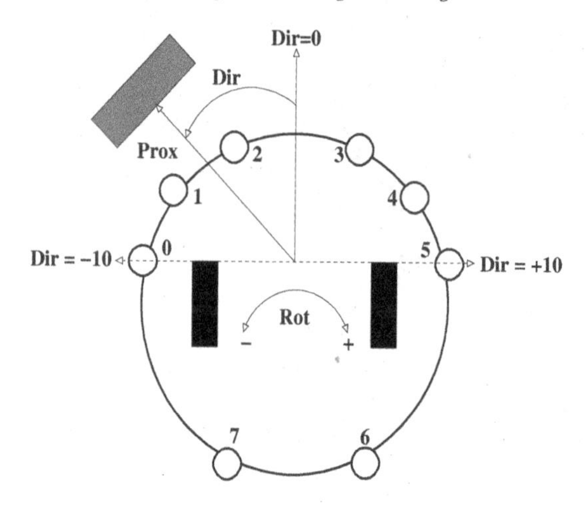

FIGURE 4.2: The sensor and motor variables of the Khepera robot.

Then, in a learning phase (identification), we drive the robot with a joystick to push objects. During that phase, the robot collects, every tenth of a second, both the values of its sensory variables and the values of its motor variable (determined by the joystick position). These data sets are then used to identify the free parameters of the parametric forms.

Finally, the robot must autonomously reproduce the behavior it has just learned. Every tenth of a second it decides the values of its motor variable, knowing the values of its sensory variables and the internal representation of the task (the description).

4.1.2.1 Specification

Having defined our goal, we describe the three steps necessary to define the preliminary knowledge.

1. Choose the pertinent variables
2. Decompose the joint distribution
3. Define the parametric forms

Variables: First, the programmer specifies which variables are pertinent for the task. To push objects it is necessary to have an idea of the position of the objects relative to the robot. The front proximeters provide this information. However, we chose to summarize the information from these six proximeters by the two variables Dir and $Prox$.

We also chose to set the translation speed to a constant and to operate the robot by its rotation speed Rot.

These three variables are all we require to push obstacles. Their definitions are summarized as follows

$$Dir \in \{-10, \ldots, 10\}, \mathbf{Card}\{Dir\} = 21 \quad (4.3)$$
$$Prox \in \{0, \ldots, 15\}, \mathbf{Card}\{Prox\} = 16 \quad (4.4)$$
$$Rot \in \{-10, \ldots, 10\}, \mathbf{Card}\{Rot\} = 21 \quad (4.5)$$

Decomposition: In the second specification step, we give a decomposition of the joint probability $P(Dir \wedge Prox \wedge Rot)$ as a product of simpler terms.

$$P(Dir \wedge Prox \wedge Rot) = P(Dir \wedge Prox)\, P(Rot|Dir \wedge Prox) \quad (4.6)$$

This equality simply results from the application of the conjunction rule (2.8).

Forms: To be able to compute the joint distribution, we must finally assign parametric forms to each of the terms appearing in the decomposition:

$$P(Dir \wedge Prox) \equiv \mathbf{Uniform} \quad (4.7)$$
$$P(Rot \mid Dir \wedge Prox) \equiv \mathbf{B}(\mu(Dir, Prox), \sigma(Dir, Prox)) \quad (4.8)$$

We have no *a priori* information about the direction or distance of the obstacles. Hence, $P(Dir \wedge Prox)$ is a uniform distribution, with all directions and proximities having the same probability. As we have 21×16 different possible values for $Dir \wedge Prox$ we get:

$$P(Dir \wedge Prox) = \frac{1}{21 \times 16} = \frac{1}{336} \quad (4.9)$$

For each sensory situation, we believe that there is one and only one rotation speed (Rot) that should be preferred. The distribution $P(Rot \mid Dir \wedge Prox)$ is thus unimodal. However, depending on the situation, the decision to be made for Rot may be more or less certain. This is presumed by assigning a bell-shaped[4] parametric form to $P(Rot \mid Dir \wedge Prox)$. For each possible position of the object relative to the robot we have a bell-shaped distribution. Consequently, we have $21 \times 16 = 336$ bell-shaped distributions and we have $2 \times 21 \times 16 = 772$ free parameters: 336 means and 336 standard deviations.

[4]Bell-shaped distributions are distributions of discrete variables that have a Gaussian shape. They are noted with the **B** symbol and defined by their means and standard deviations as regular Gaussian distributions of continuous variables.

4.1.2.2 Identification

To set the values of these free parameters we drive the robot with a joystick and collect a set of data.

Every tenth of a second, we obtain the value of Dir and $Prox$ from the proximeters and the value of Rot from the joystick. Let us call the particular set of data corresponding to this experiment δ_{push}. A datum collected at time t is a triplet $(rot^t, dir^t, prox^t)$. During the 30 seconds of learning, 300 such triplets are recorded.

From the collection δ_{push} of such data, it is very simple to compute the corresponding values of the free parameters. We first sort the data in 336 groups, each corresponding to a given position of the object and then compute the mean and standard deviations of Rot for each of these groups.

There are only 300 triplets for 336 groups. Moreover, these 300 triplets are concentrated around some particular position often observed when pushing obstacles. Consequently, it may often happen that a given position never occurs and that no data is collected for this particular situation. In that case, we set the corresponding mean to 0 and the standard deviation to 10. The bell-shaped distribution is then flat, close to a uniform distribution.

Figure 4.3 presents three of the 336 curves. The first one corresponds to an obstacle very close to the left ($[Dir = -10], [Prox = 13]$), and shows that the robot should turn to the left rapidly with average uncertainty. The second one corresponds to an obstacle right in front and in contact ($[Dir = 0], [Prox = 15]$), and shows that the robot should go straight with very low uncertainty. Finally, the last one shows an unobserved situation where the uncertainty is maximal ($[Dir = 3], [Prox = 0]$).

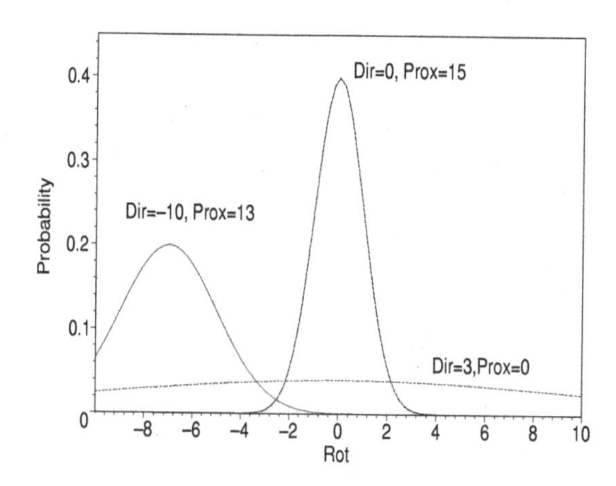

FIGURE 4.3: $P(Rot \mid Dir \wedge Prox)$ when pushing objects for different situations.

4.1.2.3 Results

To render the pushing obstacle behavior just learned, a decision on *Rot* is made every tenth of a second according to the following algorithm.

1. The sensors are read and the values of dir^t and $prox^t$ are computed.
2. The corresponding distribution $P\left(Rot \mid Dir = dir^t \wedge Prox = prox^t\right)$ is selected from the 336 distributions stored in memory.
3. A value rot^t is drawn at random according to this distribution and sent to the motors.

As shown in Movie[5] 1, the Khepera learns how to push obstacles in 30 seconds. It learns the particular dependency, corresponding to this specific behavior, between the sensory variables *Dir* and *Prox*, and the motor variable *Rot*. This dependency is largely independent of the particular characteristics of the objects (such as weight, color, balance, or nature). Therefore, as shown in Movie 2, the robot is able to push different objects. This, of course, is only true within certain limits. For instance, the robot will not be able to push an object if it is too heavy.

The file "chapter4/kepera.py" contains the code to learn and to render the appropriate behavior. Learning $P(Vrot \mid Dir \wedge Prox)$ is obtained by adding observed data to a learner.

```
for row in VrotDirProxReader:
        sample[Vrot] = int(row[0])
        sample[Dir] = int(row[1])
        sample[Prox]= int(row[2])
        # learn with this new point
        VrotDirProxLearner.add_point(sample)
```

Rendering the behavior is obtained by reading the values of $Dir \wedge Prox$ and by drawing the value of $Vrot$ according to the corresponding distribution.

```
render_question.instantiate([dir,prox]).draw(VrotValue)
```

4.1.3 Following contours

The goal of the second experiment is to teach the robot how to follow the contour of an object.

We will follow the same steps as in the previous experiment: first, a specification phase, then an identification phase where we also drive the robot with a joystick but this time to follow the contours.

[5] http:/www.probayes.com/Bayesian-Programming-Book/Movies.

We keep the exact same specification, changing only the data to be learned. The resulting description is, however, completely different: following contours of objects instead of pushing them.

4.1.3.1 Specification

Variables: To follow the contours, we must know where the object is situated relative to the robot. This is defined by the variables Dir and $Prox$, as in the previous experiment. We must also pilot the robot using its rotation speed with the variable Rot. The required variables are thus exactly the same as previously:

$$Dir \in \{-10, \ldots, 10\}, \mathbf{Card}\{Dir\} = 21 \tag{4.10}$$
$$Prox \in \{0, \ldots, 15\}, \mathbf{Card}\{Prox\} = 16 \tag{4.11}$$
$$Dir \in \{-10, \ldots, 10\}, \mathbf{Card}\{Rot\} = 21 \tag{4.12}$$

Decomposition: The decomposition does not change either:

$$P(Dir \wedge Prox \wedge Rot) = P(Dir \wedge Prox)\, P(Rot|Dir \wedge Prox) \tag{4.13}$$

Forms: Finally, the parametric forms are also the same:

$$P(Dir \wedge Prox) \equiv \mathbf{Uniform} \tag{4.14}$$
$$P(Rot \mid Dir \wedge Prox) \equiv \mathbf{B}(\mu(Dir, Prox), \sigma(Dir, Prox)) \tag{4.15}$$

4.1.3.2 Identification

In contrast, the learned data are completely different, because we are driving the robot to do some contour following (see Movie 3). The learning process is the same but the data set, called δ_{follow}, is completely different.

The collection δ_{follow} of data leads to completely different values of the 336 means and standard deviation of the bell-shaped distributions. This clearly appears in the following distributions presented for the same relative positions of the object and the robot as in the previous experiment:

- Figure 4.4 shows the two distributions obtained after learning for both experiments (pushing objects and following contours) when the object is close to the left ($[Dir = -10]$, $[Prox = 13]$). When pushing, the robot turns left to face the object; on the contrary, when following, the robot goes straight, bordering the object.

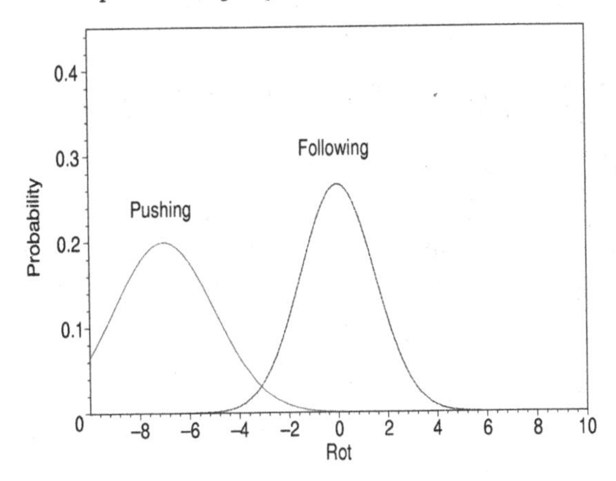

FIGURE 4.4: $P(Rot \mid Dir = -10] \wedge [Prox = 13])$ when pushing objects and when following contours.

- Figure 4.5 shows the two distributions obtained after learning for both experiments (pushing objects and following contours) when the object is in contact right in front of the robot $P(Rot \mid Dir = 0] \wedge [Prox = 15])$. When pushing, the robot goes straight. On the contrary, when following, the robot turns to the right to have the object on its left. However, the uncertainty is larger in this last case.

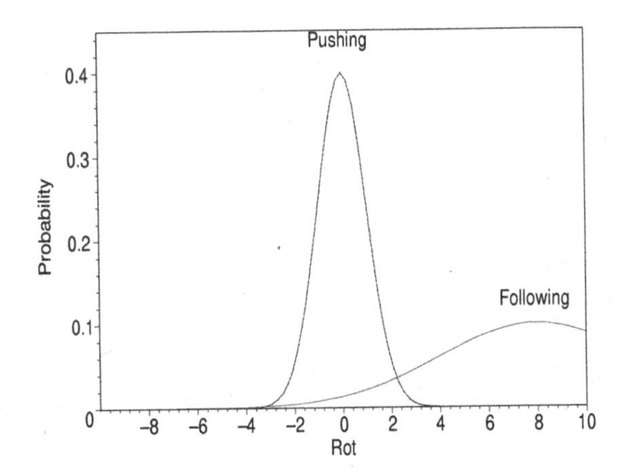

FIGURE 4.5: $P(Rot \mid Dir = 0] \wedge [Prox = 15])$ when pushing objects and when following contours.

4.1.3.3 Result

The restitution process is also the same, but as the bell-shaped distributions are different, the resulting behavior is completely different, as demonstrated by Movie 3. It should be noted that one turn around the object is enough to learn the contour following behavior.

By changing the data source in "chapter4/kepera.py" from "following.csv" to "pushing.csv" it is possible to change the behavior of the robot. These two files have been produced by the program in "chapter4/simulatefollowing.py" and "chapter4/simulatepushing.py".

4.2 Description of a water treatment unit

Let us return to the previous example of a water treatment unit, and try to build the description of this process.

4.2.1 Specification

4.2.1.1 Variables

Following our Bayesian Programming methodology, the first step in building this description is to choose the pertinent variables.

The variables to be used by our Bayesian model are obviously the following:

$$I_0, I_1, F, S, C, O \in \{0, \ldots, 10\} \tag{4.16}$$

where every one of these variables has a cardinality of 11. Of course H is missing.

4.2.1.2 Decomposition

Using the conjunction postulate (2.8) iteratively, we can write that the joint probability distribution of the six variables is equal to:

$$\begin{aligned} & P(I_0 \wedge I_1 \wedge F \wedge S \wedge C \wedge O) \\ = \; & P(I_0) \times P(I_1|I_0) \times P(F|I_0 \wedge I_1) \times P(S|I_0 \wedge I_1 \wedge F) \\ & \times P(C|I_0 \wedge I_1 \wedge F \wedge S) \times P(O|I_0 \wedge I_1 \wedge F \wedge S \wedge C) \end{aligned} \tag{4.17}$$

This is an exact mathematical expression. The designer knows more about the process than this exact form. For instance, he or she knows that:

1. The qualities of the two input streams I_0 and I_1 are independent:

$$P(I_1 \mid I_0) = P(I_1)$$

2. The state of operation of the unit is independent of both entries:

$$P(F \mid I_0 \wedge I_1) = P(F)$$

3. The reading of the sensor depends only on I_0 and F. It does not depend on I_1.

$$P(S \mid I_0 \wedge I_1 \wedge F) = P(S \mid I_0 \wedge F) \tag{4.18}$$

4. The control C may not be established without knowing the desired output O. Consequently, as long as O is unknown, C is independent of the entries and of the state of operation.

$$P(C \mid I_0 \wedge I_1 \wedge F \wedge S) = P(C) \tag{4.19}$$

   A few more thoughts may be necessary about this simplification, which is rather subtle. If you do know the desired output O, the control C will obviously depend on the entries, the state of operation, and the reading of the sensor. However, if you do not know the objective, could you think of any reason to condition the control C on these variables? If you do not know where you want to go, do you have any good reason to choose a specific direction, even knowing the map and where you are?

5. The output O depends on I_0,I_1,F,S, and C. However, because of the presence of the sensor, there is some redundancy between I_0, F, and S. If you know I_0 and F, then obviously knowing S does not bring any new information. This could be used to state: $P(O \mid I_0 \wedge I_1 \wedge F \wedge S \wedge C) = P(O \mid I_0 \wedge I_1 \wedge F \wedge C)$. Knowing I_0 and S, there is still some uncertainty about F (see Figures 3.9 and 3.10). However, as the value of F is not directly accessible for learning, we may consider as a first approximation that, knowing I_0 and S, F may be neglected:

$$P(O \mid I_0 \wedge I_1 \wedge F \wedge S \wedge C) = P(O \mid I_0 \wedge I_1 \wedge S \wedge C) \tag{4.20}$$

Finally, the decomposition of the joint probability will be specified as:

$$\begin{aligned} & P(I_0 \wedge I_1 \wedge F \wedge S \wedge C \wedge O) \\ = & P(I_0) \times P(I_1) \times P(F) \times P(S|I_0 \wedge F) \\ & \times P(C) \times P(O|I_0 \wedge I_1 \wedge S \wedge C) \end{aligned} \tag{4.21}$$

We see here a first example of the "art of decomposing" a joint distribution. The decomposition is a means to compute the joint distribution and, consequently, answer all possible questions. This decomposition has the following qualities:

- It is a better model than the basic joint distribution, because we add some useful knowledge through points 1 to 5 above.
- It is easier to compute than the basic joint distribution, because instead of working in a six-dimensional space, we will do the calculation in spaces of smaller dimension (see Chapter 5 for more on this).
- It is easier (or at least possible) to identify. The simplification of point 5 has been made for that purpose.

Figure 4.6 represents the same decomposition with a graphical model.

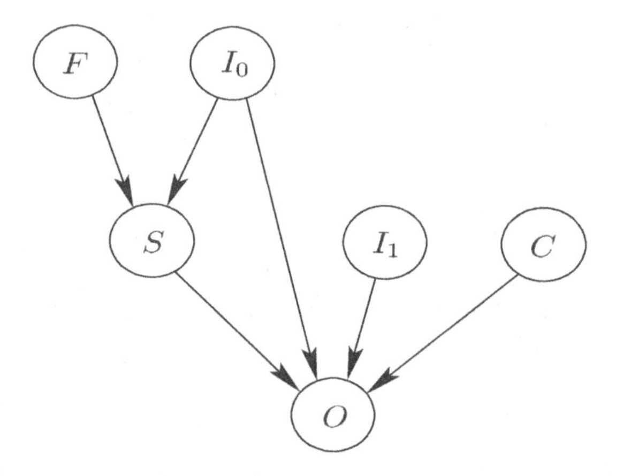

FIGURE 4.6: The graphical model of Expression 4.21: a water treatment unit.

4.2.1.3 Forms

To finish the specification task, we must still specify the parametric forms of the distribution appearing in the decomposition:

1. We have no a priori information on the entries I_0 and I_1:

$$P(I_0) \equiv \textbf{Uniform} \tag{4.22}$$
$$P(I_1) \equiv \textbf{Uniform} \tag{4.23}$$

2. Neither do we have any a priori information on F:

$$P(F) \equiv \textbf{Uniform} \tag{4.24}$$

3. We know an exact model of the sensor S:

$$P(S) = \delta_{\textbf{Int}\left(\frac{\mathbf{I_0+F}}{\mathbf{2}}\right)}(I_0 \wedge F) \tag{4.25}$$

where:

$$\delta_{\mathbf{Int}\left(\frac{\mathbf{I_0+F}}{\mathbf{2}}\right)} \tag{4.26}$$

is a Dirac distribution with probability one if and only if:

$$S = Int\left(\frac{I_0 + F}{2}\right) \tag{4.27}$$

4. Not knowing the desired output O, all possible controls are equally probable:

$$P(C) \equiv \mathbf{Uniform} \tag{4.28}$$

5. Finally, each of the 11^4 distributions $P(O \mid I_0 \wedge I_1 \wedge S \wedge C)$ (one for each possible value of $I_0 \wedge I_1 \wedge S \wedge C$) is defined as a histogram on the 11 possible values of O.

4.2.2 Identification

After the specification phase, we end up with $11^5 = 161,051$ free parameters to identify. To do this we will run the simulator described in Chapter 3, drawing at random with uniform distributions I_0, I_1, F, H, and C. For each of these draws (for instance, 10 of them), we compute the corresponding values of the sensor S and the output O. We then update the 11^4 histograms according to the values of I_0, I_1, S, C, and O.

4.2.3 Bayesian program

This may be summarized with the following Bayesian program:

$$
Pr\begin{cases}
Ds\begin{cases}
Sp(\pi)\begin{cases}
Va: \\
I_0, I_1, F, S, C, O \\
Dc: \\
\begin{cases}
& P(I_0 \wedge I_1 \wedge F \wedge S \wedge C \wedge O) \\
= & P(I_0) \times P(I_1) \times P(F) \times P(S|I_0 \wedge F) \\
& \times P(C) \times P(O|I_0 \wedge I_1 \wedge S \wedge C)
\end{cases} \\
Fo: \\
P(I_0) = Uniform \\
P(I_1) = Uniform \\
P(F) = Uniform \\
P(S|I_0 \wedge F) = \delta_{S=Int\left(\frac{I_0+F}{2}\right)} \\
P(C) = Uniform \\
P(O|I_0 \wedge I_1 \wedge S \wedge C) = Histograms
\end{cases} \\
Id
\end{cases} \\
Qu:
\end{cases}
\tag{4.29}
$$

4.2.4 Results

Such histograms have already been presented in previous chapters, for instance, in Figure 3.8 reproduced below as Figure 4.7 may be seen a collection of 11 of these histograms $P(O \mid I_0 \wedge I_1 \wedge S \wedge C)$ when $[I_0 = 2]$,$[I_1 = 8]$,$[C = 2]$, and S varies. The complete description of the elementary water treatment unit is made of 11^4 histograms, 11^3 times as much data as in Figure 4.7.

4.3 Lessons, comments, and notes

4.3.1 Description = Specification + Identification

Descriptions are the basic elements that are used, combined, composed, manipulated, computed, compiled, and questioned in different ways to build Bayesian programs.

A description is the probabilistic model of the observed phenomenon.

As such, it results from both the prior knowledge of the programmer about this phenomenon[6] and from the experimental data.

[6]We adopt here an unambiguous subjectivist epistemological position about probability. Explanation about the fundamental controversy opposing objectivism and subjectivism

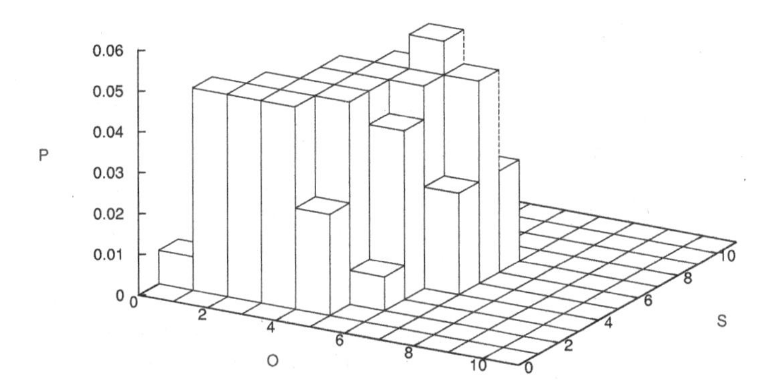

FIGURE 4.7: The 11 probability distributions of O when $S \in [0, \ldots, 10]$, the inputs and the control are set to $[I_0 = 2]$, $[I_1 = 8]$, $[C = 2]$, and the values of the external factor and the internal functioning $H \wedge F$ are not known.

The programmer's *preliminary* knowledge is expressed in a first phase called the specification phase (see Section 4.3.2 next).

The experimental data are taken into account during an identification (or learning) phase where the free parameters of the specification are given their values. The two robotics experiments described above (pushing and following contours) prove the importance of this identification phase. For a given specification but different experimental data, you can obtain completely different descriptions (i.e., different models of the phenomenon).

4.3.2 Specification = Variables + Decomposition + Forms

The preliminary knowledge of the programmer is always expressed the same way:

1. Choose the pertinent variables.

2. Decompose the joint distribution.

3. Define the parametric forms.

---

may be found in the FAQ/FAM: Objectivism vs. subjectivism controversy and the "mind projection fallacy," Section 16.13.

This strict and simple methodology and framework for the expression of the preliminary knowledge of the programmer present several fundamental advantages:

1. It is a programming baseline that guides any development of a Bayesian program.

2. It compels the programmer to express all available knowledge using this formalism, thus forcing a rigorous modeling process.

3. It warrants that no piece of knowledge stays implicit. Everything that is known about the phenomenon is expressed within this formalism. Nothing is hidden elsewhere in any other piece of code of the program.

4. It is a formal and unambiguous common language to describe models. It could be used very efficiently to discuss and compare different models.

5. As will be seen in the sequel, this formalism is generic and may be used to express a huge variety of models. Although simple, it offers a very strong power of expression.

It is important to note the huge difference between *preliminary* knowledge and *a priori* knowledge. The term *a priori* knowledge is usually restricted to the choice of the parameters of the distributions appearing in the description which are not learned. The term preliminary knowledge covers a much larger reality (choice of the variables, choice of the decomposition, and choice of the parametric forms) thus recognizing the complete role of the programmer in the modeling process.

4.3.3 Learning is a means to transform incompleteness into uncertainty

Descriptions are probabilistic models of a phenomenon. Descriptions are not complete. They do not escape the incompleteness curse of any nonprobabilistic model. For instance, the description of the elementary water treatment unit does not take into account the variable H, which stays hidden.

However, because of learning, the influence of H is nevertheless taken into account, as its effect on the phenomenon has been captured in the values of the histograms. Learning is a means to transform incompleteness (the effect of hidden variables) into uncertainty. The magic of this transformation is that after incompleteness has been transformed into uncertainty (probability distributions), then it is possible to reason with these distributions.

Furthermore, learning is also a means to estimate the importance of the hidden variable and consequently the quality of the model (description). If learning leads to flat distributions, it means that the neglected variables play a very important role and that the model should be improved. On the other hand, if learning leads to very informative (low entropy) distributions, then it confirms the quality of the model and the secondary influence of the hidden variables.

Chapter 5

The Importance of Conditional Independence

> What we call chance is, and may only be, the ignored cause of known effect[1]
>
> *Dictionaire Philosophique*
> Voltaire [1993–1764, 2005]

The goal of this chapter is both to explain the notion of conditional independence and to demonstrate its importance in actually solving and computing complex problems.

5.1 Water treatment center Bayesian model

In this chapter, we complete the construction of the Bayesian model for the water treatment center.

The complete process consists of four single units. Similarly, the complete Bayesian model is made of the four single models specified and identified in the previous chapter. Putting these four models together presupposes some strong structural knowledge that can be translated into conditional independence hypotheses.

Figure 5.1 presents the functioning diagram of the water treatment center.

[1]Ce que nous appelons le hasard n'est, et ne peut être, que la cause ignorée d'un effet connu.

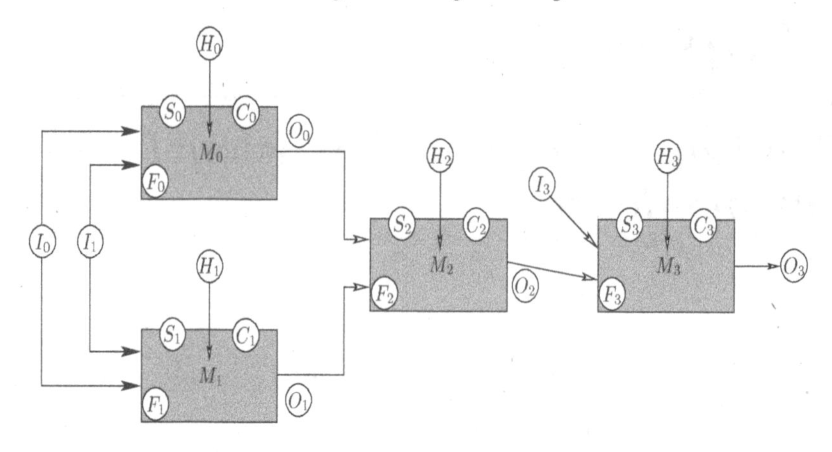

FIGURE 5.1: A complete water treatment center.

The units M_0 and M_1 take the same inputs I_0 and I_1. They respectively produce O_0 and O_1 as outputs, which in turn are used as inputs by M_2. M_3 takes I_3 and O_3 as inputs, and finally produces O_3 as output. The four water treatment units have four internal states (respectively F_0, F_1, F_2, and F_3), four sensors (respectively S_0, S_1, S_2, and S_3), four controllers (respectively C_0, C_1, C_2, and C_3), and may all be perturbed by some external factors (respectively H_0, H_1, H_2, and H_3). The production of each of these units is governed by Equations 3.3 and 3.4. The sensors take their values according to Equation 3.5.

5.2 Description of the water treatment center

As in the previous chapter, to build the Bayesian model of the whole plant, we assume that H_0, H_1, H_2, and H_3 are hidden variables not known by the designer.

5.2.1 Specification

5.2.1.1 Variables

There are now 19 variables in our global Bayesian model:

$$I_0, I_1, I_3, F_0, F_1, F_2, F_3, S_0, S_1, S_2, S_3, C_0, C_1, C_2, C_3, O_0, O_1, O_2, O_3 \in \{0, \ldots, 10\}$$

Each of these variables has a cardinality equal to 11.

5.2.1.2 Decomposition

Using the conjunction postulate (2.8) iteratively as in the previous chapter it is possible to write the joint probability on the 19 variables as:

$$
\begin{aligned}
& P(I_0 \wedge I_1 \wedge I_3 \wedge \ldots \wedge O_2 \wedge O_3) \\
= \; & P(I_0 \wedge I_1 \wedge I_3) \\
\times \; & P(F_0 \wedge S_0 \wedge C_0 \wedge O_0 \mid I_0 \wedge I_1 \wedge I_3) \\
\times \; & P(F_1 \wedge S_1 \wedge C_1 \wedge O_1 \mid I_0 \wedge I_1 \wedge I_3 \wedge F_0 \wedge S_0 \wedge C_0 \wedge O_0) \\
\times \; & P(F_2 \wedge S_2 \wedge C_2 \wedge O_2 \mid I_0 \wedge I_1 \wedge I_3 \wedge F_0 \wedge S_0 \wedge C_0 \wedge O_0 \wedge F_1 \wedge S_1 \wedge C_1 \wedge O_1) \\
\times \; & P(F_3 \wedge S_3 \wedge C_3 \wedge O_3 \mid I_0 \wedge I_1 \wedge I_3 \ldots \wedge S_1 \wedge C_1 \wedge O_1 \wedge F_2 \wedge S_2 \wedge C_2 \wedge O_2)
\end{aligned}
$$

This is an exact mathematical formula where we tried to regroup variables as they appeared in the four different units. Although it is exact, this formula should obviously be further simplified! This can be done by using some additional knowledge:

1. The operation of M_0 depends on its entries I_0 and I_1 but is independent of the water flow I_3:

$$
\begin{aligned}
& P(F_0 \wedge S_0 \wedge C_0 \wedge O_0 | I_0 \wedge I_1 \wedge I_3) \\
= \; & P(F_0 \wedge S_0 \wedge C_0 \wedge O_0 | I_0 \wedge I_1)
\end{aligned} \tag{5.1}
$$

2. The functioning of unit M_1 depends on its entries I_0 and I_1, but is obviously independent of entry I_3 and of the operation of unit M_0 (specified by variables F_0, S_0, C_0, and O_0). This leads to:

$$
\begin{aligned}
& P(F_1 \wedge S_1 \wedge C_1 \wedge O_1 | I_0 \wedge I_1 \wedge I_3 \wedge F_0 \wedge S_0 \wedge C_0 \wedge O_0) \\
= \; & P(F_1 \wedge S_1 \wedge C_1 \wedge O_1 | I_0 \wedge I_1)
\end{aligned} \tag{5.2}
$$

3. The operation of M_2 obviously depends on the operation of M_0 and M_1, which produce its inputs. For instance, changing C_0 will change O_0, which in turn will change S_2, O_2, and eventually C_2. Apparently, M_2 depends on all the previous variables except I_3, and the only obvious simplification concerns I_3:

$$
\begin{aligned}
& P(F_2 \wedge S_2 \wedge C_2 \wedge O_2 \mid I_0 \wedge I_1 \wedge I_3 \wedge F_0 \wedge S_0 \wedge C_0 \wedge O_0 \wedge F_1 \wedge S_1 \wedge C_1 \wedge O_1) \\
& \quad = P(F_2 \wedge S_2 \wedge C_2 \wedge O_2 \mid I_0 \wedge I_1 \wedge F_0 \wedge S_0 \wedge C_0 \wedge O_0 \wedge F_1 \wedge S_1 \wedge C_1 \wedge O_1)
\end{aligned}
$$

4. However, if we know the value of O_0, then we do not care anymore about the values of I_0, I_1, F_0, S_0, and C_0, which all influence the operation of M_2 only by means of the output O_0. Similarly, if we

know the value of O_1, then we do not care anymore about the values of F_1, S_1, and C_1. This is called *conditional independence* between variables and is a main tool to build interesting and efficient descriptions. One should be very careful that conditional independence has nothing in common with independence. The variable S_2 depends on C_0 ($P(S_2 \mid C_0) \neq P(S_2)$), but is conditionally independent of C_0 if O_0 is known ($P(S_2 \mid C_0 \wedge O_0) = P(S_2 \mid O_0)$).

See Section 5.3.1 in the sequel for further discussions of this point. This finally leads to:

$$\begin{aligned} & P(F_2 \wedge S_2 \wedge C_2 \wedge O_2 | I_0 \wedge I_1 \wedge I_3 \wedge F_0 \wedge S_0 \wedge \cdots \wedge C_1 \wedge O_1) \\ = \; & P(F_2 \wedge S_2 \wedge C_2 \wedge O_2 | O_0 \wedge O_1) \end{aligned} \tag{5.3}$$

5. For the same kind of reasons, we find:

$$\begin{aligned} & P(F_3 \wedge S_3 \wedge C_3 \wedge O_3 | I_0 \wedge I_1 \wedge I_3 \wedge F_0 \wedge S_0 \wedge \cdots \wedge C_2 \wedge O_2) \\ = \; & P(F_3 \wedge S_3 \wedge C_3 \wedge O_3 | I_3 \wedge O_2) \end{aligned} \tag{5.4}$$

At this stage, we have the following decomposition:

$$\begin{aligned} & P(I_0 \wedge I_1 \wedge I_3 \wedge \ldots \wedge O_2 \wedge O_3) \\ = \; & P(I_0 \wedge I_1 \wedge I_3) \\ \times \; & P(F_0 \wedge S_0 \wedge C_0 \wedge O_0 \mid I_0 \wedge I_1) \\ \times \; & P(F_1 \wedge S_1 \wedge C_1 \wedge O_1 \mid I_0 \wedge I_1) \\ \times \; & P(F_2 \wedge S_2 \wedge C_2 \wedge O_2 \mid O_0 \wedge O_1) \\ \times \; & P(F_3 \wedge S_3 \wedge C_3 \wedge O_3 \mid I_3 \wedge O_2) \end{aligned} \tag{5.5}$$

Using the same kind of assumptions as in the previous chapter, we may further simplify this expression, stating that:

1. Concerning M0:

$$\begin{aligned} & P(F_0 \wedge S_0 \wedge C_0 \wedge O_0 \mid I_0 \wedge I_1) \\ = \; & P(F_0) \\ \times \; & P(S_0 \mid I_0 \wedge F_0) \\ \times \; & P(C_0) \\ \times \; & P(O_0 \mid I_0 \wedge I_1 \wedge S_0 \wedge C_0) \end{aligned} \tag{5.6}$$

2. Concerning M1:

$$
\begin{aligned}
&P(F_1 \wedge S_1 \wedge C_1 \wedge O_1 \mid I_0 \wedge I_1) \\
&= P(F_1) \\
&\times P(S_1 \mid I_0 \wedge F_1) \\
&\times P(C_1) \\
&\times P(O_1 \mid I_0 \wedge I_1 \wedge S_1 \wedge C_1)
\end{aligned} \tag{5.7}
$$

3. Concerning M2:

$$
\begin{aligned}
&P(F_2 \wedge S_2 \wedge C_2 \wedge O_2 \mid O_0 \wedge O_1) \\
&= P(F_2) \\
&\times P(S_2 \mid O_0 \wedge F_2) \\
&\times P(C_2) \\
&\times P(O_2 \mid O_0 \wedge O_1 \wedge S_2 \wedge C_2)
\end{aligned} \tag{5.8}
$$

4. Concerning M3:

$$
\begin{aligned}
&P(F_3 \wedge S_3 \wedge C_3 \wedge O_3 \mid I_3 \wedge O_2) \\
&= P(F_3) \\
&\times P(S_3 \mid I_3 \wedge F_3) \\
&\times P(C_3) \\
&\times P(O_3 \mid I_3 \wedge O_2 \wedge S_3 \wedge C_3)
\end{aligned} \tag{5.9}
$$

After some reordering, we obtain the following final decomposition and the associated graphical model (Figure 5.2):

$$
\begin{aligned}
&P(I_0 \wedge I_1 \wedge I_3 \wedge \ldots \wedge O_2 \wedge O_3) \\
&= P(I_0) \times P(I_1) \times P(I_3) \\
&\times P(F_0) \times P(F_1) \times P(F_2) \times P(F_3) \\
&\times P(C_0) \times P(C_1) \times P(C_2) \times P(C_3) \\
&\times P(S_0 \mid I_0 \wedge F_0) \times P(S_1 \mid I_0 \wedge F_1) \times P(S_2 \mid O_0 \wedge F_2) \times P(S_3 \mid I_3 \wedge F_3) \\
&\times P(O_0 \mid I_0 \wedge I_1 \wedge S_0 \wedge C_0) \times P(O_1 \mid I_0 \wedge I_1 \wedge S_1 \wedge C_1) \\
&\times P(O_2 \mid O_0 \wedge O_1 \wedge S_2 \wedge C_2) \times P(O_3 \mid I_3 \wedge O_2 \wedge S_3 \wedge C_3)
\end{aligned} \tag{5.10}
$$

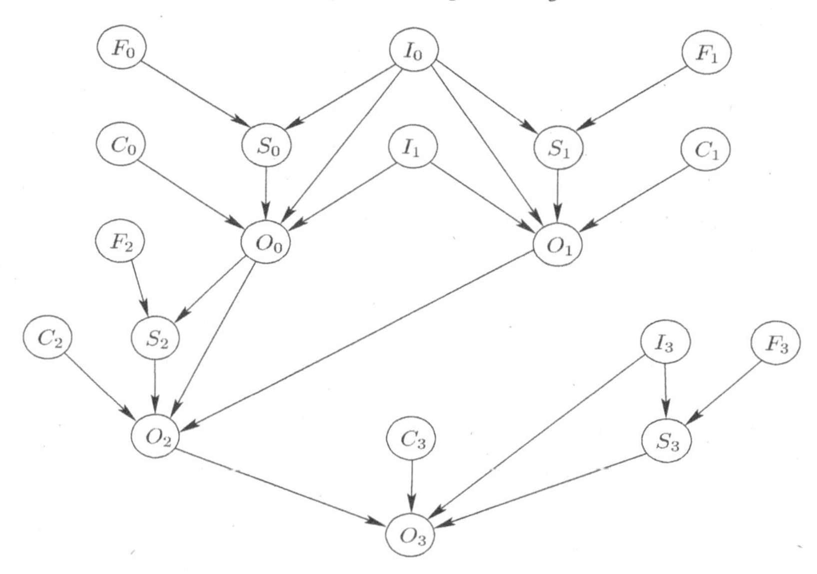

FIGURE 5.2: The graphical model of the decomposition of the joint distribution as defined by Equation 5.10.

5.2.1.3 Forms

The distributions $P(I_0)$, $P(I_1)$, $P(I_3)$, $P(F_0)$, $P(F_1)$, $P(F_2)$, $P(F_3)$, $P(C_0)$, $P(C_1)$, $P(C_2)$, and $P(C_3)$ are all assumed to be uniform distributions.

$P(S_0 \mid I_0 \wedge F_0)$, $P(S_0 \mid I_0 \wedge F_1)$, $P(S_0 \mid O_0 \wedge F_2)$, and $P(S_3 \mid I_3 \wedge F_3)$ are all Dirac distributions.

Finally, the four distributions relating the output to the inputs, the sensor, and the control are all specified as histograms as in the previous chapter.

5.2.2 Identification

We have now four series of 11^4 histograms to identify (4×11^5 free parameters).

In a real control and diagnosis problem, the four production units, even though they are identical, would most probably function slightly differently (because of incompleteness and some other hidden variables besides H). In that case, the best thing to do would be to perform four different identification campaigns to take these small differences into account by learning.

In this didactic example, as the four units are simulated and formal, they really are perfectly identical (there are no possible hidden variables but H in our formal model as specified by Equations 3.3, 3.4, and 3.5). Consequently, we use the exact same histogram for the four units as was identified in the previous chapter.

5.2.3 Bayesian program

All this may be summarized by the following Bayesian program:

$$
Pr\begin{cases}
Ds\begin{cases}
Sp(\pi)\begin{cases}
Va: I_0, I_1, I_3,, F_0, \dots C_3, O_0, O_1, O_2, O_3 \\
Dc:\begin{cases}
P(I_0 \wedge I_1 \wedge I_3 \wedge \dots \wedge O_2 \wedge O_3) \\
= P(I_0) \times P(I_1) \times P(I_3) \\
\times P(F_0) \times P(F_1) \times P(F_2) \times P(F_3) \\
\times P(C_0) \times P(C_1) \times P(C_2) \times P(C_3) \\
\times P(S_0 \mid I_0 \wedge F_0) \times P(S_1 \mid I_0 \wedge F_1) \\
\times P(S_2 \mid O_0 \wedge F_2) \times P(S_3 \mid I_3 \wedge F_3) \\
\times P(O_0 \mid I_0 \wedge I_1 \wedge S_0 \wedge C_0) \\
\times P(O_1 \mid I_0 \wedge I_1 \wedge S_1 \wedge C_1) \\
\times P(O_2 \mid O_0 \wedge O_1 \wedge S_2 \wedge C_2) \\
\times P(O_3 \mid I_3 \wedge O_2 \wedge S_3 \wedge C_3)
\end{cases} \\
Fo:\begin{cases}
P(I_0) \dots P(C_3) \equiv \textbf{Uniform} \\
P(S_0 \mid I_0 \wedge F_0), \dots, P(S_3 \mid I_3 \wedge F_3) \equiv \textbf{Dirac} \\
P(O_0 \mid I_0 \wedge I_1 \wedge S_0 \wedge C_0) \equiv \textbf{Histogram} \\
\dots \\
P(O_3 \mid I_3 \wedge 0_2 \wedge S_3 \wedge C_3) \equiv \textbf{Histogram}
\end{cases}
\end{cases} \\
Identification
\end{cases} \\
Qu :?
\end{cases}
\tag{5.11}
$$

5.3 Lessons, comments, and notes

5.3.1 Independence versus conditional independence

To build this Bayesian model of our water treatment center we intensively used independence between variables and overall conditional independence between variables. Let us return briefly to these two concepts, which should not be confused.

Two variables X and Y are *independent* from one another if and only if $P(X \wedge Y) = P(X) \times P(Y)$ which is logically equivalent because of the conjunction postulate to $P(X \mid Y) = P(X)$ and of course also to $P(Y \mid X) = P(Y)$.

Two variables X and Y are *independent conditionally* to a third one, Z, if and only if $P(X \wedge Y \mid Z) = P(X \mid Z) \times P(Y \mid Z)$, which is also logically equivalent because of the conjunction postulate to $P(X \mid Y \wedge Z) = P(X \mid Z)$ and to $P(Y \mid X \wedge Z) = P(Y \mid Z)$.

However, two variables may be independent but not conditionally independent to a third one. Two variables may also be conditionally independent but not independent.

For the first case, for production unit M_0, the first entry I_0, and the internal state F_0, are independent but are not conditionally independent knowing S_0 the reading of the sensor. Figure 5.3 shows, for instance, the corresponding probabilities for $I_0 = 5$ and $S_0 = 6$.

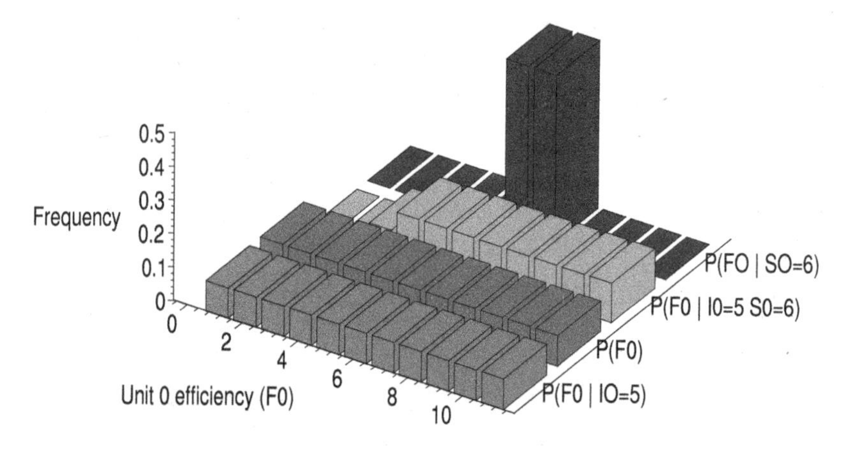

FIGURE 5.3: The distributions $P(F_0 \mid I_0)$, $P(F_0)$, $P(F_0 \mid I_0 \wedge S_0)$, and $P(F_0 \mid S_0)$ are represented. Note that $P(F_0 \mid I_0) = P(F_0)$ and $P(F_0 \mid I_0 \wedge S_0) \neq P(F_0 \mid S_0)$.

This is a very common situation where the two causes of a phenomenon are independent but are conditionally dependent on one another, knowing their common consequence (see Figure 5.4). Otherwise, no sensor measuring several factors would be of any use.

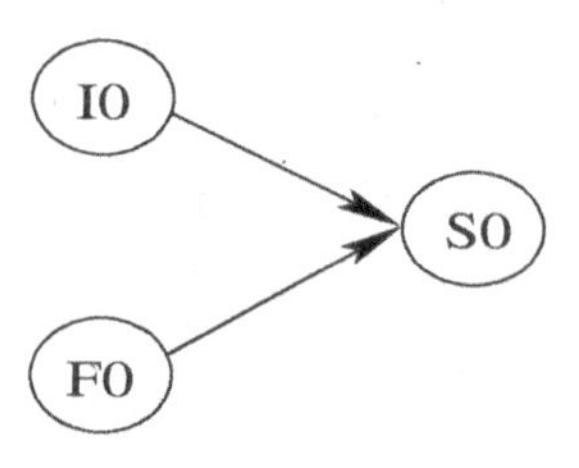

FIGURE 5.4: Two independent causes for the same phenomenon.

For the second case, as already stated, S_2 depends on C_0 $(P(S_2 \mid C_0) \neq$

$P(S_2)$) but is conditionally independent of C_0 if O_0 is known ($P(S_2 \mid C_0 \wedge O_0) = P(S_2 \mid O_0)$).

Figure 5.5 shows this for $C_0 = 10$ and $O_0 = 5$.

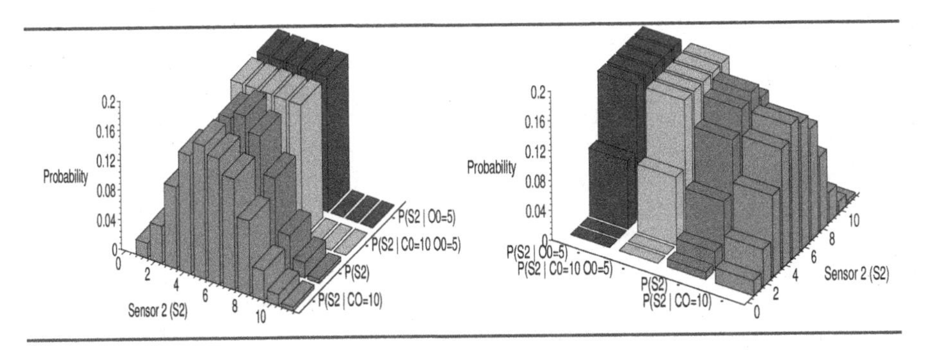

FIGURE 5.5: Two views of distributions $P(S_2 \mid C_0)$, $P(S_2)$, $P(S_2 \mid C_0 \wedge O_0)$, and $P(S_2 \mid S_0)$. Note that $P(S_2 \mid O_0) \neq P(S_2)$ and $P(S_2 \mid C_0 \wedge O_0) = P(S_2 \mid O_0)$.

This is also a very common situation where there is a causal chain between three variables (see Figure 5.6).

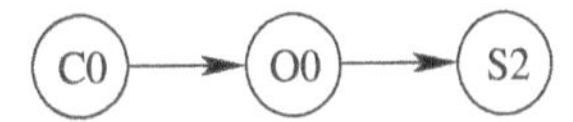

FIGURE 5.6: Causal chain.

5.3.2 The importance of conditional independence

This kind of conditional independence is of great importance for Bayesian programming for two main reasons:

- It expresses crucial designer's knowledge.
- It makes the computation tractable by breaking the curse of dimensionality.

For any model of any phenomenon, knowing what matters with, what does not influence what, and, most importantly, what bias could be neglected compared to another one is fundamental knowledge.

A model where everything depends on everything else is a very poor model, indeed. In probabilistic terms, such a model would be a joint distribution on all the relevant variables coded as a huge table containing the probabilities of all the possible cases. In our example, simple as it is, it would be a table containing the 2^{63} probability values necessary for the joint distribution

$P(I_0 \wedge I_1 \wedge I_3 \wedge \ldots \wedge O_2 \wedge O_3)$ on our 19 variables that take 11 values each:

$$11^{19} \approx 2^{63}$$

Such a table would encode all the necessary information, but in a very poor manner. Hopefully, a model does not usually code the joint distribution in this way but rather uses a decomposition and the associated conditional independencies to express the knowledge in a structured and formal way. The probabilistic model of the production models as expressed by Equation 5.10 only requires 2^{18} probability values to encode the joint distribution:

$$(11 \times 11) + (11^3 \times 4) + (11^5 \times 4) \approx 2^{18}$$

Thanks to conditional independence the curse of dimensionality has been broken! What has been shown to be true here for the required memory space is also true for the complexity of inferences. Conditional independence is the principal tool to keep the calculation tractable. Tractability of Bayesian inference computation is of course a major concern as it has been proved NP-hard [Cooper, 1990]. This subject will be developed in Chapter 14, which reviews the main algorithms for Bayesian inference and is summed up in the FAQ/FAM: "Computation complexity of Bayesian inference," Section 16.7.

Chapter 6

Bayesian Program = Description + Question

> Far better an approximate answer to the *right* question which is often vague, than an *exact* answer to the wrong question which can always be made precise.[1]
>
> *The Future of Data Analysis*
> John W. Tukey [1962]

In the two previous chapters, we built a description (Bayesian model) of our water treatment center. In this chapter, we use this description to solve different problems: prediction of the output, choice of the best control strategy, and diagnosis of failures. This shows that multiple questions may be asked about the same description to solve very different problems. This

[1] Emphasized text is in the original citation.

clear separation between the model and its use is a very important feature of Bayesian Programming.

6.1 Water treatment center Bayesian model (end)

Let us first recall the present model of the process:

$$
Pr\begin{cases}
Ds\begin{cases}
Sp(\pi)\begin{cases}
Va: I_0, I_1, I_3, , F_0, \ldots C_3, O_0, O_1, O_2, O_3 \\
Dc:\begin{cases}
P(I_0 \wedge I_1 \wedge I_3 \wedge \ldots \wedge O_2 \wedge O_3) \\
= P(I_0) \times P(I_1) \times P(I_3) \\
\times P(F_0) \times P(F_1) \times P(F_2) \times P(F_3) \\
\times P(C_0) \times P(C_1) \times P(C_2) \times P(C_3) \\
\times P(S_0 \mid I_0 \wedge F_0) \times P(S_1 \mid I_0 \wedge F_1) \\
\times P(S_2 \mid 0_0 \wedge F_2) \times P(S_3 \mid I_3 \wedge F_3) \\
\times P(O_0 \mid I_0 \wedge I_1 \wedge S_0 \wedge C_0) \\
\times P(O_1 \mid I_0 \wedge I_1 \wedge S_1 \wedge C_1) \\
\times P(O_2 \mid 0_0 \wedge 0_1 \wedge S_2 \wedge C_2) \\
\times P(O_3 \mid I_3 \wedge 0_2 \wedge S_3 \wedge C_3)
\end{cases} \\
Fo:\begin{cases}
P(I_0) \ldots P(C_3) \equiv \textbf{Uniform} \\
P(S_0 \mid I_0 \wedge F_0), \ldots, P(S_3 \mid I_3 \wedge F_3) \equiv \textbf{Dirac} \\
P(O_0 \mid I_0 \wedge I_1 \wedge S_0 \wedge C_0) \equiv \textbf{Histogram} \\
\ldots \\
P(O_3 \mid I_3 \wedge 0_2 \wedge S_3 \wedge C_3) \equiv \textbf{Histogram}
\end{cases}
\end{cases} \\
Identification
\end{cases} \\
Qu: ?
\end{cases}
\tag{6.1}
$$

Quite a simple model indeed, thanks to our knowledge of this process, even if it took some time to make all its subtleties explicit! However, the question is still unspecified. Specifying and answering different possible questions is the purpose of the sequel of this chapter.

6.2 Forward simulation of a single unit

We first want to predict the output of a single unit (for instance M_0) knowing its inputs, sensor reading, and control, or some of these values.

6.2.1 Question

For instance, we may look for the value of O_0 knowing the values of I_0, I_1, S_0, and C_0. The corresponding question will be:

$$P(O_0 \mid I_0 \wedge I_1 \wedge S_0 \wedge C_0)$$

This distribution is directly available since it is given to specify the description. This could also be inferred using the simple algorithm formulae 2.34.

Let's define $Free$, $Searched$, and $Known$ as:

$$\begin{array}{l} Free = I_3 \wedge F_0 \wedge F_1 \wedge F_2 \wedge F_3 \wedge S_1 \wedge S_2 \wedge S_3 \wedge C_1 \wedge C2 \wedge C_3 \wedge O_1 \wedge O_2 \wedge O_3 \\ Known = I_0 \wedge I_1 \wedge S_0 \wedge C_0 \\ Searched = O_0 \end{array} \tag{6.2}$$

In this particular case $P(Search \mid Known)$ appears in the decomposition and the previous Bayesian program may be rewritten as follows.

$$Pr \begin{cases} Ds \begin{cases} Sp(\pi) \begin{cases} Va: Free, Known, Searched \\ Dc: \begin{cases} P(Free \wedge Searched \wedge Known) = \\ P(Free \wedge Known) \times P(Search \mid Known) \end{cases} \\ Fo: \begin{cases} P(Free \wedge Known) \textbf{ Any Distribution} \\ P(Searched \mid Known) \textbf{ Any Distribution} \end{cases} \end{cases} \\ Identification \end{cases} \\ Qu: P(Searched \mid known) \end{cases} \tag{6.3}$$

According to Equation 2.34:

$$P(Searched \mid known) = \frac{\sum_{Free} P(Free \wedge Searched \wedge known)}{\sum_{Free} \sum_{Searched} P(Free \wedge Searched \wedge known)}$$

This can be developed using the specification of $P(Free \wedge Searched \wedge Known)$ into:

$$P(Searched \mid known) = \frac{\sum_{Free} [P(Free \wedge known) \times P(Search \mid known)]}{\sum_{Free} \sum_{Searched} [P(Free \wedge known) \times P(Search \mid known)]}$$

using the marginalization rule $\sum_{Free} P(Free \wedge known) = P(known)$:

$$
\begin{aligned}
&P(Searched \mid known) \\
&= \frac{P(known) \times P(Search \mid known)}{\sum_{Free} [P(Free \wedge known)] \sum_{Searched} [P(Search \mid known)]}
\end{aligned}
$$

Using again the marginalization rule $\sum_{Free} P(Free \wedge known) = P(known)$ and the normalization rule $\sum_{Searched} P(Search \mid known) = 1$ we obtain:

$$
P(Searched \mid known) = P(Searched \mid [Known = known])
$$

Which could have been obtained directly since it appears in the decomposition.

6.2.2 Results

The corresponding results have already been presented in Chapter 4 for $[I_0 = 2]$, $[I_1 = 8]$, $[C_0 = 2]$, and for all the values of S_0 (see Figure 4.7).

6.3 Forward simulation of the water treatment center

6.3.1 Question

We may now want to predict the output of the whole water treatment center, knowing the three inputs I_0, I_1, and I_3, the four sensor readings S_0, S_1, S_2, and S_3, and the four control values C_0, C_1, C_2, and C_3.

The corresponding question is:

$$
P(O_3 \mid I_0 \wedge I_1 \wedge I_3 \wedge S_0 \wedge S_1 \wedge S_3 \wedge S_4 \wedge C_0 \wedge C_1 \wedge C_2 \wedge C_3) \tag{6.4}
$$

It can be computed using the inference algorithms described in Chapter 14 or by invoking an inference engine as in the example below. However, it is always useful to have a feeling of the complexity of a given inference. One way is to apply the formulae 2.34 with:

$$
\begin{aligned}
&Searched = \{O_3\} \\
&Known = \{I_0, I_1, I_3, S_0, S_1, S_2, S_3, S_4, C_0, C_1, C_2, C_3\} \\
&Free = \{F_0, F_1, F_2, F_3, O_0, O_1, O_2\}
\end{aligned}
\tag{6.5}
$$

$$P(Searched \mid known) = \frac{\sum_{Free} P(Searched \wedge known \wedge Free)}{P(known)}$$

As the distributions for the entries, the sensors, and the controllers are uniform distributions $P(known)$ is a constant for any value $\{i_0, i_1, i_3, s_0, s_1, s_2, s_3, s_4, c_0, c_1, c_2, c_3\}$ of the $Known$ variables.

$$P(O_3 \mid i_0 \wedge i_1 \wedge i_3 \wedge s_0 \wedge s_1 \wedge s_2 \wedge s_3 \wedge s_4 \wedge c_0 \wedge c_1 \wedge c_2 \wedge c_3) =$$

$$\frac{1}{Z} \times \sum_{F_0,F_1,F_2,F_3,O_0,O_1,O_2} \left[\begin{array}{l} P(i_0) \times P(i_1) \times P(i_3) \\ \times P(F_0) \times P(F_1) \times P(F_2) \times P(F_3) \\ \times P(c_0) \times P(c_1) \times P(c_2) \times P(c_3) \\ \times P(s_0 \mid i_0 \wedge F_0) \times P(s_1 \mid i_0 \wedge F_1) \\ \times P(s_2 \mid O_0 \wedge F_2) \times P(s_3 \mid i_3 \wedge F_3) \\ \times P(O_0 \mid i_0 \wedge i_1 \wedge s_0 \wedge c_0) \\ \times P(O_1 \mid i_0 \wedge i_1 \wedge s_1 \wedge c_1) \\ \times P(O_2 \mid O_0 \wedge O_1 \wedge s_2 \wedge c_2) \\ \times P(O_3 \mid i_3 \wedge O_2 \wedge s_3 \wedge c_3) \end{array} \right]$$

Uniform distributions can be further simplified by incorporating them in the normalization constant to yield:

$$P(O_3 \mid i_0 \wedge i_1 \wedge i_3 \wedge s_0 \wedge s_1 \wedge s_2 \wedge s_3 \wedge s_4 \wedge c_0 \wedge c_1 \wedge c_2 \wedge c_3) =$$

$$\frac{1}{Z} \times \sum_{F_0,F_1,F_2,F_3,O_0,O_1,O_2} \left[\begin{array}{l} \times P(s_0 \mid i_0 \wedge F_0) \times P(s_1 \mid i_0 \wedge F_1) \\ \times P(s_2 \mid O_0 \wedge F_2) \times P(s_3 \mid i_3 \wedge F_3) \\ \times P(O_0 \mid i_0 \wedge i_1 \wedge s_0 \wedge c_0) \\ \times P(O_1 \mid i_0 \wedge i_1 \wedge s_1 \wedge c_1) \\ \times P(O_2 \mid O_0 \wedge O_1 \wedge s_2 \wedge c_2) \\ \times P(O_3 \mid i_3 \wedge O_2 \wedge s_3 \wedge c_3) \end{array} \right]$$

Finally, after some reordering (see Chapter 14) of the sums to minimize the amount of computation, we obtain:

$$P(O_3 \mid i_0 \wedge i_1 \wedge i_3 \wedge s_0 \wedge s_1 \wedge s_2 \wedge s_3 \wedge s_4 \wedge c_0 \wedge c_1 \wedge c_2 \wedge c_3) =$$
$$\frac{1}{Z} \times \sum_{F_0} P(s_0 \mid i_0 \wedge F_0) \times \sum_{F_1} P(s_0 \mid i_0 \wedge F_1) \times \sum_{F_3} P(s_3 \mid i_3 \wedge F_3) \times$$
$$\sum_{O_0} \left[\begin{array}{l} P(O_0 \mid i_0 \wedge i_1 \wedge s_0 \wedge c_0) \times \\ \sum_{F_2} P(s_2 \mid O_0 \wedge F_2) \times \\ \sum_{O_1} \left[\begin{array}{l} P(O_1 \mid i_0 \wedge i_1 \wedge s_1 \wedge c_1) \times \\ \sum_{O_2} \left[\begin{array}{l} P(O_2 \mid O_0 \wedge O_1 \wedge s_2 \wedge c_2) \times \\ P(O_3 \mid i_3 \wedge O_2 \wedge s_3 \wedge c_3) \end{array} \right] \end{array} \right] \end{array} \right]$$

6.3.2 Results

The file "chapter6/treatmentcenter.py" contains the model of the water treatment center. The conditional probability distributions on O_0, O_1, O_2, O_3 are set using a file "calibration.txt" containing the 11^4 distributions. This file has been generated using the program found in "chapter6/calibration.py" which contains a model of a water treatment unit with the hidden variable H to generate uncertainties.

It is possible to obtain a symbolic representation of the joint distribution with:

```
print model
```

As well as a symbolic representation of the simplification made by the interpreter to compute the forward model.

$$P(O_3 \mid i_0 \wedge i_1 \wedge i_3 \wedge s_0 \wedge s_1 \wedge s_2 \wedge s_3 \wedge s_4 \wedge c_0 \wedge c_1 \wedge c_2 \wedge c_3) \tag{6.6}$$

```
print model.ask(O[3],I0^I1^I3^C^S)
```

The inference engine may provide other and extra simplifications:

```
Sum_{O1} {P(O1|I(0) I(1) S1 C1)
          Sum_{O0} {P(O0|I(0) I(1) S0 C0)
                    Sum_{F2} {P(F2)P(S2|O0 F2) }
                    sum_{O2} {P(O2|O0 O1 S2 C2)
                              P(O3|I(3) O2 S3 C3)
                              }
                    }
          }
```

Some of the results obtained for the forward simulation of the water treatment center are presented in Figure 6.1. For the same three inputs ($i_0 = 1, i_1 =$

$8, i_3 = 10$), with the same four sensor readings ($s_0 = 5, s_1 = 5, s_2 = 7, s_3 = 9$) but with three different kinds of control we obtain completely different forecasts for the output. These three curves show clearly that with these specific inputs and readings, an average control ($c_0 = c_1 = c_2 = c_3 = 5$) is more efficient than no control or an overstated control.

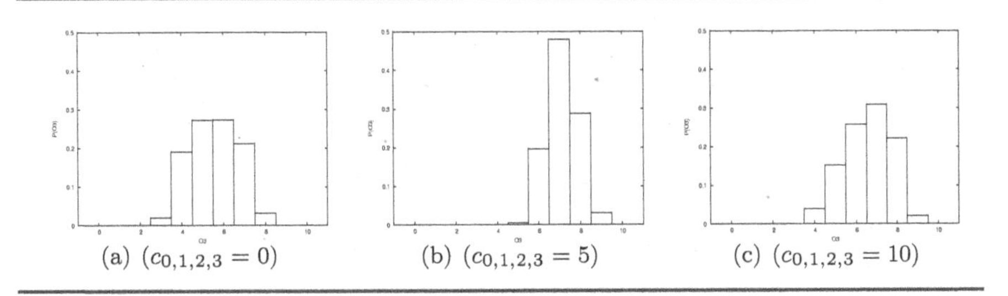

(a) ($c_{0,1,2,3} = 0$) (b) ($c_{0,1,2,3} = 5$) (c) ($c_{0,1,2,3} = 10$)

FIGURE 6.1: Direct distributions for the output O_3 for three different controls.

6.4 Control of the water treatment center

Forecasting the output O_3, as in the previous section, is certainly a valuable tool to choose an appropriate control policy for the water treatment center. However, there are $11^4 = 14,641$ such possible different policies. Testing all of them, one after another, would not be very practical.

6.4.1 Question (1)

Bayesian Programming endeavors to offer a direct solution to the control problem when asked the question:

$$P(C_0 \wedge C_1 \wedge C_2 \wedge C_3 \mid i_0 \wedge i_1 \wedge i_3 \wedge s_0 \wedge s_1 \wedge s_2 \wedge s_3 \wedge s_4 \wedge o_3) \qquad (6.7)$$

6.4.2 Results (1)

If we fix the objective O_3 to nine given $i_0 = 2, i_1 = 8, i_3 = 10, s_0 = 5, s_1 = 5, s_2 = 7, s_3 = 9$ the most probable values of the control are $c_0 = 5, c_1 = 5, c_2 = 4, c_3 = 4$.

However, many controls have the same or almost the same probability. This means that the water treatment process is robust and that the sensitivity to the control is low.

The forecast for this control choice is presented in Figure 6.2. It shows that even if this is the best control choice, the probability of obtaining $o_3 = 9$ is only 3% when the probability of obtaining $o_3 = 6$ is around 25%.

This suggests that searching the controls for a given value of o_3 may still not be the best question. Indeed this question could lead to a control choice that ensures the highest probability for $o_3 = 9$, but at the price of very high probabilities for much worse outputs.

The same model defined in file "chapter6/treatmentcenter.py" may be used to compute the optimal value C of

$$P(C_0 \wedge C_1 \wedge C_2 \wedge C_3 \mid i_0 \wedge i_1 \wedge i_3 \wedge s_0 \wedge s_1 \wedge s_2 \wedge s_3 \wedge s_4 \wedge o_3)$$

with $i_0 = 2, i_1 = 8, i_3 = 10, s_0 = 5, s_1 = 5, s_2 = 7, s_3 = 9$, and $o_3 = 9$

```
question1=model.ask(C,I0^I1^I3^S^O[3])
resultD=question1.instantiate([2,8,10,5,5,7,9,9])
```

The optimization can be done with:

```
val_opt = resultD.best()
```

In this case the system chose to use a genetic algorithm to perform optimization. An exact solution may be obtained with:

```
compiled_resultD= resultD.compile()
compiled_val_opt =  compiled_resultD.best()
```

The following commands print several probability values for different values of C reaching the maximum.

```
print compiled_resultD.compute(compiled_val_opt)
print compiled_resultD.compute(val_opt)
print compiled_resultD.compute([6,6,4,9])
```

6.4.3 Question (2)

We might rather try to keep O_3 superior to a minimum value $VALMIN$ ($O_3 \geq VALMIN$). That is, we need to compute controls C which leads as much as we can to a water quality better than $VALMIN$. One way is to introduce a constraint within the initial model by defining two new variables $H \in \{0, 1\}$ and $VALMIN$, as well as a new Dirac distribution:

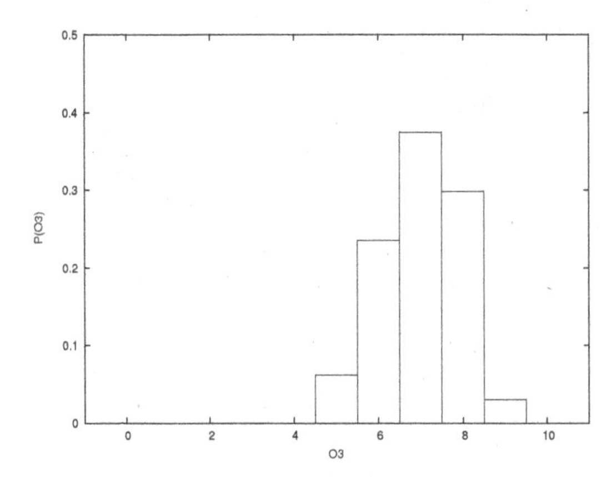

FIGURE 6.2: $P(O_3 \mid i_0 \wedge i_1 \wedge i_3 \wedge s_0 \wedge s_1 \wedge s_2 \wedge s_3 \wedge s_4 \wedge c_0 \wedge c_1 \wedge c_3)$ with $i_0 = 2, i_1 = 8, i_3 = 10, s_0 = 5, s_1 = 5, s_2 = 7, s_3 = 9$, and $c_0 = 5, c_1 = 5, c_2 = 4, c_3 = 4$.

$$
Pr \begin{cases} Ds \begin{cases} Sp(\pi) \begin{cases} Va: I_0, \ldots O_3, H, VALMIN \\ Dc: \begin{cases} P(I_0 \wedge I_1 \wedge I_3 \wedge \ldots \wedge O_2 \wedge O_3) \\ = P(VALMIN) \times P(H) \\ \times P(I_0) \times P(I_1) \times P(I_3) \\ \times P(F_0) \times P(F_1) \times P(F_2) \times P(F_3) \\ \times P(C_0) \times P(C_1) \times P(C_2) \times P(C_3) \\ \times P(S_0 \mid I_0 \wedge F_0) \times P(S_1 \mid I_0 \wedge F_1) \\ \times P(S_2 \mid O_0 \wedge F_2) \times P(S_3 \mid I_3 \wedge F_3) \\ \times P(O_0 \mid I_0 \wedge I_1 \wedge S_0 \wedge C_0) \\ \times P(O_1 \mid I_0 \wedge I_1 \wedge S_1 \wedge C_1) \\ \times P(O_2 \mid O_0 \wedge O_1 \wedge S_2 \wedge C_2) \\ \times P(O_3 \mid I_3 \wedge O_2 \wedge S_3 \wedge C_3) \end{cases} \\ Fo: \begin{cases} P(VALMIN)\ \textbf{Uniform} \\ P(I_0) \ldots P(C_3) \equiv \textbf{Uniform} \\ P(S_0 \mid I_0 \wedge F_0), \ldots, P(S_3 \mid I_3 \wedge F_3) \equiv \textbf{Dirac} \\ P(O_0 \mid I_0 \wedge I_1 \wedge S_0 \wedge C_0) \equiv \textbf{Histogram} \\ \ldots \\ P(O_3 \mid I_3 \wedge O_2 \wedge S_3 \wedge C_3) \equiv \textbf{Histogram} \\ P(H \mid VALMIN \wedge O_3) \equiv \textbf{Dirac:} \\ \quad if(O_3 \geq VALMIN) : H = 1 else H = 0 \end{cases} \end{cases} \\ Identification \end{cases} \\ Qu: P(C_{0,1,2,3} \mid i_{0,1,3} \wedge s_{0,1,2,3} \wedge h \wedge valmin) \end{cases}
$$

6.4.4 Results (2)

Let us, for example, analyze the cases where $VALMIN \in \{5, 6, 7\}$.

For example, $P(C_{0,1,2,3} \mid i_{0,1,3} \wedge s_{0,1,2,3} \wedge h = 1 \wedge valmin = 5)$ will provide a distribution on controls $C_{0,1,2,3}$ which maximizes the chances for O_3 to be greater than five.

Figure 6.3 presents, for the three considered values of $VALMIN$, the probability distribution

$$P\left(O_3 \mid i_{0,1,3} = 2, 8, 10 \wedge s_{0,1,2,3} = 5, 5, 7, 9 \wedge c_{0,1,2,3} = c^i_{0,1,2,3}\right) \tag{6.8}$$

for the most probable obtained controls:

$$c^i_{0,1,2,3} = \max_{c \in C_{O,1,2,3}} P\left(C_{0,1,2,3} \mid i_{0,1,3} \wedge s_{0,1,2,3} \wedge h = 1 \wedge valmin = i\right) \tag{6.9}$$

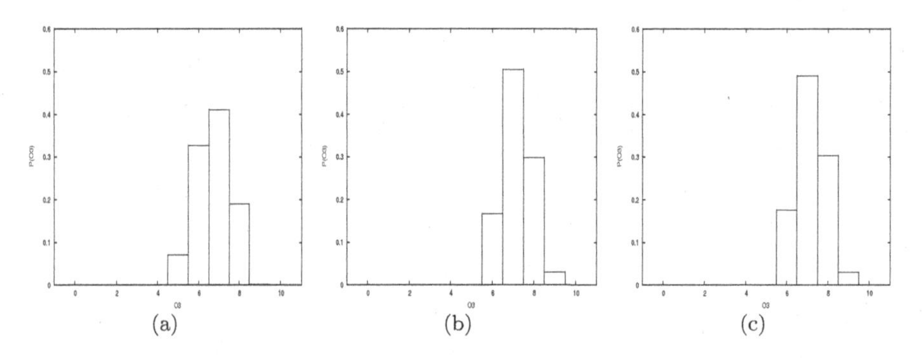

(a) (b) (c)

FIGURE 6.3: Distributions on O_3 obtained with different controls maximizing $P(C_{0,1,2,3} \mid i_{0,1,3} = 2, 8, 10 \wedge s_{0,1,2,3} = 5, 5, 7, 9 \wedge h = 1 \wedge valmin = i)$:
(a): $VALMIN = 5$ which implies control: $c^*_0 = 5, c^*_1 = 0, c^*_2 = 0, c^*_3 = 5$
(b): $VALMIN = 6$ which implies control: $c^*_0 = 5, c^*_1 = 5, c^*_2 = 5, c^*_3 = 6$
(c): $VALMIN = 7$ which implies control: $c^*_0 = 5, c^*_1 = 5, c^*_2 = 4, c^*_3 = 7$

Note that for $VALMIN = 7$ (Figure 6.3c) the probability that $O_3 = 6$ is not 0. Indeed, the question asked here is "Find the best control to maximize the probability for the output O_3 to be more than $VALMIN$?" It does not impose that O_3 should absolutely be more than $VALMIN$.

Yet another question could be "Find the best control that warrants that the output will not be less than $VALMIN$?"

The file "chapter6/treatmentcenter.py" contains the code used to draw Figure 6.3. The constraint H is defined with a Python function:

```
def constraint(O_,I_):
        if I_[O[3]].to_float() >= I_[VALMIN].to_float():
            O_[H]=1
        else:
            O_[H]=0
```

The corresponding Dirac distribution and a uniform distribution on $VALMIN$ are the decomposition initially defined for the program 6.1.

```
NewJointDistributionList=JointDistributionList
NewJointDistributionList.push_back(plUniform(VALMIN))
constraintmodel=plPythonExternalFunction(H, \
                                        VALMIN^O[3], \
                                        constraint)
NewJointDistributionList.push_back(\
            plFunctionalDirac(H,VALMIN^O[3],constraintmodel))
```

For $VALMIN = 5$ the $C*$ is obtained with:

```
newquestion1=newmodel.ask(C,I0^I1^I3^S^H^VALMIN)
newresultD=newquestion1.instantiate([2,8,10,5,5,7,9,1,5])
compiled_newresultD= newresultD.compile()
new_opt_val = compiled_newresultD.best()
```

Finally we may combine the readings with the desired control and use the initial model to compute the distribution on O_3:

```
opt_known_val = known_val^new_opt_val
newresultA = question.instantiate(opt_known_val)
```

6.5 Diagnosis

We may also use our Bayesian model to diagnose failures. Let us suppose that the output is only seven. This means that at least one of the four units is in poor working condition. We want to identify these defective units so we can fix them.

6.5.1 Question

The question is: "What is going wrong?" We must look for the values of F_0, F_1, F_2, and F_3 knowing the entries, the sensor values, the control, and the final output. The corresponding question is:

$$P(F_0 \wedge F_1 \wedge F_2 \wedge F_3 \mid i_0 \wedge i_1 \wedge i_3 \wedge s_0 \wedge s_1 \wedge s_2 \wedge s_3 \wedge c_0 \wedge c_1 \wedge c_2 \wedge c_3 \wedge o_3)$$

6.5.2 Results

For $O_3 = 7$, the usual entries $I_0 = 2, I_1 = 8, I_3 = 10$, sensor readings equal to $S_0 = 5, S_1 = 5, S_2 = 4, S_3 = 9$, and controls all set to 5. We obtain a maximum of probabilities for $F_0 = 8, F_1 = 8, F_2 = 3, F_3 = 8$.

If we consider the values of F_0, F_1, F_2, F_3 making the 90% quantile we obtain values for F_2 below 5 with an average of 3 with the values of F_0, F_1, F_3 always above 8, indicating a greater chance of failure for unit 2.

We can confirm this hypothesis by asking the question (see Figure 6.4):

$$P(F_2 \mid i_0 \wedge i_1 \wedge i_3 \wedge s_0 \wedge s_1 \wedge s_2 \wedge s_3 \wedge c_0 \wedge c_1 \wedge c_2 \wedge c_3 \wedge o_3)$$

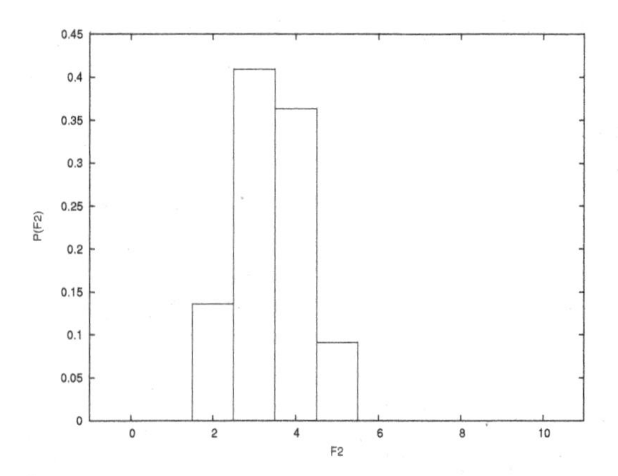

FIGURE 6.4: $P(F_2 \mid i_0 \wedge i_1 \wedge i_3 \wedge s_0 \wedge s_1 \wedge s_2 \wedge s_3 \wedge s_4 \wedge c_0 \wedge c_1 \wedge c_3 \wedge o_3)$ with $i_0 = 2, i_1 = 8, i_3 = 10, s_0 = 5, s_1 = 5, s_2 = 7, s_3 = 9$, and $c_0 = 5, c_1 = 5, c_2 = 5, c_3 = 5, o_3 = 7$.

In the file "chapter6/treatmentcenter.py," Python is used to sort the values of F according to their probabilities.

```
#Diagnosis
#question
diagnosisquestion = model.ask(F,I0^I1^I3^S^C^O[3])
# for a given value of the Known variables
resultdiagnosis = diagnosisquestion.instantiate( \
                         [2,8,10,5,5,4,9,5,5,5,5,7])
#allow faster access
compiled_resultdiagnosis = resultdiagnosis.compile()
vF = plValues(F)
indexed_proba_table = []
#One way to store the result in a list (probal,
#                                       value1,
#                                       value2,....)
for v in vF:
    indexed_proba_table.append(\
                  [compiled_resultdiagnosis.compute(v)] \
                  + [vF[F[x]] for x in range(4)])
#Using Python to sort this list
sorted_indexed_prob_table = sorted(indexed_proba_table, \
                             key=lambda el: el[0], \
                             reverse = True)
```

6.6 Lessons, comments, and notes

6.6.1 Bayesian Program = Description + Question

A Bayesian program consists of two parts: a *description*, which is the probabilistic model of the observed phenomenon, and a *question* used to interrogate this model.

Any partition of the set of relevant variables in three subsets: the set of searched variables (which should not be empty), the set of known variables, and the complementary set of free variables, defines a valid question.

If we call $Searched$ the conjunction of variables of the searched set, $Known$ the conjunction of variables with known values, and $Free$ the conjunction of variables in the complementary set, a question corresponds to the bundle of distributions:

$$P(Search \mid Known) \tag{6.10}$$

One distribution may be obtained by setting the values of the $Known$ variables a to particular value $known_i$

$$P(Search \mid known_i) \tag{6.11}$$

This may be summarized by:

"Bayesian Program = Description + Question."

6.6.2 The essence of Bayesian inference

The essence of Bayesian inference is to be able to compute:

$$P(Search \mid known)$$

This computation is always made by applying the same steps:

1. Utilization of the marginalization rule (2.12):

$$P(Search \mid known) = \sum_{Free} P(Search \wedge Free \mid known) \tag{6.12}$$

2. Utilization of the conjunction rule (2.8):

$$P(Search \mid known) = \sum_{Free} \frac{P(Search \wedge Free \wedge known)}{P(known)} \tag{6.13}$$

3. Replacement of the denominator by a normalization constant:

$$P(Search \mid Known) = \frac{1}{Z} \times \sum_{Free} P(Search \wedge Free \wedge known) \tag{6.14}$$

As the *decomposition* gives us a means to compute the joint distribution $P(Search \wedge Free \wedge Known)$ as a product of simpler distributions, we are always able to compute the wanted distribution.

However, the number of possible values for the variable *Searched* may be huge. Exhaustive and explicit computation of $P(Search \wedge Free \wedge Known)$ is then impossible even for a single value of $Known = known_i$. It must be either approximated, sampled, or used to find the most probable values. In any of these three cases, the regions of interest in the searched space are the areas of high probability, which most often cover a tiny portion of the whole space. The fundamental problem is to find them, which is a very difficult optimization problem, indeed.

Worse, for each single value $search_j$ of *Search*, we need to sum on *Free* (see Equation 6.14) to compute $P(search_j \mid known_i)$. The number of possible values for *Free* may also be huge and the integration problem itself may be a heavy computational burden. An approximation of the sum must be made, which means that it must effectively be computed on a sample of the whole space. Where should we sample? Obviously, to find a good estimation of the sum we should sample in areas where the probability of $P(search_j \wedge Free \wedge known_i)$ is high. This is yet another optimization problem where we must search the high-probability areas of a distribution.

All the algorithms of Bayesian inference try to deal with these two overlapped optimization problems using various methods. A survey of these algorithms will be presented in Chapter 14.

6.6.3 No inverse or direct problem

When using a functional model defined as:

$$Y = \mathbf{F}(X)$$

computing Y knowing X is a direct problem. A search for X knowing Y is an inverse problem.

Most of the time, the direct problem is easy to solve, as we know the function. Often the inverse one is very difficult because $\mathbf{F}^{-1}$, the inverse function, is at best unknown and even, sometimes, nonanalytic. Unfortunately, most of the time, the inverse problem is much more interesting than the direct one.

For instance, if $\mathbf{F}$ is the function that predicts the performance of a racing yacht, based on a knowledge of her characteristics, what is really of interest is to find the characteristics that ensure the best performance.

When using a probabilistic model defined as:

$$P(X \wedge Y)$$

the difference between direct and inverse problems vanishes.

In the joint distribution, indeed, all the variables play the exact same mathematical role. Whatever the decomposition, any of the variables can in turn, in different questions, be considered as either searched, known, or free. There is no difference in the nature of the computation to calculate either $P(Y \mid X)$ or $P(X \mid Y)$. However, even if any question can be asked of the probabilistic model, we stressed in the previous section that some of them can be time and resource consuming to answer.

There may be no more direct or inverse problems, but there are still some difficult problems characterized by either a huge integration space ($Free$), or a huge search space ($Searched$), or both.

6.6.4 No ill-posed problem

When using functional models, inverse problems, even when they are solvable, may be ill-posed in the sense that $X = \mathbf{F}^{-1}(Y)$ may have several solutions. This is a very common situation in control problems. For instance, if you are given some values of the control variables, you can predict the outputs exactly, but if you are given the goal, you have numerous control solutions to achieve it.

In these cases, functional models do not have enough information to give you any hint to help you choose between the different values of X satisfying $X = \mathbf{F}^{-1}(Y)$.

In contrast, $P(X \mid Y)$ gives much more information, because it allows you to find the relative probabilities of the different solutions.

Part II

Bayesian Programming Cookbook

Chapter 7

Information Fusion

Based on your pupil dilation, skin temperature, and motor functions, I calculate an 83% probability that you will not pull the trigger.

Terminator 3: Rise of the Machines

7.1 "Naive" Bayes sensor fusion

7.1.1 Statement of the problem

The most common application of Bayesian techniques is the fusion of information coming from different sensors to estimate the state of a given phenomenon.

The situation is always the same: you want information about a given phenomenon, this phenomenon influences sensors that you can read, and from these readings you try to estimate the phenomenon.

Usually the readings are neither completely informative about the phenomenon, nor completely consistent with one another. Consequently, you are compelled to a probabilistic approach and the question you want to answer is:

$$P(S|r_1 \wedge \ldots \wedge r_N \wedge \pi) \tag{7.1}$$

where S is the state and $\{r_1, \ldots, r_N\}$ the readings.

A very common difficulty is the profusion of sensors which leads to a very high dimensionality state space for the joint distribution $P(S \wedge R_1 \wedge \ldots \wedge R_N|\pi)$.

A very common solution to break this curse of dimensionality is to make the very strong assumption that knowing the phenomenon the sensors may be considered to provide independent readings. Knowing the common cause, the different consequences are considered independent:

$$P(S \wedge R_1 \wedge \ldots \wedge R_N|\pi) = P(S|\pi) \times \prod_{n=1}^{N} [P(R_n|S \wedge \pi)] \tag{7.2}$$

7.1.2 Bayesian program

The corresponding Bayesian program is the following:

$$Pr \begin{cases} Ds \begin{cases} Sp(\pi) \begin{cases} Va: \\ S, R_1, \ldots, R_N \\ Dc: \\ \begin{cases} & P(S \wedge R_1 \wedge \ldots \wedge R_N | \pi) \\ = & P(S|\pi) \times \prod_{n=1}^{N} [P(R_n | S \wedge \pi)] \end{cases} \\ Fo: \\ any \end{cases} \\ Id \end{cases} \\ Qu: \\ P(S | r_1 \wedge \ldots \wedge r_N \wedge \pi) \end{cases} \tag{7.3}$$

$P(S|r_1 \wedge \ldots \wedge r_N \wedge \pi)$ can be computed very efficiently as it is simply equal to:

$$P(S|r_1 \wedge \ldots \wedge r_N \wedge \pi) = \frac{1}{Z} \times P(S|\pi) \times \prod_{n=1}^{N} [P(r_n|S \wedge \pi)] \tag{7.4}$$

where Z is a normalization constant.

The major interest of this decomposition is that it dispels the curse of complexity. Indeed, when you need potentially $card(R)^N \times card(S)$ parameters to store $P(S \wedge R_1 \wedge \ldots \wedge R_N|\pi)$, you need only $(card(R) - 1) \times N \times card(S)$ parameters to store $P(S|\pi) \times \prod_{n=1}^{N} [P(R_n|S \wedge \pi)]$.

Furthermore, $P(R_n|S \wedge \pi)$ is usually much easier to learn than $P(S|r_1 \wedge \ldots \wedge r_N \wedge \pi)$ and needs much less data.

Finally, another interesting property of this decomposition is that if one (or several) of the observations is missing it just vanishes from the computation:

$$\begin{aligned} & P(S|r_1 \wedge \ldots \wedge r_{N-1} \wedge \pi) \\ = & \frac{1}{Z} \times \sum_{R_N} \left[P(S|\pi) \times \prod_{n=1}^{N} [P(R_n|S \wedge \pi)] \right] \\ = & \frac{1}{Z} \times P(S|\pi) \times \prod_{n=1}^{N-1} [P(r_n|S \wedge \pi)] \times \sum_{R_N} [P(R_N|S \wedge \pi)] \\ = & \frac{1}{Z} \times P(S|\pi) \times \prod_{n=1}^{N-1} [P(r_n|S \wedge \pi)] \end{aligned} \tag{7.5}$$

7.1.3 Instance and results

There are very numerous applications of this principle.

We saw a first instance in the second chapter of this book, where the phenomenon was spam, and the sensors the presence or absence of words in the analyzed text.

Credit card fraud detection is another example, where the phenomenon is fraud and the sensors are about a hundred characteristics of the transaction such as, for instance, the amount, the hour, the activity of the seller, the country where the transaction took place, or the number of transactions with this card in the previous 24 hours.

Localization relative to landmarks is yet another, where the phenomenon is the position of the observer, and the observations, the measures of distances, and the bearings[1] of known landmarks. Let us develop a simple version of this last example in some detail for illustration.

Before the GPS era, landmarks were used for locations by doing triangulation either with optical (for instance, bearing compass), electromagnetic (for instance, radio-goniometry or radar landmarks) or, even, sound information (for instance, using bells mounted on buoys or lighthouses when low visibility weather conditions prevented the use of optical devices). These different devices provided bearings of landmarks and possibly distances with various precision.

Let us take an example inspired from this situation (see Figure 7.1).

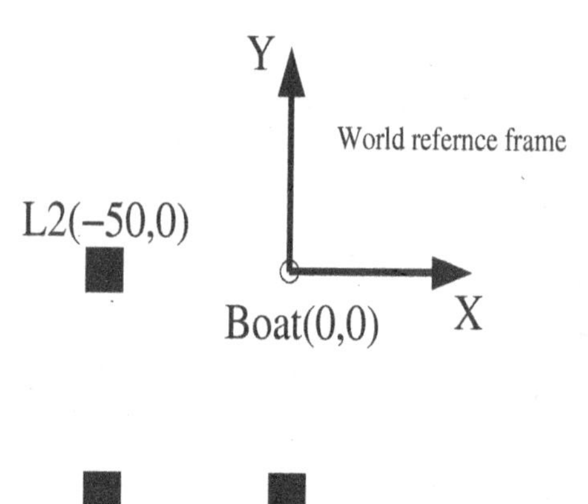

FIGURE 7.1: In this section, the boat is located at $X = Y = 0$ and the three landmarks are located at $(-50, -50)$, $(-50, 0)$, and $(0, -50)$.

In this instance, $S \equiv X \wedge Y$, where X and Y are spatial coordinates, both

[1]In the sequel of this text we use "bearing" for "compass bearing" or "azimuth": the angle between the direction of a target and north counted in degrees in a clockwise direction.

in the range $\{-50, 50\}$. $R_1 \equiv D_1$, $R_2 \equiv D_2$, $R_3 \equiv D_3$ (in the range $\{0, 141\}$) are the measured distances to three known and identifiable landmarks located at coordinates $(-50, -50)$, $(-50, 0)$, and $(0, -50)$. $R_4 \equiv B_1$, $R_5 \equiv B_2$, and $R_6 \equiv B_3$ are the measured bearings in degrees of these three same landmarks, the three of them in the range $\{0, 359\}$.[2]

The Bayesian program used is:

$$
Pr\begin{cases}
Ds\begin{cases}
Sp(\pi)\begin{cases}
Va: \\
X, Y, D_1, D_2, D_3, B_1, B_2, B_3 \\
Dc: \\
\begin{cases}
P(X \wedge Y \wedge \ldots \wedge B_3|\pi) \\
= P(X \wedge Y|\pi) \times \prod_{n=1}^{3}\begin{bmatrix} & P(D_n|X \wedge Y \wedge \pi) \\ \times & P(B_n|X \wedge Y \wedge \pi)\end{bmatrix}
\end{cases} \\
Fo: \\
\begin{cases}
P(X \wedge Y|\pi) = Uniform \\
P(D_n|X \wedge Y \wedge \pi) = \\
B\left(\left[\mu = f_d^n(X, Y)\right], \left[\sigma = g_d^n(X, Y)\right]\right) \\
P(B_n|X \wedge Y \wedge \pi) = \\
B\left(\left[\mu = f_b^n(X, Y)\right], \left[\sigma = 10\right]\right)
\end{cases}
\end{cases} \\
Id
\end{cases} \\
Qu: \\
P(X \wedge Y|d_1 \wedge d_2 \wedge d_3 \wedge b_1 \wedge b_2 \wedge b_3 \wedge \pi)
\end{cases}
\tag{7.6}
$$

The central hypothesis is that knowing the position of the observer, the distances and the bearings of the three landmarks are independent from one another.

The measured distances are supposed to be distributed according to normal laws with the readings as means, and with varying standard deviations increasing with distance. However, as distances can take only discrete values we use "bell-shaped" distributions (denoted $B(\mu, \sigma)$) which are approximations of continuous Gaussians for discrete variables.

The measured bearings are supposed to be distributed according to normal laws with the readings as means, and with a fixed standard deviation of 10°.

The specification includes several conditional distributions which are defined with functions: f_d^n, g_d^n, f_b^n. These functions are the sensor models used to model the sensor readings and their uncertainties according to the location of the boat. For example $f_d^1(X, Y)$ computes the distance to the first landmark knowing the location of the boat:

$$f_d^1(X, Y) = \sqrt{(X + 50)^2 + (Y + 50)^2}$$

[2]These angles are in degrees, and do not follow the navigation conventions: north = 90 and bearings are measured counterclockwise.

We assume the uncertainty gets bigger as the boat is further away from the landmark and becomes:

$$g_d^1(X,Y) = \frac{f_d^1(X,Y)}{10} + \alpha$$

where α is some minimal uncertainty on the reading of the distance. Similarly we define the bearing knowing the location of the boat:

$$f_b^1 = \arctan \frac{Y-50}{X-50}$$

The above Bayesian program is implemented in "chapter7/fusion.py". In these programs, sensor models are python functions. For example, to define $P(D_1|X \wedge Y \wedge \pi) = B\left(\left[\mu = f_d^1(X,Y)\right)\right], \left[\sigma = g_d^1(X,Y)\right)\right]\right)$ we first describe the sensor model with two functions.

```
def f_d_1(Output_,Input_):
    Output_[0] =  hypot( Input_[X]+50.0,Input_[Y]+50.0)

def g_d_1(Output_,Input_):
    Output_[0]= hypot( Input_[X]+50.0,Input_[Y]+50.0)/10.0 + 5
```

These functions are used to define a conditional probability distribution on D1:

```
plCndNormal(D1,X^Y, \
    plPythonExternalFunction(X^Y,f_d_1), \
    plPythonExternalFunction(X^Y,g_d_1)))
```

This distribution is added to the list (JointDistributionList) of all the distributions defining the specification. The joint distribution is then defined as:

```
localisation_model=plJointDistribution(X^Y^D1^D2^D3^B1^B2^B3,\
                                        JointDistributionList)
```

We can ask a lot of different questions to this description as, for instance, for a supposed observer in position $(0,0)$:

1. The distribution on the position knowing the three distances and the three bearings (see Figure 7.2a):

$$P(X \wedge Y|d_1 \wedge d_2 \wedge d_3 \wedge b_1 \wedge b_2 \wedge b_3 \wedge \pi)$$

2. The distribution on the position knowing only the bearings of the three landmarks (the most common situation in our boat navigation scenario — see Figure 7.2b):

$$P(X \wedge Y|b_1 \wedge b_2 \wedge b_3 \wedge \pi)$$

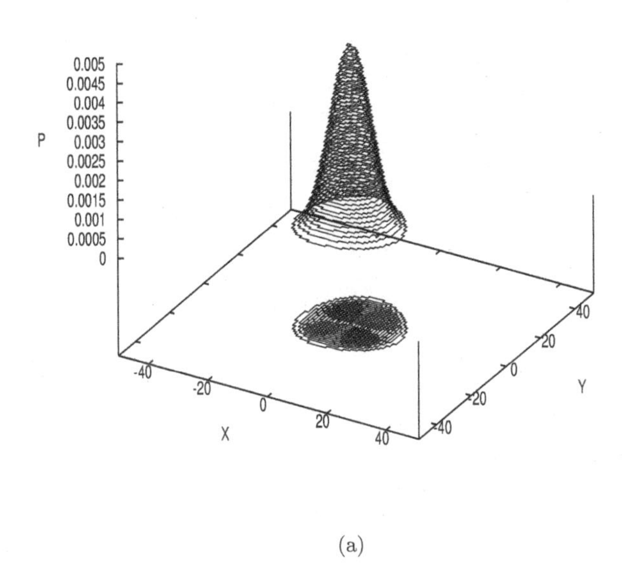

(a)

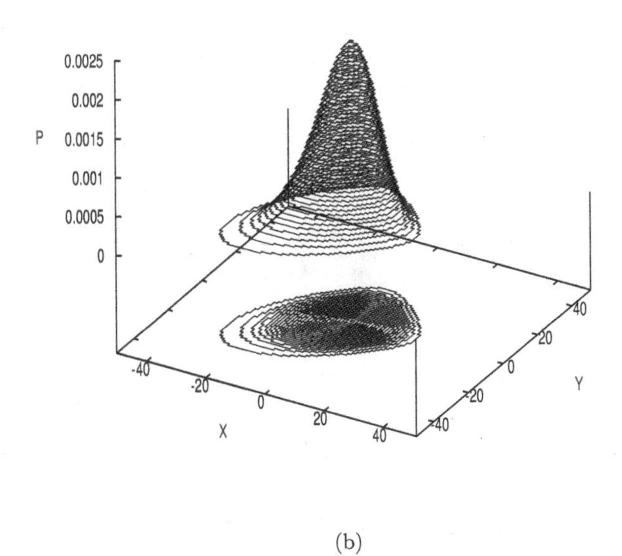

(b)

FIGURE 7.2: The probability of being at a given location given the readings: (a): $P(X \wedge Y|d_1 = 70 \wedge d_2 = 50 \wedge d_3 = 50 \wedge b_1 = 225 \wedge b_2 = 180 \wedge b_3 = 270 \wedge \pi)$; (b): $P(X \wedge Y|b_1 = 225 \wedge b_2 = 180 \wedge b_3 = 270 \wedge \pi)$.

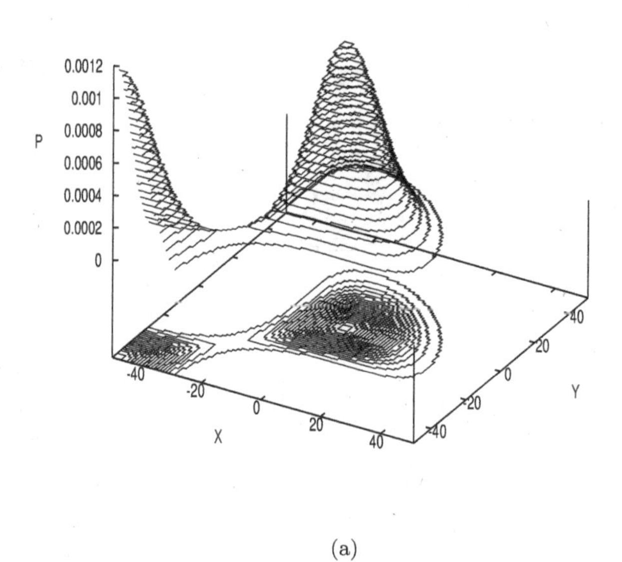

(a)

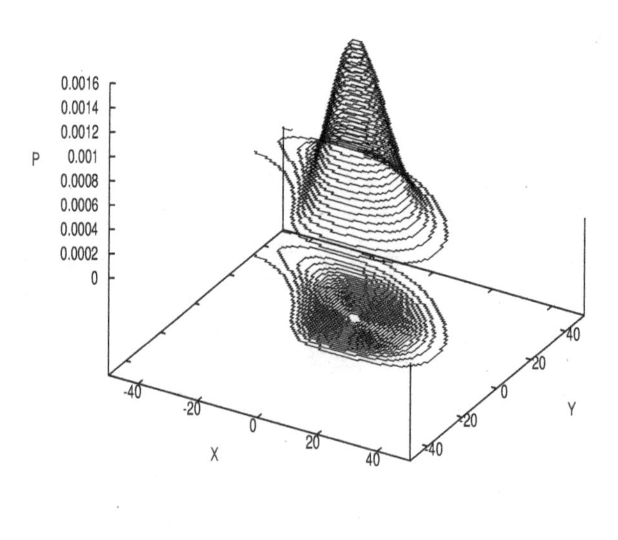

(b)

FIGURE 7.3: The probability of being at a given location given the readings:
(a): $P(X \wedge Y | d_2 = 50 \wedge d_3 = 50 \wedge \pi)$
(b): $P(X \wedge Y | d_1 = 70 \wedge d_2 = 50 \wedge d_3 = 70 \wedge \pi)$

3. The distribution on the position knowing only the distance of the landmarks (L_2 and L_3) (see Figure 7.3a).

$$P(X \wedge Y | d_2 \wedge d_3 \wedge \pi)$$

4. A question with the three distances but with a wrong value ($d_3 = 70$) for the reading d_3 (see Figure 7.3b).

Finally, a different kind of question where we look for the distribution on the bearing of the third landmark knowing the distances and bearings to the two first ones (see Figure 7.4):

$$P(B_3 | b_1 \wedge b_2 \wedge d_1 \wedge d_2 \wedge \pi)$$

The result of this question could be either used to search for the third landmark if you do not know which direction to look for it (most probable direction 270) or, even, to detect a potential problem with your measure of B_3 if it is not coherent with the other readings (i.e., it has a very low probability, for instance for $b_3 = 150$).

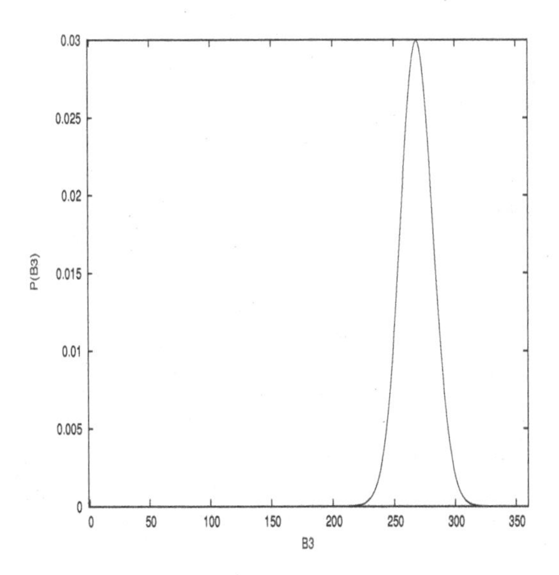

FIGURE 7.4: The probability distribution on the location of the third landmark knowing $b_1 = 225$, $b_2 = 180$, $d_1 = 70$, $d_2 = 50$.

The questions 1, 3, and 4 are defined as follows in the file "chapter7/fusion.py":

```
PXY_K_D1D2D3B1B2B3=localisation_model.ask(X^Y,D1^D2^D3^B1^B2^B3)
PXY_K_D2D3=localisation_model.ask(X^Y,D2^D3)
PXY_K_D1D2D3=localisation_model.ask(X^Y,D1^D2^D3)
```

All the questions use the same description "localization_model".

7.2 Relaxing the conditional independence fundamental hypothesis

7.2.1 Statement of the problem

The "naive" sensor fusion model assumes the complete conditional independence of the sensors knowing the phenomenon. It is obviously a very strong hypothesis.

This hypothesis is always wrong, or rather never totally true, as there are always in the physical world some factors (hidden variables) other than the observed phenomenon that correlate the readings of the sensors.

Deciding whether these correlations may be neglected is a choice of modeling usually made based on two contradictory criteria: (i) Does this simplification lead to results of a sufficient quality for the problem we are dealing with? (ii) Does it lead to a tractable model in terms of computing time and resources?

For instance, in the spam example of the second chapter, it is obvious that knowing that a text is spam or not is far from sufficient to explain all the correlations between the words appearing in this text. But even with a model assuming such a wrong hypothesis, as the obtained results are satisfying and as this hypothesis leads to very fast computation, the "naive" model can be used satisfactorily.

However, in some cases the "naive" hypothesis is clearly not sufficient. In such cases the obvious solution is to complicate the model by adding more variables explaining these correlations, as will be shown in the sequel. Yet, a first possibility to take into account these correlations does not suppose to introduce new variables. It consists in relaxing the "naive" hypothesis assuming, not the conditional independence of single variables, but of tuples of variables knowing the phenomenon:

$$P(S \wedge R_1 \wedge \ldots \wedge R_N|\pi) = P(S|\pi) \times \prod_{k=1}^{K} \left[P\left(R_k^1 \wedge \ldots \wedge R_k^{M_k}|S \wedge \pi\right)\right] \quad (7.7)$$

with $\sum_{k=1}^{K} [M_k] = N$.

7.2.2 Bayesian program

The corresponding Bayesian program is the following:

$$Pr \begin{cases} Ds \begin{cases} Sp(\pi) \begin{cases} Va: \\ S, R_1, \ldots, R_N \\ Dc: \\ \begin{cases} & P(S \wedge R_1 \wedge \ldots \wedge R_N | \pi) \\ = & P(S|\pi) \times \prod_{k=1}^{K} \left[P\left(R_k^1 \wedge \ldots \wedge R_k^{M_k} | S \wedge \pi\right)\right] \end{cases} \\ Fo: \\ any \end{cases} \\ Id \end{cases} \\ Qu: \\ P(S|r_1 \wedge \ldots \wedge r_N \wedge \pi) \end{cases} \tag{7.8}$$

$P(S|r_1 \wedge \ldots \wedge r_N \wedge \pi)$ can still be computed very efficiently as it is simply equal to:

$$P(S|r_1 \wedge \ldots \wedge r_N \wedge \pi) = \frac{1}{Z} \times P(S|\pi) \times \prod_{k=1}^{K} \left[P\left(r_k^1 \wedge \ldots \wedge r_k^{M_k} | S \wedge \pi\right)\right] \tag{7.9}$$

However, if the tuples are too large, storing the distributions $P\left(R_k^1 \wedge \ldots \wedge R_k^{M_k} | S \wedge \pi\right)$ may very quickly either become very costly in terms of memory or complicated if a parametric representation is used.

7.2.3 Instance and results

Coming back to our previous example, if we are using radar, it may be difficult to assume that the reading of distance and bearing are independent. Indeed, the propagation of radar waves is organized in "lobes" that introduce dependencies between them.

These dependencies may be difficult to model but the joint distributions $P(D_n \wedge B_n | X \wedge Y \wedge \pi)$ could eventually be learned by experience.

To keep things simple for didactic purposes, let say that:

$$P(D_n \wedge B_n | X \wedge Y \wedge \pi) = P(D_n | X \wedge Y \wedge \pi) \times P(B_n | D_n \wedge X \wedge Y \wedge \pi) \tag{7.10}$$

We can express in this way that the measure of the bearing depends on the distance of the landmark. For instance, by having a decreasing function $g_b^n(D_n)$ we may say that the further the landmark the more precise the measure of the bearing:

$$P(B_n|D_n \wedge X \wedge Y \wedge \pi) = B\left([\mu = f_b^n(X,Y)], [\sigma = g_b^n(D_n)]\right)$$

For this example, we choose to set the standard deviation to vary from a minimum of 5 to a maximum of 20:

$$g_b^n = \max(5, (20 - \frac{500}{10 + d_n})$$

This leads to the following Bayesian program:

$$
Pr\left\{\begin{array}{l}
Ds\left\{\begin{array}{l}
Sp(\pi)\left\{\begin{array}{l}
Va:\\
X,Y,D_1,D_2,D_3,B_1,B_2,B_3\\
Dc:\\
\left\{\begin{array}{ll}
& P(X\wedge Y\wedge\ldots\wedge B_3|\pi)\\
= & P(X\wedge Y|\pi)\times\prod_{n=1}^{3}\left[\begin{array}{c} P(D_n|X\wedge Y\wedge\pi)\\ \times \\ P(B_n|D_n\wedge X\wedge Y\wedge\pi)\end{array}\right]
\end{array}\right.\\
Fo:\\
\left\{\begin{array}{ll}
& P(X\wedge Y|\pi)=Uniform\\
& P(D_n|X\wedge Y\wedge\pi)\\
= & B([\mu=f_d^n(X,Y)],[\sigma=g_d^n(X,Y)])\\
& P(B_n|D_n\wedge X\wedge Y\wedge\pi)\\
= & B([\mu=f_b^n(X,Y)\,b_n],[\sigma=g_b^n(D_n)])
\end{array}\right.
\end{array}\right.\\
Id
\end{array}\right.\\
Qu:\\
P(X\wedge Y|b_1\wedge b_2\wedge b_3\wedge\pi)
\end{array}\right.
\tag{7.11}
$$

We may use this description to recompute the distribution on the location of the boat using only the bearing readings:

$$P(X \wedge Y|b_1 \wedge b_2 \wedge b_3 \wedge \pi)$$

As expected, the result is quite different (see Figure 7.5, to be compared with Figure 7.2b).

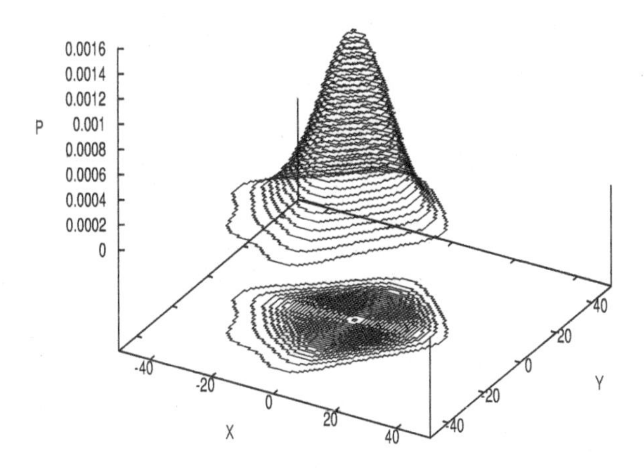

FIGURE 7.5: The probability distribution on the location of the boat assuming the precision of the bearings depends on the distance with the following readings: $b_1 = 225, b_2 = 180, b_3 = 270$ (to be compared with Figure 7.2b).

7.3 Classification

7.3.1 Statement of the problem

It may happen quite often that we are not interested in knowing the state in all its details. It may also happen that the information provided by the sensors is so imprecise that we can only access a rough evaluation of this state.

In both cases, it is possible to introduce a variable C to classify the states into categories that are either considered sufficient for the task or are imposed as the best that can be achieved.

Rather than $P(S|r_1 \wedge \ldots \wedge r_N \wedge \pi)$, the question asked to this model is:

$$P(C|r_1 \wedge \ldots \wedge r_N \wedge \pi) \tag{7.12}$$

7.3.2 Bayesian program

The classification Bayesian program is as follows:

$$
Pr\begin{cases}
Ds\begin{cases}
Sp(\pi)\begin{cases}
Va: \\
S, C, R_1, \ldots, R_N \\
Dc: \\
\begin{cases}
& P(S \wedge C \wedge R_1 \wedge \ldots \wedge R_N|\pi) \\
= & P(S|\pi) \times P(C|S \wedge \pi) \times \prod_{n=1}^{N} [P(R_n|S \wedge \pi)]
\end{cases} \\
Fo: \\
any
\end{cases} \\
Id
\end{cases} \\
Qu: \\
P(C|r_1 \wedge \ldots \wedge r_N \wedge \pi)
\end{cases}
\tag{7.13}
$$

$P(C|S \wedge \pi)$ encodes the definition of the classes.

$P(S|\pi) \times P(C|S \wedge \pi)$ may be eventually replaced by $P(C|\pi) \times P(S|C \wedge \pi)$ if more convenient to defined the classes by specifying the constraints on the states knowing the classes.

$P(C|r_1 \wedge \ldots \wedge r_N \wedge \pi)$ can be computed by the following formula:

$$
\begin{aligned}
& P(C|r_1 \wedge \ldots \wedge r_N \wedge \pi) \\
= & \frac{1}{Z} \times \sum_{S} \left[P(S|\pi) \times P(C|S \wedge \pi) \times \prod_{n=1}^{N} [P(r_n|S \wedge \pi)] \right]
\end{aligned}
\tag{7.14}
$$

The sum on S may be costly but often, hopefully, reduced as the classes impose by definition that $P(C|S \wedge \pi)$ is zero for large ranges of S.

7.3.3 Instance and results

Returning to our boat navigation example, if the weather conditions (thick fog) prevent you from using any visual landmarks and if no modern localization devices (radar, GPS, etc.) are available, the minimum you may try to ensure is to stay in a safe zone (no rocks, no shallows) based, for instance, only on very rough bearing information obtained from the sound of the bells.

Assuming the landmark L1 is above a shipwreck we may define the status of our boat as "safe" or "unsafe." C takes these two values and $P(C|X \wedge Y \wedge \pi)$ may be defined as follows:

$$
P(C = unsafe \mid X \wedge Y \wedge \pi) = Min(1.0, \frac{30}{\sqrt{(Y+50)^2 + (X+50)^2}}) \tag{7.15}
$$

If we are very close ($d < 30$) the danger is certain and it becomes less and less probable when we get further from the shipwreck.

One feature of ProBT (see file "chapter7/classification.py") is to allow the user to define his own conditional probability distribution.

```
def danger_prob_function(Input_):
    proba_for_danger = \
       min(1.0,30.0 / \
              (1+sqrt(pow((Input_[2]+50),2)+ \
                     pow((Input_[1]+50),2))))
    if Input_[0] == 1:
        return  proba_for_danger
    else:
        return 1-proba_for_danger
```

This function is the probability distribution associated with $P(C \mid X \wedge Y)$. An order is assumed for the values of the probabilistic variables:

```
C -> Input_[0]
X -> Input_[1]
Y -> Input_[2]
```

This order is given when declaring this function as a conditional probability distribution.

```
plPythonExternalProbFunction(C^X^Y,danger_prob_function)
```

We will assume we can approximatively relate the direction of the bells to one of the following four cardinal directions: "NE," "NW," "SE," and "SW."

Knowing the location of the boat we may compute $P(B_n|X \wedge Y \wedge \pi)$ as follows (essentially saying in which quarter of space you are hearing a given bell):

$$
\begin{array}{ll}
P(B_n \mid X \wedge Y \wedge \pi) :: & \\
\theta = atan2(-(Y + L_n^Y), -(X + L_n^X)) & \\
if(0 < \theta < \frac{\pi}{2}: & P(B_n = NE) = 0.5; P(B_n = SE) = 0.2 \\
 & P(B_n = NW) = 0.2; P(B_n = SW) = 0.1 \\
if(\frac{\pi}{2} < \theta < \pi): & P(B_n = NE) = 0.2; P(B_n = SE) = 0.1 \\
 & P(B_n = NW) = 0.5; P(B_n = SW) = 0.2; \\
if(-\frac{\pi}{2} < \theta < 0): & P(B_n = NE) = 0.2; P(B_n = SE) = 0.5 \\
 & P(B_n = NW) = 0.1; P(B_n = SW) = 0.2; \\
if(0 < \theta < -\frac{\pi}{2}): & P(B_n = NE) = 0.1; P(B_n = SE) = 0.2 \\
 & P(B_n = NW) = 0.2; P(B_n = SW) = 0.5;
\end{array}
\tag{7.16}
$$

The corresponding Bayesian program is now:

$$
Pr\begin{cases}
Ds\begin{cases}
Sp(\pi)\begin{cases}
Va: \\
X, Y, C, B_1, B_2, B_3 \\
Dc: \\
\begin{cases}
 & P(X \wedge Y \wedge \ldots \wedge B_3|\pi) \\
= & P(X \wedge Y|\pi) \times P(C|X \wedge Y \wedge \pi) \\
 & \times \prod_{n=1}^{3} \left[P(B_n|X \wedge Y \wedge \pi)\right]
\end{cases} \\
Fo: \\
\begin{cases}
P(X \wedge Y|\pi) = Uniform \\
P(C|X \wedge Y \wedge \pi) : \text{equation}(7.15) \\
P(B_n|X \wedge Y \wedge \pi) = \text{equation}(7.16)
\end{cases}
\end{cases} \\
Id
\end{cases} \\
Qu: \\
P(C|B_1 \wedge B_2 \wedge B_3 \wedge \pi)
\end{cases}
\tag{7.17}
$$

We can divide the space into four regions labeled with the true reading of the bearings in that region: If we use these readings as examples, we can obtain the probability of being "unsafe" with the corresponding reading (see Figure 7.6). Note this probability does not correspond to the probability of being in that region but to the probability of being in danger if we have this reading.

7.4 Ancillary clues

7.4.1 Statement of the problem

Ancillary clues (variable A) may help you in your prediction. These clues are not, strictly speaking, part of the state S of the phenomenon but it is important to take them into account in the sensor models. Consequently, each sensor model $P(R_n|S \wedge \pi)$ should be replaced by:

$$
P(R_n|S \wedge A \wedge \pi) \tag{7.18}
$$

7.4.2 Bayesian program

The Bayesian program including ancillary clues has the following structure:

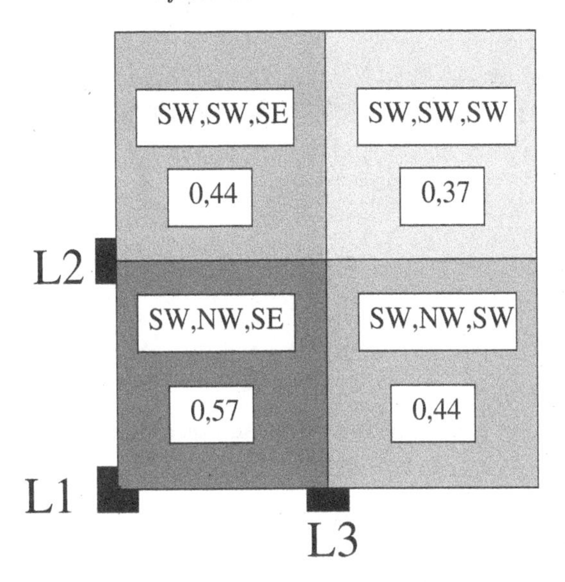

FIGURE 7.6: The probability of being "unsafe" if the readings correspond to one of the labels of the danger regions.

$$
Pr\begin{cases}
Ds\begin{cases}
Sp(\pi)\begin{cases}
Va:\\
S, A, R_1, \ldots, R_N\\
Dc:\\
\begin{cases}
& P(S \wedge A \wedge R_1 \wedge \ldots \wedge R_N|\pi)\\
= & P(A \wedge S|\pi) \times \prod_{n=1}^{N}\left[P(R_n|S \wedge A \wedge \pi)\right]
\end{cases}\\
Fo:\\
any
\end{cases}\\
Id
\end{cases}\\
Qu:\\
\begin{cases}
P(S|r_1 \wedge \ldots \wedge r_N \wedge \pi)\\
P(S|a \wedge r_1 \wedge \ldots \wedge r_N \wedge \pi)
\end{cases}
\end{cases}
\tag{7.19}
$$

$P(A \wedge S|\pi)$ encodes the eventual relations between the state of the phenomenon and the ancillary clues.

There could be none. The ancillary clue could be independent of the state: $P(A \wedge S|\pi) = P(A|\pi) \times P(S|\pi)$.

But if there are some relations that deserve to be taken into account, as in the classification case, $P(A \wedge S|\pi)$ could be either defined as $P(A|\pi) \times P(S|A \wedge \pi)$ or as $P(S|\pi) \times P(A|S \wedge \pi)$.

The real important innovation here is that the sensor models depend on both S and A. Either A is used to refine the value of the parameters of the sensor models or it can even completely change the very nature and mathematical form of this sensor model.

This model may be used to estimate the state either knowing the value a of the ancillary clue or ignoring it.

If the value of A is known, then the computation of $P(S|a \wedge r_1 \wedge \ldots \wedge r_N \wedge \pi)$ is straightforward.

If the value of A is unknown, the computation of $P(S|r_1 \wedge \ldots \wedge r_N \wedge \pi)$ supposes to marginalize on A. In that case, $P(A|\pi)$ can be seen as "soft evidence" (see the next chapter for details on "soft evidence"), which is useful for a better estimation of S.

7.4.3 Instance and results

Returning to the boat localization problem in presence of fog, we may introduce a new Boolean variable V $(A \equiv V)$ with the semantic that if V is true, the visibility is perfect, and if V is false, there is no visibility at all.

If the visibility is perfect ($V = true$), the sensor model used is the same as the one used in the Bayesian program 7.6 at the beginning of this chapter. If there is no visibility ($V = false$), the sensor model assumes a larger uncertainty. For example, the sensor readings with no visibility will have a standard deviation of 30, which will be reduced to 10 when there is good visibility.

The resulting Bayesian program is:

$$
Pr\begin{cases}
Ds\begin{cases}
Sp(\pi)\begin{cases}
a: \\
X,Y,V,B_1,B_2,B_3 \\
Dc: \\
\begin{cases}
 & P(X\wedge Y\wedge\ldots\wedge B_3|\pi) \\
= & P(X\wedge Y|\pi)\times P(V|\pi) \\
 & \times\prod_{n=1}^{3}\left[P(B_n|V\wedge X\wedge Y\wedge\pi)\right]
\end{cases} \\
Fo: \\
\begin{cases}
 & P(X\wedge Y|\pi)=Uniform \\
 & P(V|\pi)=Soft-evidence \\
 & P(B_n|\left[V=false\right]\wedge X\wedge Y\wedge\pi) \\
= & B(\left[\mu=f_n(X,Y)\right],\left[\sigma=30\right]) \\
 & P(B_n|\left[V=true\right]\wedge X\wedge Y\wedge\pi) \\
= & B(\left[\mu=f_n(X,Y)\right],\left[\sigma=10\right])
\end{cases}
\end{cases} \\
Id
\end{cases} \\
Qu: \\
P(X\wedge Y|b_1\wedge b_2\wedge b_3\wedge\pi)
\end{cases}
\tag{7.20}
$$

Computing $P(X\wedge Y|b_1\wedge b_2\wedge b_3\wedge\pi)$ supposes to marginalize on V:

$$
\begin{aligned}
& P(X\wedge Y|b_1\wedge b_2\wedge b_3\wedge\pi) \\
= & \frac{1}{Z}\times\sum_V\left[P(V|\pi)\times\prod_{n=1}^{3}\left[P(b_n|V\wedge X\wedge Y\wedge\pi)\right]\right] \\
= & \frac{1}{Z}\times\left[\begin{array}{cc}
 & P(\left[V=0\right]|\pi)\times\prod_{n=1}^{3}\left[P(b_n|\left[V=0\right]\wedge X\wedge Y\wedge\pi)\right] \\
+ & P(\left[V=1\right]|\pi)\times\prod_{n=1}^{3}\left[P(b_n|\left[V=1\right]\wedge X\wedge Y\wedge\pi)\right]
\end{array}\right]
\end{aligned}
\tag{7.21}
$$

$P(V|\pi)$ may be considered as "soft evidence." Indeed, the weather forecast gives you an estimation of the visibility in percent that can be used as the value for this soft evidence.

The computation above appears as a weighting sum between the two models, the weights being the estimation of the visibility.

We may compute $P(X\wedge Y|b_1\wedge b_2\wedge b_3\wedge\pi)$ for different values of this soft evidence:

1. $P(\left[V=false\right]|\pi)=1$ (no visibility) see Figure 7.7a.
2. $P(\left[V=false\right]|\pi)=0.9$ (almost no visibility) see Figure 7.7b.

3. $P([V = true]|\pi) = 0.5$ (partial visibility) see Figure 7.7c.
4. $P([V = true]|\pi) = 1$ (clear weather) see Figure 7.7d (identical to Figure 7.2b).

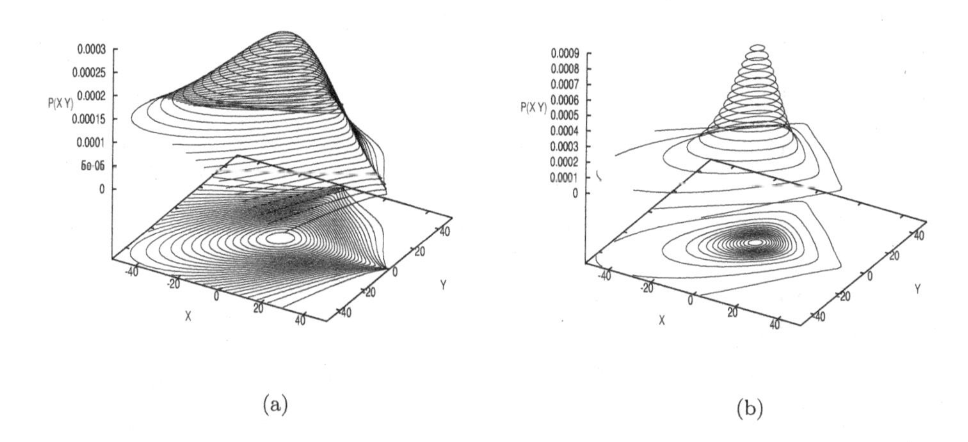

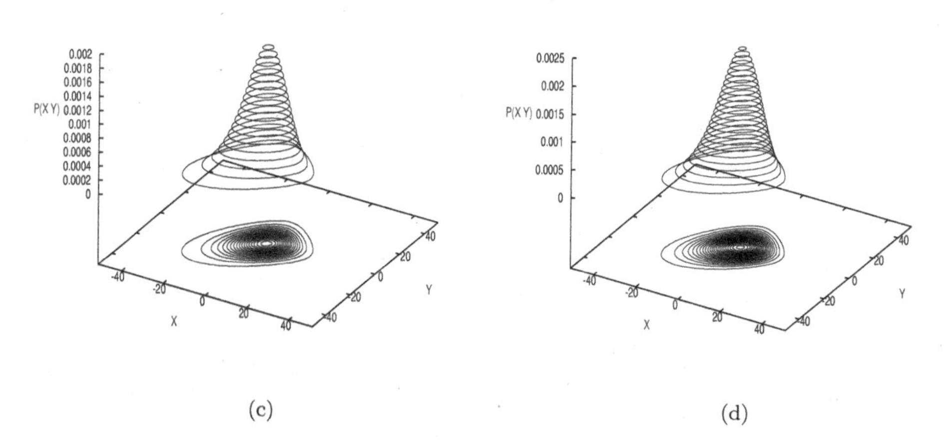

FIGURE 7.7: Distributions with several soft evidences:
(a): $P(V = false) = 1, b_1 = 225, b_2 = 180, b_3 = 270$
(b): $P(V = false) = 0.9, b_1 = 225, b_2 = 180, b_3 = 270$
(c): $P(V = false) = 0.5, b_1 = 225, b_2 = 180, b_3 = 270$
(d): $P(V = true) = 1, b_1 = 225, b_2 = 180, b_3 = 270$

The file "chapter7/ancillary.py" provides an example of how to change soft evidence by replacing one distribution in a decomposition.

```
#estimating the location with poor visibility
localisation_model.replace(V,plProbTable(V,[0.1,0.9]))
PXY_K_B1B2B3=localisation_model.ask(X^Y,B1^B2^B3)
PXY=PXY_K_B1B2B3.instantiate(sensor_reading_values)
```

The initial distribution on V $(P(V=1)=0)$ is replaced with another distribution $P(V=1)=0.1$, changing the inference on the location.

7.5 Sensor fusion with false alarm

7.5.1 Statement of the problem

Sensors are never completely reliable.

The readings they provide are never exactly the ground truth of what they are supposed to measure. Hidden variables are always the explanation for these discrepancies. For instance, a sensor usually does not measure solely what it is supposed to measure.[3] Thus, an infrared proximeter measures the distance of the target but it measures also its colors: red targets reflect more infrared light than blue ones, consequently they return more energy to the sensor and, finally, they appear closer. These discrepancies are taken into account by the probabilistic sensor model $P(R|S \wedge \pi)$.

Sensors may also miss certain detections. In these cases, the sensor does not give any reading even when this information should be available (this happens frequently when doing fusion on information coming from a database where some fields may not always be filled). This is also, most of the time, taken into account by the parametric form of $P(R|S \wedge \pi)$.

Finally, the sensor may return some "false alarm." In these cases, the sensor provides a reading as if a "target" was present even if there is none. To deal with this problem a common solution is to introduce a binary variable F, meaning that there is no false alarm if $F=0$ and that there is one if $F=1$. The sensor model then gets a little bit more complicated as $P(R|S \wedge \pi)$ is replaced by $P(R|S \wedge F \wedge \pi)$.

If $F=0$ there is no false alarm and we use the original sensor model:

[3]This is a main concern in physical experiments where considerable effort is made to build the setup that will warrant that the sensors measure only the search quantity. For instance, neutrino detectors are buried several thousand meters under mountains so as to be free from cosmic radiation.

$$P(R|S \wedge [F=0] \wedge \pi) = P(R|S \wedge \pi) \tag{7.22}$$

On the contrary, is $F = 1$ there is a false alarm and whatever the state of the phenomenon, we have no information on the reading:

$$P(R|S \wedge [F=1] \wedge \pi) = Uniform \tag{7.23}$$

7.5.2 Bayesian program

This may be summarized as the following Bayesian program:

$$Pr \begin{cases} Ds \begin{cases} Sp(\pi) \begin{cases} Va: \\ S, F_1, \ldots, F_N, R_1, \ldots, R_N \\ Dc: \\ \begin{cases} & P(S \wedge F_1 \wedge \ldots \wedge R_N|\pi) \\ = & P(S \wedge F_1 \wedge \ldots \wedge F_N|\pi) \times \prod_{n=1}^{N} [P(R_n|S \wedge F_n \wedge \pi)] \end{cases} \\ Fo: \\ P(R_n|S \wedge [F_n=0] \wedge \pi) = P(R|S \wedge \pi) \\ P(R_n|S \wedge [F_n=1] \wedge \pi) = Uniform \end{cases} \\ Id \end{cases} \\ Qu: \\ P(S|r_1 \wedge \ldots \wedge r_N \wedge \pi) \end{cases} \tag{7.24}$$

which can be compared to the Bayesian program 7.19. False alarms appear as a special case of ancillary clues.

7.5.3 Instance and results

Radar often observes "ghost" targets.

This is especially true for onboard radar due to the lack of stability of the platform and to echoes generated by the waves (especially in rough sea conditions).

We can revisit the example of Figure 7.3b, but this time taking into account a probability of false alarm $P(F=1|\pi) = 0.3$.

$$
Pr\begin{cases}
Ds\begin{cases}
Sp(\pi)\begin{cases}
Va: \\
X, Y, F_1, F_2, F_3, D_1, D_2, D_3 \\
Dc: \\
\begin{cases}
& P(X \wedge Y \wedge \ldots \wedge D_3|\pi) \\
= & P(X \wedge Y|\pi) \times \prod_{n=1}^{3}[P(F_n|\pi)] \\
& \times \prod_{n=1}^{3}[P(D_n|X \wedge Y \wedge F_n \wedge \pi)]
\end{cases} \\
Fo: \\
\begin{cases}
& P(X \wedge Y|\pi) = Uniform \\
& P(F_n = true|\pi) = 0.3 \\
& P(D_n|X \wedge Y \wedge [F_n = 0] \wedge \pi) \\
= & G\left([\mu = d_n], \left[\sigma = 1 + \frac{d_n}{10}\right]\right) \\
& P(D_n|X \wedge Y \wedge [F_n = 1] \wedge \pi) = Uniform
\end{cases}
\end{cases} \\
Id
\end{cases} \\
Qu: \\
P(X \wedge Y|d_1 \wedge d_2 \wedge d_3 \wedge \pi)
\end{cases}
\tag{7.25}
$$

$P(X \wedge Y|d_1 \wedge d_2 \wedge d_3 \wedge \pi)$ is a weighted sum between the different variations of the models according to the different combinations of false alarms for the different targets. The weights are given by the product of the probabilities of false alarms:

$$
\begin{aligned}
& P(X \wedge Y|d_1 \wedge d_2 \wedge d_3 \wedge \pi) \\
= & \frac{1}{Z} \times \sum_{F_1 \wedge F_2 \wedge F_3} \left[\begin{array}{c} P(X \wedge Y|\pi) \times \prod_{n=1}^{3}[P(F_n|\pi)] \\ \times \prod_{n=1}^{3}[\quad P(D_n|X \wedge Y \wedge F_n \wedge \pi)\ ] \end{array} \right]
\end{aligned}
\tag{7.26}
$$

We observe in Figure 7.8 (to be compared to Figure 7.3b) that this new model with a false alarm is more robust than the one without one when we have a wrong reading on $D3$.

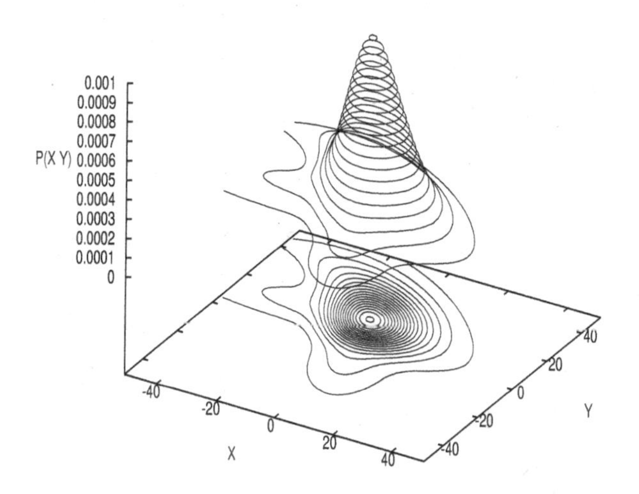

FIGURE 7.8: $P(X \wedge Y \mid d_1 = 50 \wedge d_2 = 70 \wedge d_3 = 70)$ with $P(F_n = 1) = 0.3$.

7.6 Inverse programming

7.6.1 Statement of the problem

Often, some "action" (any kind of computer process with or without consequences on a physical device) has to be triggered by a given situation.

This is very commonly implemented using rules that state that "if condition r_1 & condition r_2 & ... & condition r_N are true then action a should be executed" (see, for instance, finite state automata scripting languages, or expert systems).

This approach has both the advantages of being intuitive and very efficient in terms of computing time.

However, one fundamental drawback it has is that the number of rules may very rapidly become huge when the complexity of the system augments. Indeed, you potentially need one rule for each possible tuple of the N conditions even if some rules may group ranges of values having the same effect. As a consequence, the readability and the maintenance of these systems may become cumbersome and even intractable.

A possible solution, inspired from the above information fusion models, is "inverse programming." The symptom of the curse of dimensionality is the same as for information fusion: the number of cases to be taken into account for the conjunction of conditions $R_1 \wedge \ldots \wedge R_N$ grows exponentially with N. The solution to break this curse of dimensionality is of the same nature as in Section 7.1.1: that makes the assumption that by knowing the action A (instead of the state S) the sensors provide independent readings:

$$P(A \wedge R_1 \wedge \ldots \wedge R_N|\pi) = P(A|\pi) \times \prod_{n=1}^{N} [P(R_n|A \wedge \pi)] \tag{7.27}$$

This is called "inverse" programming because instead of having to specify rules of the form $R_1 \wedge \ldots \wedge R_N \rightarrow A$, we need to specify probability distributions of the form $P(R_n|A \wedge \pi)$.

7.6.2 Bayesian program

The Bayesian program summarizing the inverse programming approach is the following:

$$Pr \begin{cases} Ds \begin{cases} Sp(\pi) \begin{cases} Va: \\ A, R_1, \ldots, R_N \\ Dc: \\ \begin{cases} & P(A \wedge R_1 \wedge \ldots \wedge R_N|\pi) \\ = & P(A|\pi) \times \prod_{n=1}^{N} [P(R_n|A \wedge \pi)] \end{cases} \\ Fo: \\ any \end{cases} \\ Id \end{cases} \\ Qu: \\ P(A|r_1 \wedge \ldots \wedge r_N \wedge \pi) \end{cases} \tag{7.28}$$

As this Bayesian program is mathematically identical to 7.3, it has the same appealing properties to answer the question: $P(A|r_1 \wedge \ldots \wedge r_N \wedge \pi)$. The difference between the two is in the semantics of the question because instead of looking for the state of the phenomenon, we are searching for the appropriate action.

It also has the same weakness and may be extended in parallel ways to overcome them. For instance, in inverse programming you often need to relax the conditional independence hypothesis (as in Section 7.2) to express that a specific conjunction of conditions is necessary to trigger the action.

7.6.3 Instance and results

Inverse programming has been, for instance, successfully used to program and to train video game avatars to play autonomously (see Le Hy et al. [2004], Le Hy [2007], and Le Hy and Bessière [2008]).

Coming back to our navigation example, we would like to decide the heading direction to go toward landmark 1, avoiding a new landmark 0 located at the center of the map. The action is the heading direction H and the sensors are two bearings B_0 and B_1.

Knowing the heading direction, whatever the position of the boat, the bearing of the goal landmark should be approximately the same, as we want to go toward this landmark:

$$P(B_1|H) = B(\mu = H, \sigma) \tag{7.29}$$

Knowing the heading direction, whatever the position of the boat the bearing of the obstacle should be different, as we want to avoid it.

$$P(B_0|H) = \frac{1}{Z}\left[1 - \lambda B(\mu = H, \sigma')\right]$$

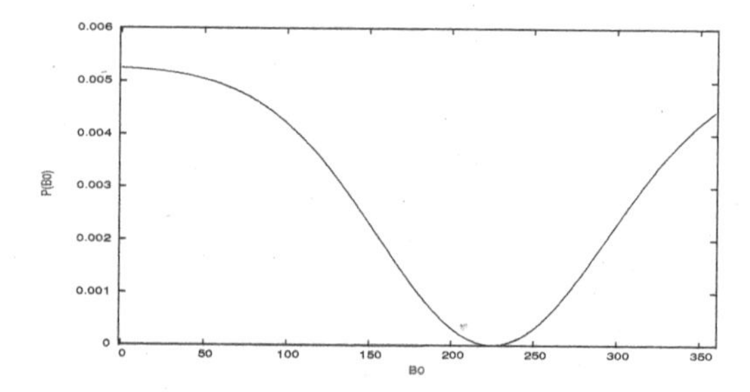

FIGURE 7.9: The probability of going toward a given direction (225) is lower than all other directions. It is a way to avoid certain directions making them much less probable.

The Bayesian program is the following:

$$
Pr\begin{cases}Ds\begin{cases}Sp(\pi)\begin{cases}Va: \\ H, B_1, B_0 \\ Dc: \\ \begin{cases} & P(H \wedge B_0 \wedge B_1 \wedge B_3|\pi) \\ = & P(H|\pi) \times \prod_{n=0}^{3}[P(B_n|H \wedge \pi)] \end{cases} \\ Fo: \\ \begin{cases} P(H|\pi) = Uniform \\ P(B_0|H) = \frac{1}{Z}[1 - \lambda B(\mu = H, \sigma')] \\ P(B_1|H) = B(B_1, \mu = H, \sigma) \end{cases}\end{cases} \\ Id: \end{cases} \\ Qu: \\ P(H|b_0 \wedge b_1 \wedge \pi)\end{cases} \tag{7.30}
$$

$P(H|b_0 \wedge b_1 \wedge \pi)$ can be computed at any point in space. The resulting vector field is displayed in Figure 7.10.

Figure 7.10 has been generated using the program "chapter7/invpgm.py". The following instruction allows us to get to the most probable value for the heading H given the readings.

```
PH=PHkB0B1.instantiate(sensor_reading_values)
best=PH.compile().best()
```

best is a vector: the angle value can be obtained with

```
best[0]
```

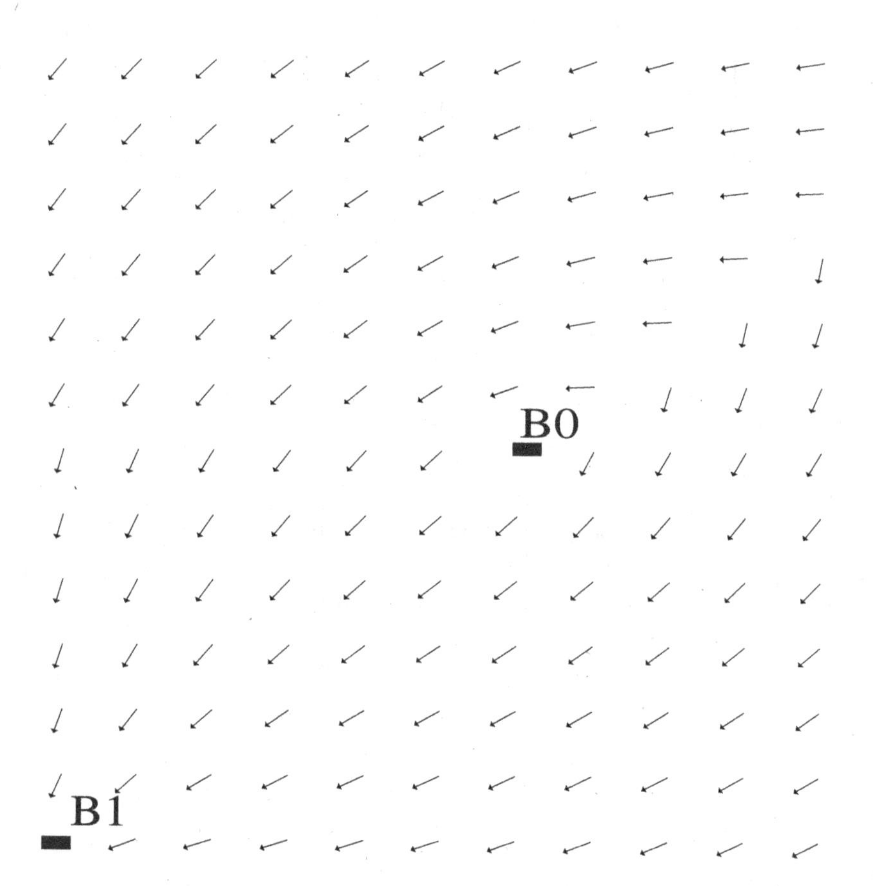

FIGURE 7.10: The vector field corresponding to $\max_h P(H = h|b_0 \wedge b_1 \wedge \pi)$.

Chapter 8

Bayesian Programming with Coherence Variables

Today's posterior distribution is tomorrow's prior

Bayesian Analysis in Regression Problems
David Lindley [1970]

What means "equality" for Bayesian variables?

Two different calculi may lead to the same result. This is the case if you try to compute the same thing with two different methods in a "consistent" or "coherent"[1] calculus system.

You can impose it as a constraint of your model by specifying that a given equation should be respected. Solving the equation then consists of finding the conditions of the two terms of the equation in order to make them "equal."[2]

It can finally be used as a programming notion when you "assign"[3] the result of a calculus to a given variable in order to use it in a subsequent calculus.[4]

However, for all these fundamental notions of logic, mathematic, and computing the results of the calculus are always values either Boolean, numeric, or symbolic.

In probabilistic computing, the basic objects that are manipulated are not values but rather probability distributions on variables. In this context, the "equality" has a different meaning as it should say that two variables have the same probability distribution.

To realize this, we introduce in this chapter the notion of coherence variables. A coherence variable is a Boolean variable. If the coherence variable is equal to 1 (or "true") it imposes that the two variables are "coherent" which means that they should share the same probability distribution knowing the same premises.

8.1 Basic example with Boolean variables

8.1.1 Statement of the problem

To begin, let us examine the simplest case with two Boolean variables (A and B) and one coherence variable (Λ). The decomposition of the joint distribution is the following:

$$P(A \wedge B \wedge \Lambda) = P(A) \times P(B) \times P(\Lambda | A \wedge B) \tag{8.1}$$

A and B are assumed to be independent.

---

[1] "Consistent" and "coherent" are equivalent notions. "Consistent" is usually used as a semantic notion for "satisfiability" when "coherence" is usually used as a syntactic notion for "noncontradictory."

[2] "Equal" here stands for "=" if the terms have numeric values, "⇔" if they have Boolean values, "unifiable" if they are more complex logical expressions.

[3] Often denoted in computer languages by "=" or ":=".

[4] This case is quite different from the two first as the two terms do not play symmetric roles, the right term being computed first in order to assign its result to the left one.

$P([\Lambda=1]|A\wedge B)$ is a Dirac distribution with value one if and only if $A=B$:

$$P([\Lambda=1]|A\wedge B)=\delta_{A=B} \tag{8.2}$$

8.1.2 Bayesian program

This may be summarized by the following Bayesian program:

$$Pr\begin{cases} Ds\begin{cases} Sp(\pi)\begin{cases} Va: \\ A,B,\Lambda \\ Dc: \\ \begin{cases} & P(A\wedge B\wedge \Lambda|\pi) \\ = & P(A|\pi)\times P(B|\pi)\times P(\Lambda|A\wedge B\wedge\pi) \end{cases} \\ Fo: \\ P([\Lambda=1]|A\wedge B\wedge\pi)=\delta_{A=B} \end{cases} \\ Id \end{cases} \\ Qu: \\ P(A|[\Lambda=1]\wedge\pi) \\ P(B|[\Lambda=1]\wedge\pi) \end{cases} \tag{8.3}$$

The interesting question is $P(A|[\Lambda=1]\wedge\pi)$:

$$\begin{aligned} & P(a|\lambda\wedge\pi) \\ = & \frac{1}{Z}\times\sum_B\left[P(a|\pi)\times P(B|\pi)\times P(\lambda|a\wedge B\wedge\pi)\right] \\ = & \frac{1}{Z}\times\begin{bmatrix} P(a|\pi)\times P(\bar{b}|\pi)\times P(\lambda|a\wedge\bar{b}\wedge\pi) \\ P(a|\pi)\times P(b|\pi)\times P(\lambda|a\wedge b\wedge\pi) \end{bmatrix} \\ = & \frac{1}{Z}\times P(a|\pi)\times P(b|\pi) \end{aligned} \tag{8.4}$$

where we use the more compact logical notation: $a\equiv[A=1]$ and $\bar{a}\equiv[A=0]$.

If we compute the normalization constant Z we obtain:

$$\begin{aligned} & P(a|\lambda\wedge\pi) \\ = & \frac{P(a|\pi)\times P(b|\pi)}{P(a|\pi)\times P(b|\pi)+P(\bar{a}|\pi)\times P(\bar{b}|\pi)} \end{aligned} \tag{8.5}$$

We check that $P(A|[\Lambda=1]\wedge\pi)=P(B|[\Lambda=1]\wedge\pi)$ which means that the semantic of the coherence variable is respected.

8.1.3 Instance and results

8.1.3.1 Logical and algebraic interpretation

If we know that B is true (i.e., $P(b|\pi) = 1$), then whatever the prior on A we get: $P(a|\lambda \wedge \pi) = 1$ meaning that A is necessarily true. In logical terms, if Λ is true then $b \Rightarrow a$. We also obtain the same result asking the question $P(a|b \wedge \lambda \wedge \pi)$ which, indeed, is the exact same semantic.

If we know that B is false (i.e. $P(b|\pi) = 0$), then whatever the prior on A we get: $P(a|\lambda \wedge \pi) = 0$ meaning that A is necessarily false. In logical terms, if Λ is true then $\bar{b} \Rightarrow \bar{a}$.

Consequently, if Λ is true then in logical terms $a \Leftrightarrow b$ or in algebraic terms $A = B$. If we have certainty on B then $\Lambda = 1$ has the semantic of a classical equality.

The only exception is when we have at the same time: $[\Lambda = 1]$, $P(b|\pi) = 1$, and $P(a|\pi) = 0$. In that case, $P(a|\lambda \wedge \pi)$ is not determined, which may be interpreted as a signal that we are trying to reason with contradictory hypotheses and a way to detect the incoherence of our hypotheses.[5]

8.1.3.2 Probabilistic interpretation

If we have no certainty on B (i.e., $P(b|\pi) \neq 1$ and $P(b|\pi) \neq 0$), but we know a probability distribution on B (i.e., $P(b|\pi) = x$) then there are two cases:

1. If we have a uniform prior on A ($P(a|\pi) = \frac{1}{2}$) then we have:

$$P(a|\lambda \wedge \pi) = P(b|\pi) = x \tag{8.6}$$

   If we have no certainty on B and a noninformative prior on A then $\Lambda = 1$ has the semantic of transmitting the probability distribution on B to A, a kind of probabilistic "assignment."

2. If we have a nonuniform prior on A (i.e., $P(a|\pi) = y$) then we get:

$$P(a|\lambda \wedge \pi) = \frac{y \times x}{y \times x + (1-y) \times (1-x)} \tag{8.7}$$

   If we have both constraints at the same time (as expressed by $P(a|\pi) = y$ and $P(b|\pi) = x$) then $P(a|\lambda \wedge \pi)$ (and $P(b|\lambda \wedge \pi)$ as they are equal) is a compromise between these two constraints expressed by Equation 8.7.

[5]See Section 2.6.2 titled "Godel's theorem" of Jaynes' book [2003] (pages 45–47) for a very stimulating discussion on this subject and about the perspectives it opens relatively to the meaning of Godel's theorem in probability.

8.1.3.3 $\mathbf{P(a|\bar{\lambda} \wedge \pi)}$

What happens if Λ is set to false?

In that case we get:

$$\begin{aligned} & P\left(a|\bar{\lambda} \wedge \pi\right) \\ = & \frac{1}{Z} \times P\left(a|\pi\right) \times P\left(\bar{b}|\pi\right) \end{aligned} \tag{8.8}$$

If B is true, we get that A is false and if B is false, we get that A is true. The logical interpretation is that $a \Leftrightarrow \bar{b}$ and the algebraic interpretation is that the value of $\neg B$ is equal to A.

If we have no certainty on B and a uniform prior on A then we get:

$$P\left(a|\bar{\lambda} \wedge \pi\right) = P\left(\bar{b}|\pi\right) = 1 - P\left(b|\pi\right) \tag{8.9}$$

8.2 Basic example with discrete variables

8.2.1 Statement of the problem

A common need in Bayesian programming is to express that a given discrete variable A should have the same probability distribution as another discrete variable B.

The same approach as for Boolean variables, using coherence variables, can be used with discrete variables. If we have two discrete variables A and B and one Boolean coherence variable Λ, the decomposition of the joint distribution is the same as that in the case of Boolean variables:

$$P\left(A \wedge B \wedge \Lambda\right) = P\left(A\right) \times P\left(B\right) \times P\left(\Lambda|A \wedge B\right) \tag{8.10}$$

A and B are still assumed independent and $P\left(\left[\Lambda = 1\right]|A \wedge B\right)$ is again a Dirac distribution with a value of one if and only if $A = B$:

$$P\left(\left[\Lambda = 1\right]|A \wedge B\right) = \delta_{A=B} \tag{8.11}$$

We check that $P\left(A|\left[\Lambda = 1\right] \wedge \pi\right)$ and $P\left(B|\left[\Lambda = 1\right] \wedge \pi\right)$ are proportional which means that the semantic of the coherence variable is respected. Indeed, we can not impose a strict equality as the range of A and B may be different (see details in the sequel of this section).

If we want to "assign" A we also have to assume that $P\left(A\right)$ is uniform.

Different assumptions than this noninformative prior will be treated further in this chapter, particularly in Section 8.5 titled: "Reasoning with soft evidence."

8.2.2 Bayesian program

The corresponding Bayesian program is exactly the same as in the case of Boolean variables:

$$
Pr\begin{cases}
Ds\begin{cases}
Sp(\pi)\begin{cases}
Va: \\
A, B, \Lambda \\
Dc: \\
\begin{cases}
& P(A \wedge B \wedge \Lambda|\pi) \\
= & P(A|\pi) \times P(B|\pi) \times P(\Lambda|A \wedge B \wedge \pi)
\end{cases} \\
Fo: \\
P([\Lambda = 1]|A \wedge B \wedge \pi) = \delta_{A=B}
\end{cases} \\
Id
\end{cases} \\
Qu: \\
P(A|[\Lambda = 1] \wedge \pi) \\
P(B|[\Lambda = 1] \wedge \pi)
\end{cases}
\tag{8.12}
$$

For $P(A|[\Lambda = 1] \wedge \pi)$ we get:

$$
\begin{aligned}
& P(A|\lambda \wedge \pi) \\
= & \frac{1}{Z} \times \sum_{B} [P(A|\pi) \times P(B|\pi) \times P(\lambda|A \wedge B \wedge \pi)] \\
= & \frac{1}{Z} \times P(A|\pi) \times P([B = A]|\pi)
\end{aligned}
\tag{8.13}
$$

And, if we further assume that $P(A)$ is uniform, we get for all possible values of A:

$$
P([A = x]|\lambda \wedge \pi) \propto P([B = x]|\pi) \tag{8.14}
$$

8.2.3 Instance and results

8.2.3.1 *B* has a known value

If B has a known value b, then we get:

$$
P(A|b \wedge \lambda \wedge \pi) = \delta_{A=b} \tag{8.15}
$$

which is the semantic of the classical (nonprobabilistic) assignment.

8.2.3.2 *A* and *B* with the same range

If A and B have the same range, Equation 8.14 turns to be an equality:

$$
P([A = x]|\lambda \wedge \pi) = P([B = x]|\pi) \tag{8.16}
$$

8.2.3.3 Range of B included in range A

If the range of B is included in the range of A, we can reduce to the case where they have the same range assuming that $P(B|\pi) = 0$ for the missing values of B.

If A is out of the range of B, we get:

$$P(A|\lambda \wedge \pi) = 0 \tag{8.17}$$

If A is in the range of B, we get:

$$P([A = x] | \lambda \wedge \pi) = \frac{P([B = x] | \pi)}{\sum_{a \in A} [P([B = a] | \pi)]} \tag{8.18}$$

where the sum on A is made only for the values of A that are in the range of B (see Figure 8.1).

8.2.3.4 Range of A included in range B

Similarly, if the range of A is included in the range of B, we can reduce to the previous case assuming that A and B have the same range and that $P(A|\pi) = 0$ for the missing values of A. We then get for all possible values of A:

$$P([A = x] | \lambda \wedge \pi) = \frac{P([B = x] | \pi)}{\sum_{a \in A} [P([B = a] | \pi)]} \tag{8.19}$$

8.2.3.5 Range of A and range of B intersect

The same can be done if the range of A and the range of B intersect.

If A is out of the range of B we get:

$$P(A|\lambda \wedge \pi) = 0 \tag{8.20}$$

If A is in the range of B, we get:

$$P([A = x] | \lambda \wedge \pi) = \frac{P([B = x] | \pi)}{\sum_{a \in A} [P([B = a] | \pi)]} \tag{8.21}$$

where the sum on A is made only for the values of A that are in the range of B.

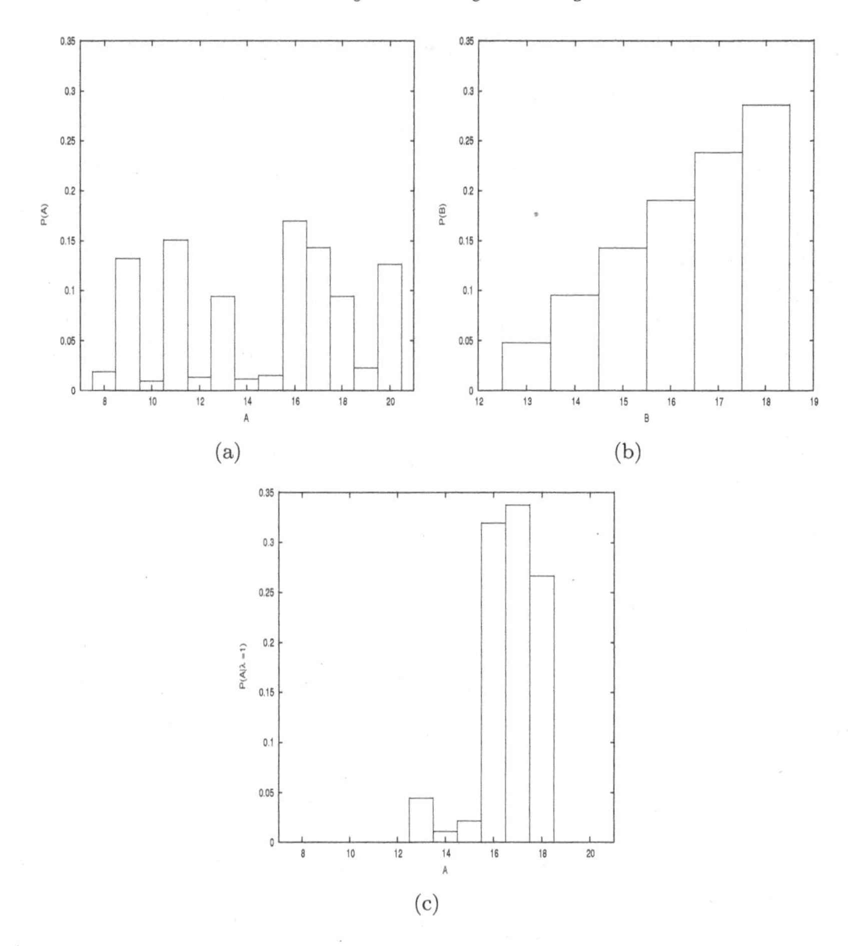

FIGURE 8.1: The assignment operator when the range of B is included in the range of A:
(a): $P(A)$
(b): $P(B)$
(c): $P(A \mid \lambda = 1)$

8.2.3.6 More complicated mapping of B to A

The coherence variable Λ may also be used to encode more complicated mapping from B to A.

In such a case $P([\Lambda = 1] \mid A \wedge B) = 1$ if and only if $A = f(B)$ where f is the function encoding the mapping from B to A.

$$P([\Lambda = 1] \mid A \wedge B) = \delta_{A=f(B)} \tag{8.22}$$

A coherence variable can be used, for instance, to change the discretization of the variable (which is always a delicate question) or even for a nonlinear mapping such as, for instance, a logarithmic mapping where $f(B) = int(log_2(B))$. We get:

$$P([A = x] | \lambda \wedge \pi) = \frac{\sum_{B} \left[P(B|\pi) \times \delta_{x=int(log_2(B))} \right]}{\sum_{a \in A} \left[\sum_{B} \left[P(B|\pi) \times \delta_{a=int(log_2(B))} \right] \right]} \tag{8.23}$$

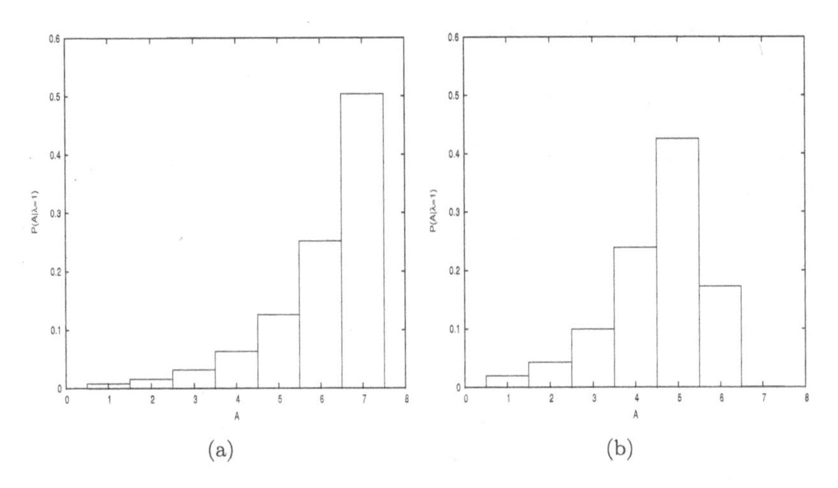

(a) (b)

FIGURE 8.2: The variable $B \in [1, 255]$ is mapped into the variable $A \in [1, 7]$ using the log_2 function: (a) $P(A|\lambda \wedge \pi)$ with a uniform prior on B, (b) $P(A|\lambda \wedge \pi)$ with a Gaussian prior on B.

The program "chapter8/logcoding.py" has been used to produce Figure 8.2. A Python function was used to define the value of λ from A and B.

```
def LogCoding(Output_, Input_):
    if Input_[A] == math.floor(math.log(Input_[B],2)) :
        Output_[LAMBDA]=1
    else:
        Output_[LAMBDA]=0
    return
```

The following statement produced a Dirac distribution based on the previous function.

```
diracDistrib = plFunctionalDirac(LAMBDA,A^B,\
                         plPythonExternalFunction(LAMBDA,A^B, \
                                                  LogCoding))
```

8.3 Checking the semantic of Λ

8.3.1 Statement of the problem

As stated at the beginning of this chapter, the semantic of the coherence variable should be that if the coherence variable is equal to 1 (or "true") it imposes that the two bound variables are "coherent" which means that they should share the same probability distribution knowing the same premises.

Let us check that this can effectively be realized with a generic model of the form:

$$\begin{aligned} & P(A \wedge B \wedge C \wedge D \wedge E \wedge \Lambda) \\ = \; & P(A \wedge C \wedge E) \times P(\Lambda | A \wedge B) \times P(B \wedge D \wedge E) \end{aligned} \tag{8.24}$$

It is generic in the sense that: (i) A and B are bound by the coherence variable λ, (ii) A is part of a model including C not shared with B and E shared with B, and (iii) symmetrically B is part of a model including D not shared with A and E shared with A.

Our objective is to prove that whatever the premises p: $P(A|p \wedge \lambda \wedge \pi) \prec P(B|p \wedge \lambda \wedge \pi)$. We will establish this property for five different cases as all the other possibilities can be reduced to these five cases by symmetry considerations.

8.3.2 Bayesian program

The corresponding Bayesian program and questions are the following:

$$Pr\begin{cases}Ds\begin{cases}Sp(\pi)\begin{cases}Va:\\A,B,C,D,E,\Lambda\\Dc:\\\begin{cases}&P(A\wedge B\wedge\wedge\wedge\wedge\Lambda|\pi)\\=&P(A\wedge C\wedge E|\pi)\times P(\Lambda|A\wedge B\wedge\pi)\\&\times P(B\wedge D\wedge E|\pi)\end{cases}\\Fo:\\any\end{cases}\\Id\end{cases}\\Qu:\\P(A|\lambda\wedge\pi)\\P(A|c\wedge\lambda\wedge\pi)\\P(A|e\wedge\lambda\wedge\pi)\\P(A|c\wedge e\wedge\lambda\wedge\pi)\\P(A|d\wedge c\wedge e\wedge\lambda\wedge\pi)\end{cases} \tag{8.25}$$

8.3.3 Instance and results

For any value x of the variable A, we have:

$$\begin{aligned}&P([A=x]|\lambda\wedge\pi)\\\prec\ &\sum_{B\wedge C\wedge D\wedge E}[P([A=x]\wedge C\wedge E|\pi)\times P(\lambda|A\wedge B\wedge\pi)\times P(B\wedge D\wedge E|\pi)]\\\prec\ &\sum_{B\wedge E}[P([A=x]\wedge E|\pi)\times P(\lambda|A\wedge B\wedge\pi)\times P(B\wedge E|\pi)]\\\prec\ &\sum_{E}[P([A=x]\wedge E|\pi)\times P([B=x]\wedge E|\pi)]\\\prec\ &P([B=x]|\lambda\wedge\pi)\end{aligned} \tag{8.26}$$

We have also:

$$\begin{aligned}&P([A=x]|c\wedge\lambda\wedge\pi)\\\prec\ &\sum_{B\wedge D\wedge E}[P([A=x]\wedge c\wedge E|\pi)\times P(\lambda|A\wedge B\wedge\pi)\times P(B\wedge D\wedge E|\pi)]\\\prec\ &\sum_{E}[P([A=x]\wedge c\wedge E|\pi)\times P([B=x]\wedge E|\pi)]\\\prec\ &P([B=x]|c\wedge\lambda\wedge\pi)\end{aligned} \tag{8.27}$$

and:

$$\begin{array}{ll} & P([A=x]\,|e\wedge\lambda\wedge\pi) \\ \prec & \sum\limits_{B\wedge C\wedge D}\left[P([A=x]\wedge C\wedge e|\pi)\times P(\lambda|A\wedge B\wedge\pi)\times P(B\wedge D\wedge e|\pi)\right] \\ \prec & P([A=x]\wedge e|\pi)\times P([B=x]\wedge e|\pi) \\ \prec & P([B=x]\,|e\wedge\lambda\wedge\pi) \end{array} \tag{8.28}$$

and also:

$$\begin{array}{ll} & P([A=x]\,|c\wedge e\wedge\lambda\wedge\pi) \\ \prec & \sum\limits_{B\wedge D}\left[P([A=x]\wedge c\wedge e|\pi)\times P(\lambda|A\wedge B\wedge\pi)\times P(B\wedge D\wedge e|\pi)\right] \\ \prec & P([A=x]\wedge c\wedge e|\pi)\times P([B=x]\wedge e|\pi) \\ \prec & P([B=x]\,|c\wedge e\wedge\lambda\wedge\pi) \end{array} \tag{8.29}$$

and, finally:

$$\begin{array}{ll} & P([A=x]\,|d\wedge c\wedge e\wedge\lambda\wedge\pi) \\ \prec & \sum\limits_{B}\left[P([A=x]\wedge c\wedge e|\pi)\times P(\lambda|A\wedge B\wedge\pi)\times P(B\wedge d\wedge e|\pi)\right] \\ \prec & P([A=x]\wedge c\wedge e|\pi)\times P([B=x]\wedge d\wedge e|\pi) \\ \prec & P([B=x]\,|d\wedge c\wedge e\wedge\lambda\wedge\pi) \end{array} \tag{8.30}$$

which proves that the semantic of $[\Lambda = 1]$ is respected.

8.4 Information fusion revisited using coherence variables

8.4.1 Statement of the problems

8.4.1.1 Expression of ignorance in sensor fusion models

The basic sensor model $P(S \wedge R_n) = P(S) \times P(R_n|S)$, as proposed in the previous chapter, encounters some difficulties with the expression of ignorance.

For instance, if we have no informative prior on the phenomenon then we assume that $P(S)$ is uniform. At the same time we may want to assume that we also have a noninformative prior on the reading: $P(R_n)$ is uniform. These two assumptions are not compatible, beginning with the first one we can compute $P(R_n)$:

$$P(R_n) = \sum_S [P(S) \times P(R_n|S)] \tag{8.31}$$

which, most of the time, is not uniform.

You may not be convinced that expressing both these noninformative priors is a practical necessity. Let us get back to the "false alarm" example of Section 7.5 of Chapter 7. The sensor model is the following:

$$P(S \wedge F \wedge R_n) = P(S) \times P(F) \times P(R_n|S \wedge F) \tag{8.32}$$

We use the regular sensor model if there is no false alarm ($F = 0$) but we have no information on R_n if there is a false alarm ($F = 1$). In that last case, we would like to state that $P(R_n|[F = 1])$ is uniform. This may be obtained if we state that $P(R_n|S \wedge [F = 1])$ is uniform. However, if we do so, the value of $P(R_n|S \wedge [F = 1])$ is the inverse of the cardinality of the variable R_n. This presents the drawback that a false alarm has consequences that depend on this cardinality, which is not desirable. Indeed, a false alarm on two different sensors with different ranges should have the same consequences on the fusion process.

8.4.1.2 Expert knowledge fusion

Often you want to do the fusion between information about a variable of interest S coming from N different "experts," each forming his opinion according to his own information, which can be summarized as a variable R_n.

The opinion of each expert is formalized by $P(S|R_n)$.

We are interested by the synthesis $P(S|r_1 \wedge \ldots \wedge r_N \wedge \pi)$.

A tempting approach is to mimic the naive Bayesian fusion of the previous chapter by using a decomposition of the kind:

$$P(S \wedge R_1 \wedge \ldots \wedge R_N|\pi) = \prod_{n=1}^{N} [P(R_n|\pi) \times P(S|R_n \wedge \pi)] \tag{8.33}$$

This is, of course, an invalid approach as the variable S appears several times on the left of a distribution in this decomposition.

Another possible track is to use the following decomposition:

$$\begin{aligned} & P(S \wedge S_1 \wedge \ldots \wedge S_N \wedge R_1 \wedge \ldots \wedge R_N|\pi) \\ = \; & P(S|S_1 \wedge \ldots \wedge S_N) \times \prod_{n=1}^{N} [P(R_n|\pi) \times P(S_n|R_n \wedge \pi)] \end{aligned} \tag{8.34}$$

where each expert expresses his own opinion S_n and where the distribution $P(S|S_1 \wedge \ldots \wedge S_N)$ is in charge of the synthesis of these diverging opinions. It has two essential shortcomings as (i) $P(S|S_1 \wedge \ldots \wedge S_N)$ is a very big distribution, most of the time very difficult to formalize and (ii) computing $P(S|r_1 \wedge \ldots \wedge r_N \wedge \pi)$ supposes to marginalize out the N variables S_n which is a very cumbersome computation.

Yet another approach could be to say that we use the "regular" fusion model:

$$P(S \wedge R_1 \wedge \ldots \wedge R_N|\pi) = P(S|\pi) \times \prod_{n=1}^{N} [P(R_n|S \wedge \pi)] \tag{8.35}$$

But we add that $P(R_n|S \wedge \pi)$ is obtained from a nonregular submodel π_n:

$$P(R_n|S \wedge \pi) = \frac{P(R_n|\pi_n) \times P(S|R_n \wedge \pi_n)}{\sum_{R_n} [P(R_n|\pi_n) \times P(S|R_n \wedge \pi_n)]} \tag{8.36}$$

Introducing this expression in Equation 8.35, we get:

$$\begin{aligned} & P(S \wedge R_1 \wedge \ldots \wedge R_N|\pi) \\ = \; & P(S|\pi) \times \prod_{n=1}^{N} \left[\frac{P(R_n|\pi_n) \times P(S|R_n \wedge \pi_n)}{\sum_{R_n} [P(R_n|\pi_n) \times P(S|R_n \wedge \pi_n)]} \right] \end{aligned} \tag{8.37}$$

Computing $P(S|r_1 \wedge \ldots \wedge r_N \wedge \pi)$ gives:

$$\begin{aligned} & P(S|r_1 \wedge \ldots \wedge r_N \wedge \pi) \\ = \; & \frac{P(S|\pi)}{P(r_1 \wedge \ldots \wedge r_N|\pi)} \times \prod_{n=1}^{N} \left[\frac{P(r_n|\pi_n) \times P(S|r_n \wedge \pi_n)}{\sum_{R_n} [P(R_n|\pi_n) \times P(S|R_n \wedge \pi_n)]} \right] \end{aligned} \tag{8.38}$$

Contrary to the regular sensor fusion, where $P(S|r_1 \wedge \ldots \wedge r_N \wedge \pi)$ is proportional to the product of the $P(R_n|S \wedge \pi)$ if $P(S|\pi)$ is uniform, here, $P(S|r_1 \wedge \ldots \wedge r_N \wedge \pi)$ is not proportional to the product of the $P(S|R_n \wedge \pi_n)$ terms. We lose one of the main advantages of the fusion process: its computation efficiency. Indeed, we need now to compute all the normalization terms $(\sum_{R_n} [P(R_n|\pi_n) \times P(S|R_n \wedge \pi_n)])$, which may be very cumbersome.

8.4.1.3 Coherence variable fusion

All these difficulties result from the asymmetry of the the sensor model as expressed by: $P(S \wedge R_n) = P(S) \times P(R_n|S)$.

The semantic of this expression is that the readings depend on the state.

We would rather express that the readings and the state should be "coherent." To do so, we propose a slightly different model:

$$P(S \wedge R_n \wedge \Lambda_n|\pi) = P(S|\pi) \times P(R_n|\pi) \times P(\Lambda_n|R_n \wedge S \wedge \pi) \tag{8.39}$$

where Λ_n is a coherence variable.

8.4.2 Bayesian program

This coherence variable fusion may be summarized by the following Bayesian program:

$$
Pr \begin{cases} Ds \begin{cases} Sp(\pi) \begin{cases} Va: \\ S, R_1, \ldots, R_N, \Lambda_1, \ldots, \Lambda_N \\ Dc: \\ \begin{cases} & P(S \wedge R_1 \wedge \ldots \wedge \Lambda_N|\pi) \\ = & P(S|\pi) \times \prod_{n=1}^{N} [P(R_n|\pi) \times P(\Lambda_n|S \wedge R_n \wedge \pi)] \end{cases} \\ Fo: \\ see - text \end{cases} \\ Id \end{cases} \\ Qu: \\ P(S|r_1 \wedge \ldots r_N \wedge \lambda_1 \wedge \ldots \wedge \lambda_N \wedge \pi) \\ P(S|\lambda_1 \wedge \ldots \wedge \lambda_N \wedge \pi) \end{cases} \tag{8.40}
$$

The answer to the first question is the following:

$$
\begin{aligned} & P(S|r_1 \wedge \ldots r_N \wedge \lambda_1 \wedge \ldots \wedge \lambda_N \wedge \pi) \\ = & \frac{1}{Z} \times P(S|\pi) \times \prod_{n=1}^{N} [P(\lambda_n|S \wedge r_n \wedge \pi)] \end{aligned} \tag{8.41}
$$

when the answer to the second is:

$$
\begin{aligned} & P(S|\lambda_1 \wedge \ldots \wedge \lambda_N \wedge \pi) \\ = & \frac{1}{Z'} \times P(S|\pi) \times \prod_{n=1}^{N} \left[\sum_{R_n} [P(R_n|\pi) \times P(\lambda_n|S \wedge R_n \wedge \pi)] \right] \end{aligned} \tag{8.42}
$$

Both will lead to different results depending on how $P(\lambda_n|S \wedge R_n \wedge \pi)$ is specified.

8.4.3 Instance and results

8.4.3.1 Value assignment: $P(\lambda_n|S \wedge R_n \wedge \pi) = \delta_{S=R_n}$

In that case, for the first question, either all the values r_n are equal and we get that $P([S = r_1]|r_1 \wedge \ldots r_N \wedge \lambda_1 \wedge \ldots \wedge \lambda_N \wedge \pi) = 1$, which means that we assigned this common value to S, or, they are different and the value of $P(S|r_1 \wedge \ldots r_N \wedge \lambda_1 \wedge \ldots \wedge \lambda_N \wedge \pi)$ is undefined (the normalization constant Z in Equation 8.41 is null).

This undefined value is revelatory in that we ask a question with contradictory hypotheses.[6]

For the second question we get:

$$\begin{aligned} & P(S|\lambda_1 \wedge \ldots \wedge \lambda_N \wedge \pi) \\ = & \frac{1}{Z} \times P(S|\pi) \times \prod_{n=1}^{N} [P([R_n = S] \,|\pi)] \end{aligned} \tag{8.43}$$

$P(S|\lambda_1 \wedge \ldots \wedge \lambda_N \wedge \pi)$ is the normalized product of all the priors.

8.4.3.2 Distance assignment: $P(\lambda_n|S \wedge R_n \wedge \pi) \propto e^{-d_n(S,R_n)}$

Let us suppose that we have a distance d_n that can be applied to S and R_n.

We may want to express that the closer S is to R_n, the more coherent they are.

This can easily be done by stating:

$$P(\lambda_n|S \wedge R_n \wedge \pi) \propto e^{-d_n(S,R_n)} \tag{8.44}$$

In that case we get for the first question that:

$$\begin{aligned} & P(S|r_1 \wedge \ldots r_N \wedge \lambda_1 \wedge \ldots \wedge \lambda_N \wedge \pi) \\ = & \frac{1}{Z} \times P(S|\pi) \times \prod_{n=1}^{N} \left[e^{-d_n(S,R_n)}\right] \end{aligned} \tag{8.45}$$

If, for instance, S, R_1, and R_2 are three integer variables varying between 1 and 100 and $d_n(S, R_n) = \dfrac{abs(S - R_n)}{\sigma_n}$, we can compute the distribution on S $P(S \mid R_0 = 50 \wedge R_1 = 70 \wedge \lambda_0 \wedge \lambda_1)$ when we have two different readings $R_0 = 50$ and $R_1 = 70$. Figure 8.3 shows two cases: one with an identical precision for the reading $\sigma_0 = \sigma_1 = 10$ and the other for two different precisions: $\sigma_0 = 10$ and $\sigma_1 = 20$.

For the second question, we have:

$$\begin{aligned} & P(S|\lambda_1 \wedge \ldots \wedge \lambda_N \wedge \pi) \\ = & \frac{1}{Z} \times P(S|\pi) \times \prod_{n=1}^{N} \left[\sum_{R_n} \left[P(R_n|\pi) \times e^{-d_n(S,R_n)}\right]\right] \end{aligned} \tag{8.46}$$

[6]The same remark as above. See Section 2.6.2 titled "Godel's theorem" of Jaynes' book [2003] (pages 45–47) for a very stimulating discussion on this subject and about the perspectives it opens relative to the meaning of Godel's theorem in probability.

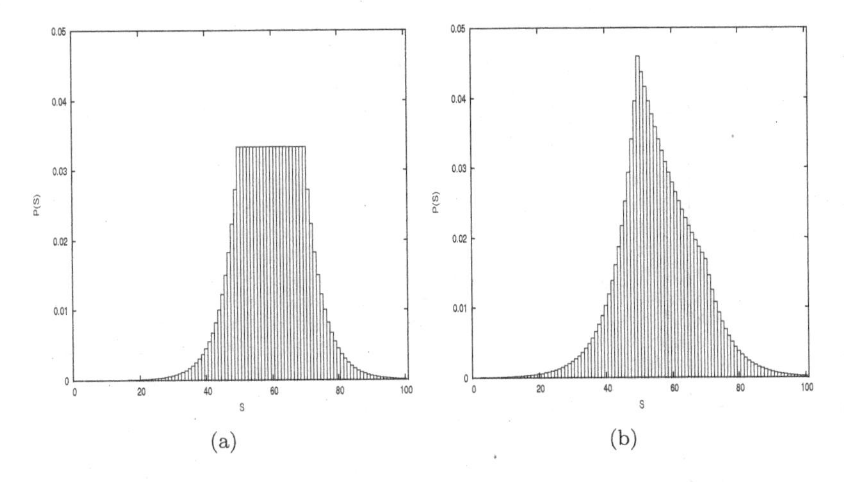

FIGURE 8.3: Distance assignment with two precisions: (a) $\sigma_0 = \sigma_1 = 10$ and (b) $\sigma_0 = 10, \sigma_1 = 20$.

8.4.3.3 Sensor fusion: $P(\lambda_n|S \wedge R_n \wedge \pi) = P(R_n|S \wedge \pi_n)$

If we state that $P(\lambda_n|S \wedge R_n \wedge \pi) = P(R_n|S \wedge \pi_n)$ we get for the first question:

$$\begin{aligned} & P(S|r_1 \wedge \ldots r_N \wedge \lambda_1 \wedge \ldots \wedge \lambda_N \wedge \pi) \\ = & \frac{1}{Z} \times P(S|\pi) \times \prod_{n=1}^{N} \left[P(r_n|S \wedge \pi_n) \right] \end{aligned} \tag{8.47}$$

which is the exact same expression as for the naive Bayesian fusion (see Equation 7.4).

However, we now have the freedom to specify priors for the readings as the $P(R_n|\pi)$ appears in the decomposition of the Bayesian program (Equation 8.40). The answer to the second question is then:

$$\begin{aligned} & P(S|\lambda_1 \wedge \ldots \wedge \lambda_N \wedge \pi) \\ = & \frac{1}{Z'} \times P(S|\pi) \times \prod_{n=1}^{N} \left[\sum_{R_n} \left[P(R_n|\pi) \times P(R_n|S \wedge \pi_n) \right] \right] \end{aligned} \tag{8.48}$$

8.4.3.4 Sensor fusion with false alarm revisited

Let us revisit, using coherence variables, the sensor fusion with the false alarm example in Section 7.5 of Chapter 7.

The Bayesian program (7.25) may be transformed into:

$$Pr\begin{cases} Ds\begin{cases} Sp(\pi)\begin{cases} Va: \\ X,Y,F_2,F_3,D_2,D_3,\Lambda_2,\Lambda_3 \\ Dc: \\ \begin{cases} & P(X\wedge Y\wedge\ldots\wedge\Lambda_3|\pi) \\ = & P(X\wedge Y|\pi)\times\prod_{n=2}^{3}\left[P(D_n|\pi)\times P(F_n|\pi)\right] \\ & \times\prod_{n=2}^{3}\left[P(\Lambda_n|X\wedge Y\wedge D_n\wedge F_n\wedge\pi)\right] \end{cases} \\ Fo: \\ \begin{cases} & P(X\wedge Y|\pi)=Uniform \\ & P([F_2=1]|\pi)=0.3 \\ & P([F_3=1]|\pi)=0.3 \\ & P(D_n|\pi)=Uniform \\ & P(\lambda_n|X\wedge Y\wedge D_n\wedge[F_n=0]\wedge\pi) \\ = & G\left([\mu=d_n],\left[\sigma=1+\frac{d_n}{10}\right]\right) \\ & P(\lambda_n|X\wedge Y\wedge D_n\wedge[F_n=1]\wedge\pi)=1/2 \end{cases} \end{cases} \\ Id \end{cases} \\ Qu: \\ P(X\wedge Y|d_2\wedge d_3\wedge\lambda_2\wedge\lambda_3\wedge\pi) \end{cases} \tag{8.49}$$

If there is no false alarm, the position of the boat ($X\wedge Y$) and the different distances should be coherent and, on the contrary, if there is a false alarm there is no reason to justify any relation between the position and the observed readings.

When there is no false alarm a good measure of the coherence is given by the regular sensor model. This can be encoded as:

$$\begin{aligned} & P(\lambda_n|X\wedge Y\wedge D_n\wedge[F_n=0]\wedge\pi) \\ = & P(D_n|X\wedge Y\wedge\pi) \\ = & G\left([\mu=d_n],\left[\sigma=1+\frac{d_n}{10}\right]\right) \end{aligned} \tag{8.50}$$

When there is a false alarm, we do not know if the position of the boat and the distances are coherent:

$$P(\lambda_n|X\wedge Y\wedge D_n\wedge[F_n=1]\wedge\pi)=1/2 \tag{8.51}$$

$P(X \wedge Y | d_2 \wedge d_3 \wedge \lambda_2 \wedge \lambda_3 \wedge \pi)$ is given by:

$$
\begin{aligned}
&P(X \wedge Y | d_2 \wedge d_3 \wedge \lambda_2 \wedge \lambda_3 \wedge \pi) \\
= & \frac{1}{Z} \times \sum_{F_2 \wedge F_3} \left[\begin{array}{c} P(X \wedge Y | \pi) \times \prod_{n=2}^{3} [P(F_n | \pi)] \\ \times \prod_{n=2}^{3} [P(\lambda_n | X \wedge Y \wedge d_n \wedge F_n \wedge \pi)] \end{array} \right]
\end{aligned} \tag{8.52}
$$

which leads to the same results as in Chapter 7.

8.4.3.5 Expert fusion: $P(\lambda_n | S \wedge R_n \wedge \pi) = P(S | R_n \wedge \pi_n)$

If we state that $P(\lambda_n | S \wedge R_n \wedge \pi) = P(S | R_n \wedge \pi_n)$ then we solve correctly the expert fusion problem.

The answer to the first question is:

$$
\begin{aligned}
&P(S | r_1 \wedge \ldots r_N \wedge \lambda_1 \wedge \ldots \wedge \lambda_N \wedge \pi) \\
= & \frac{1}{Z} \times P(S | \pi) \times \prod_{n=1}^{N} [P(S | r_n \wedge \pi_n)]
\end{aligned} \tag{8.53}
$$

which has the appealing property of being again a simple product of probability distributions as in the sensor fusion case.

The answer to the second question is:

$$
\begin{aligned}
&P(S | \lambda_1 \wedge \ldots \wedge \lambda_N \wedge \pi) \\
= & \frac{1}{Z'} \times P(S | \pi) \times \prod_{n=1}^{N} \left[\sum_{R_n} [P(R_n | \pi) \times P(S | R_n \wedge \pi_n)] \right] \\
= & \frac{1}{Z'} \times P(S | \pi) \times \prod_{n=1}^{N} [P(S | \pi_n)]
\end{aligned} \tag{8.54}
$$

It is the normalized product of all the priors on S.

8.4.3.6 Proscriptive versus prescriptive assignment

Instead of specifying that we want a given variable to have a given value, we often desire to specify that we do not want this variable to take this value. We want a "proscriptive" specification instead of a "prescriptive" one.

For instance, as we saw in the inverse programming example in Section 7.6 in Chapter 7, going toward the nth landmark may be expressed as:

$$
P(B_n | H \wedge \pi_n) = B(\mu = H, \sigma) \tag{8.55}
$$

where B_n is the bearing of the landmark and H is the heading direction. We state in this expression that when going toward a landmark, the bearing of this landmark should be approximately the same as the heading direction.

A commonly used approach to avoid an obstacle is to specify what should be done to avoid it. For instance, we could state that $P(B_n|H \wedge \pi_n) = B(\mu = H + 90, \sigma)$ to express that avoiding a landmark consists in heading 90 to its left. This has two drawbacks: (i) by prescribing a direction where to go we provide more information than necessary, and this extra information may be a handicap in certain situations and (ii) fusion of prescriptive information may convey very poor information when evaluating a solution in the queues of the distributions.

A more interesting approach is to say that avoiding a landmark consists in not heading toward this landmark. This is what we did in Section 7.6 by specifying:

$$P(B_n|H) = \frac{1}{Z}\left[1 - B(\mu = H, \sigma')\right] \tag{8.56}$$

for the landmark to avoid.

However, there is an even more simple solution using the values of coherence variables to decide if we want B_n and H to be coherent ($[\Lambda = 1]$ meaning heading toward this landmark) or to be incoherent ($[\Lambda = 0]$ meaning not heading toward this landmark).

$$Pr\begin{cases} Ds\begin{cases} Sp(\pi)\begin{cases} Va: \\ H, B_1, B_2, B_3, \Lambda_1, \Lambda_2, \Lambda_3 \\ Dc: \\ \begin{cases} P(H \wedge B_1 \wedge \ldots \wedge \Lambda_3|\pi) \\ = P(H|\pi) \times \prod_{n=1}^{3}\left[P(B_n|\pi) \times P(\Lambda_n|H \wedge B_n \wedge \pi)\right] \end{cases} \\ Fo: \\ \begin{cases} P(H|\pi) = Uniform \\ P(B_n|\pi) = Uniform \\ P(\lambda_n|H \wedge B_n \wedge \pi) = B(B_n, \mu = H, \sigma) \end{cases} \end{cases} \\ Id \end{cases} \\ Qu: \\ P\left(H|b_1 \wedge \lambda_1 \wedge b_2 \wedge \bar{\lambda_2} \wedge b_3 \wedge \bar{\lambda_3} \wedge \pi\right) \end{cases} \tag{8.57}$$

This way we obtain for the question $P\left(H|b_1 \wedge \lambda_1 \wedge b_2 \wedge \bar{\lambda_2} \wedge b_3 \wedge \bar{\lambda_3} \wedge \pi\right)$:

$$\begin{aligned} & P\left(H|b_1 \wedge \lambda_1 \wedge b_2 \wedge \bar{\lambda_2} \wedge b_3 \wedge \bar{\lambda_3} \wedge \pi\right) \\ = & \frac{1}{Z} \times B(B_1, \mu = H, \sigma) \times \left[1 - B(B_2, \mu = H, \sigma)\right] \times \left[1 - B(B_2, \mu = H, \sigma)\right] \end{aligned} \tag{8.58}$$

which is the same expression as in Section 7.6.3 and, of course leads to the same result (see Figure 7.10 in Chapter 7).

The advantage is that with the same model, we can, by changing the question, decide which landmarks are attractive and which are repulsive.

8.5 Reasoning with soft evidence

8.5.1 Statement of the problem

The information at our disposal to reason with is often probability distributions on variables instead of the values of these variables. In the literature, this is often called: "reasoning with soft evidence."

For instance, if we have the most simple model we can imagine:

$$P(A \wedge B|\pi) = P(A|\pi) \times P(B|A \wedge \pi) \tag{8.59}$$

Instead of being interested by the classical questions $P(B|a \wedge \pi)$ or $P(A|b \wedge \pi)$, we may want to infer the probability of B knowing a distribution on A $P(A|\pi')$ or the probability of A knowing a distribution $P(B|\pi')$.

A tempting notation for both these questions is $P(B|P(A|\pi') \wedge \pi)$ and $P(A|P(B|\pi') \wedge \pi)$. However, these are not valid notations of the formalism for two reasons: (i) there is no variable in the model to encode $P(A|\pi')$ and $P(B|\pi')$ and (ii) $P(A|\pi')$ and $P(B|\pi')$ are not values but probability distributions and, as such, are not supposed to appear on the right part of a question.

The first question "$P(B|P(A|\pi') \wedge \pi)$" is often solved by replacing the prior $P(A|\pi)$ by the soft evidence $P(A|\pi')$ and by computing $P(B|\pi)$:

$$P(B|\pi) = \sum_{A} [P(A|\pi') \times P(B|A \wedge \pi)] \tag{8.60}$$

Even if it performs something which intuitively is close to what we would like to do, it is not completely satisfactory because we have to use two tricks, on one hand, by replacing the prior $P(A|\pi)$ by $P(A|\pi')$ and, on the other hand, by replacing the question "$P(B|P(A|\pi') \wedge \pi)$" by $P(B|\pi)$. Furthermore, when replacing $P(A|\pi)$ by $P(A|\pi')$ we lose the freedom to express a prior on A.

Most of the time, no practical solution is proposed for the second question: "$P(A|P(B|\pi') \wedge \pi)$" as the two previous tricks do not work.

Coherence variables propose a generic solution for the formal treatment of soft evidence reasoning using the following decomposition:

$$\begin{aligned} & P(\Pi_{A'} \wedge A' \wedge \Lambda_A \wedge A \wedge B|\pi) \\ = \; & P(\Pi_{A'}|\pi)\, P(A'|\Pi_{A'} \wedge \pi) \times P(\Lambda_A|A' \wedge A \wedge \pi) \times P(A \wedge B|\pi) \end{aligned} \tag{8.61}$$

where A' is a "mirror" variable of A, $\Pi_{A'}$ stands for the parameter of the distribution on A', and where Λ_A is a coherence variable used to constrain the distribution on A' and A to be proportional for their common values.

8.5.2 Bayesian program

More details about this solution are provided by the complete Bayesian program:

$$
Pr\begin{cases}
Ds\begin{cases}
Sp(\pi)\begin{cases}
Va:\\
\Pi_{A'}, A', \Lambda_A, A, B\\
Dc:\\
\begin{cases}
 & P(\Pi_{A'} \wedge A' \wedge \Lambda_A \wedge A \wedge B|\pi)\\
= & P(\Pi_{A'}|\pi) \times P(A'|\Pi_{A'} \wedge \pi)\\
 & \times P(\Lambda_A|A' \wedge A \wedge \pi) \times P(A \wedge B|\pi)
\end{cases}\\
Fo:\\
\begin{cases}
P(\Pi_{A'}|\pi) = Uniform\\
P(A'|\Pi_{A'} \wedge \pi) = f(A', \Pi_{A'})\\
P(\lambda_A|A' \wedge A \wedge \pi) = \delta_{A=A'}
\end{cases}
\end{cases}\\
Id
\end{cases}\\
Qu:\\
P(A|\pi_{A'} \wedge \lambda_A \wedge \pi)\\
P(B|\pi_{A'} \wedge \lambda_A \wedge \pi)
\end{cases}
\tag{8.62}
$$

Our goal is to propose a clearly formalized form to transcript the semantic of the question loosely stated as "$P(B|P(A|\pi') \wedge \pi)$."

The proposed form is:

$$
P(B|\pi_{A'} \wedge \lambda_A \wedge \pi) \tag{8.63}
$$

As already mentioned, the first difficulty is to assign the variable A with a probability distribution imposed as known for a given inference. This can be easily solved using coherence variables which have been designed especially to solve this kind of problem. It leads to the following decomposition:

$$
\begin{aligned}
& P(A' \wedge \Lambda_A \wedge A \wedge B|\pi)\\
= \; & P(A'|\pi) \times P(\Lambda_A|A' \wedge A \wedge \pi) \times P(A|\pi) \times P(B|A \wedge \pi)
\end{aligned}
\tag{8.64}
$$

where we add a variable A' to mirror A and a coherence variable Λ_A to ensure that the probability distributions on A' and A are bound.

The second difficulty is the need to represent probability distribution by values. This can be accomplished by adding a parameter variable $\Pi_{A'}$ such that if $\Pi_{A'}$ has a known value $\pi_{A'}$ then $P(A'|\pi_{A'} \wedge \pi)$ is completely determined. We first assume that we have a uniform prior on the parameter: $P(\Pi_{A'}|\pi) = Uniform$. Second, we state that if the parameters are known then the value of the probability may be obtained by a function f (usually called the "parametric form"): $P(A'|\pi_{A'} \wedge \pi) = f(A', \pi_{A'})$. For instance, if we want to have a Gaussian then $P(A'|\pi_{A'} \wedge \pi) = G(A', \mu_{A'}, \sigma_{A'})$ where we have two parameters $\mu_{A'}$ the mean and $\sigma_{A'}$ the standard deviation and where $G(x,y,z) = \frac{1}{\sqrt{2\pi} \times z} \times e^{-\frac{(x-y)^2}{z^2}}$.

The answer to the question $P(B|\pi_{A'} \wedge \lambda_A \wedge \pi)$ is obtained as:

$$\begin{aligned} & P(B|\pi_{A'} \wedge \lambda_A \wedge \pi) \\ = & \frac{1}{Z} \times \sum_{A \wedge A'} \left[\begin{array}{l} P(\pi_{A'}|\pi) \times P(A'|\pi_{A'} \wedge \pi) \\ \times P(\lambda_A|A' \wedge A \wedge \pi) \times P(A \wedge B|\pi) \end{array} \right] \\ = & \frac{1}{Z'} \times \sum_{A} \left[P([A' = A]|\pi_{A'} \wedge \pi) \times P(A \wedge B|\pi) \right] \end{aligned} \quad (8.65)$$

If $P(A \wedge B|\pi) = P(A|\pi) \times P(B|A \wedge \pi)$ we get:

$$\begin{aligned} & P(B|\pi_{A'} \wedge \lambda_A \wedge \pi) \\ = & \frac{1}{Z'} \times \sum_{A} \left[P([A' = A]|\pi_{A'} \wedge \pi) \times P(A|\pi) \times P(B|A \wedge \pi) \right] \end{aligned} \quad (8.66)$$

which is similar to Equation 8.60 but where both appear as the imposed probability distribution $P([A' = A]|\pi_{A'} \wedge \pi)$ and the prior $P(A|\pi)$.

If $P(A \wedge B|\pi) = P(B|\pi) \times P(A|B \wedge \pi)$ we get:

$$\begin{aligned} & P(B|\pi_{A'} \wedge \lambda_A \wedge \pi) \\ = & \frac{1}{Z'} \times \sum_{A} \left[P([A' = A]|\pi_{A'} \wedge \pi) \times P(B|\pi) \times P(A|B \wedge \pi) \right] \end{aligned} \quad (8.67)$$

which is an answer to the second question usually not treated.

8.5.3 Instance and results

Modern radar, especially military radar, provides not only readings for the distance and bearing of a target, but also some evaluations of the confidence in these readings. Typically, they give the two parameters of a Gaussian, the mean being the reading and the standard deviation being this confidence measure.

If our goal is to fusion information of this kind about different targets to localize as in the previous chapter, then we are exactly in the situation of reasoning with soft evidence just described. The entries are the parameters of one probability distribution by target and the question is to localize knowing these parameters.

The Bayesian program (Section 7.6)[7] corresponding to naive sensor fusion then becomes (we omit the bearings to simplify the notation):

[7]Section 7.1 of Chapter 7.

$$
Pr\begin{cases} Ds\begin{cases} Sp(\pi)\begin{cases} Va: \\ X, Y, D_1, D_2, D_3, D'_1, D'_2 D'_3, \Lambda_1, \Lambda_2, \Lambda_3, \\ M_1, M_2, M_3, \Sigma_1, \Sigma_2, \Sigma_3 \\ Dc: \\ \begin{cases} & P(X \wedge Y \wedge \ldots \wedge \Sigma_3|\pi) \\ = & \prod_{n=1}^{3}[P(M_n \wedge \Sigma_n|\pi) \times P(D'_n|M_n \wedge \Sigma_n \wedge \pi)] \\ & \times P(X \wedge Y|\pi) \times \prod_{n=1}^{3}[P(D_n|X \wedge Y \wedge \pi)] \\ & \times \prod_{n=1}^{3}[P(\Lambda_n|D_n \wedge D'_n \wedge \pi)] \end{cases} \\ Fo: \\ \begin{cases} P(D'_n|\mu_n \wedge \sigma_n \wedge \pi) = B(D'_n, \mu_n, \sigma_n) \\ P(X \wedge Y|\pi) = Uniform \\ P(D_n|X \wedge Y \wedge \pi) = \delta_{D_n = d-n(X,Y)} \\ P(\Lambda_n|D_n \wedge D'_n \wedge \pi) = \delta_{D_n = D'_n} \end{cases} \end{cases} \\ Id \end{cases} \\ Qu: \\ P(X \wedge Y|\mu_1 \wedge \mu_2 \wedge \mu_3 \wedge \sigma_1 \wedge \sigma_2 \wedge \sigma_3 \wedge \lambda_1 \wedge \lambda_2 \wedge \lambda_3 \wedge \pi) \end{cases} \tag{8.68}
$$

$P(D'_n|\mu_n \wedge \sigma_n \wedge \pi)$ is a bell-shaped distribution of parameters μ_n and σ_n.

$P(D_n|X \wedge Y \wedge \pi)$, the previous sensor model, does not necessarily need to encode uncertainty anymore as this uncertainty is provided by the sensor itself with the soft evidence μ_n and σ_n. It may be chosen as a Dirac distribution, taking a value of one when D_n is equal to the distance $d-n(X,Y)$ of the boat from the nth target.

$P(\Lambda_n|D_n \wedge D'_n \wedge \pi)$ is the coherence variable Dirac distribution used to bound the soft evidence distribution from D' to D.

The answer to the localization question is:

$$
\begin{aligned} & P(X \wedge Y|\mu_1 \wedge \mu_2 \wedge \mu_3 \wedge \sigma_1 \wedge \sigma_2 \wedge \sigma_3 \wedge \lambda_1 \wedge \lambda_2 \wedge \lambda_3 \wedge \pi) \\ \propto & \prod_{n=1}^{3}[B(d-n(X,Y), \mu_n, \sigma_n)] \end{aligned} \tag{8.69}
$$

It may seem a very complicated model to finally obtain the same result as with the naive fusion model.

However, we have now a complete separation between the soft evidence modeling the sensors and the internal model describing the geometrical relations. Especially, we are completely free to specify this internal model as we want; all dependencies and all priors are acceptable.

8.6 Switch

8.6.1 Statement of the problem

A complex model is most of the time made of several submodels, partially independent from one another, only connected by well-defined interfaces.

An interesting feature is to be able to switch on or off some part of the model when needed. This feature could be implemented with coherence variables.

8.6.2 Bayesian program

Let us take an example in which we have three submodels, each of them sharing one variable used as an interface.

We have the following Bayesian program:

$$
Pr\begin{cases}
Ds\begin{cases}
Sp(\pi)\begin{cases}
Va: \\
A, I_1, B, I_2, C, I_3, \Lambda_{12}, \Lambda_{13}, \Lambda_{23} \\
Dc: \\
\begin{cases}
 & P(A\wedge\ldots\wedge\Lambda_{23}|\pi) \\
= & P(A\wedge I_1|\pi)\times P(B\wedge I_2|\pi) \\
 & \times P(C\wedge I_3|\pi) \\
 & \times P(\Lambda_{12}|I_1\wedge I_2\wedge\pi)\times P(\Lambda_{13}|I_1\wedge I_3\wedge\pi) \\
 & \times P(\Lambda_{23}|I_2\wedge I_3\wedge\pi)
\end{cases} \\
Fo: \\
\begin{cases} P(\Lambda_{ij}|I_i\wedge I_j\wedge\pi)=\delta_{I_i=I_j}\end{cases}
\end{cases} \\
Id
\end{cases} \\
Qu: \\
P(A|b\wedge\lambda_{12}\wedge\pi) \\
P(A|b\wedge c\wedge\lambda_{12}\wedge\lambda_{13}\wedge\pi) \\
P(C|b\wedge\lambda_{12}\wedge\lambda_{13}\wedge\pi) \\
P(C|b\wedge\lambda_{23}\wedge\pi)
\end{cases}
\tag{8.70}
$$

This decomposition is a good example to prove that algebraic notation is more powerful than graphical representation. Indeed, a graphical representation of such a simple decomposition is quite difficult to present and cannot be proposed without augmenting the model with some supplementary variables.[8]

We have three submodels: $P(A \wedge I_1)$, $P(B \wedge I_2)$, and $P(C \wedge I_3)$ and three coherence variables Λ_{12} between the interface variables I_1 and I_2, Λ_{13} between I_1 and I_3, and Λ_{23} between I_2 and I_3.

Among the possible different questions let us examine four of them:

$$\begin{aligned} & P(A|b \wedge \lambda_{12} \wedge \pi) \\ = \; & \frac{1}{Z} \times \sum_{I_1} \left[P(A \wedge I_1|\pi) \times P(b \wedge [I_2 = I_1]|\pi) \right] \end{aligned} \tag{8.71}$$

where two submodels are activated to search for the probability distribution on A knowing b.

$$\begin{aligned} & P(A|b \wedge c \wedge \lambda_{12} \wedge \lambda_{13} \wedge \pi) \\ = \; & \frac{1}{Z} \times \sum_{I_1} \left[P(A \wedge I_1|\pi) \times P(b \wedge [I_2 = I_1]|\pi) \times P(c \wedge [I_3 = I_1]|\pi) \right] \end{aligned} \tag{8.72}$$

where the three submodels are used to compute the probability on A knowing b and c.

$$\begin{aligned} & P(C|b \wedge \lambda_{12} \wedge \lambda_{13} \wedge \pi) \\ = \; & \frac{1}{Z} \times \sum_{A \wedge I_1} \left[P(A \wedge I_1|\pi) \times P(b \wedge [I_2 = I_1]|\pi) \times P(c \wedge [I_3 = I_1]|\pi) \right] \end{aligned} \tag{8.73}$$

where the three submodels are used to find the probability on C knowing b.

$$\begin{aligned} & P(C|b \wedge \lambda_{23} \wedge \pi) \\ = \; & \frac{1}{Z} \times \sum_{I_2} \left[P(b \wedge I_2|\pi) \times P(C \wedge [I_3 = I_2]|\pi) \right] \end{aligned} \tag{8.74}$$

where only the two submodels $P(B \wedge I_2)$ and $P(C \wedge I_3)$ are used to compute the probability on C knowing b.

8.6.3 Instance and results

For instance, such an approach has been used in the PhD thesis of Estelle Gilet (see Gilet [2009] and Gilet et al. [2011]) titled "Bayesian Action–Perception Computational Model: Interaction of Production and Recognition of Cursive Letters."

The purpose of this work is to study the complete perception–action loop

[8]This is one of the arguments in favor of using Bayesian programming instead of Bayesian networks, see the FAQ-FAM, "Bayesian programming versus Bayesian networks" in Section 16.3.

involved in handwriting. It proposes a mathematical formulation for the whole loop, based on a probabilistic model called the Bayesian Action–Perception (BAP) model. Six cognitive tasks are solved using Bayesian inference: (i) letter recognition (purely sensory), (ii) writer recognition, (iii) letter production (with different effectors), (iv) copying of trajectories, (v) copying of letters, and (vi) letter recognition (with internal simulation of movements).

One of the main interrogation is the relative role of different submodels, essentially the perception submodel, the motor one and the internal representation submodel. Coherence variables have been introduced to be able to solve the different cognitive tasks choosing which submodel to activate to compute the answer.

In this work, each I_1, I_2, and I_3 is a conjunction of about 40 scalar variables. These variables have been selected to best discriminate between letters. For example, some are a temporal discretization of the key features found in the geometrical representation of letters such as curvature.

The model of $P(A \wedge I_1)$ is learned. It captures the model of letters for a given writer. $P(B \wedge I_2)$ is a model of the perceptive system (reading) to pass visual information B to the internal encoding I_2. Finally, $P(C \wedge I_3)$ is a model of the motor system (writing) describing how to control an effector C knowing the internal representation of a letter I_2.

The meaning of the first question ($P(A|b \wedge \lambda_{12} \wedge \pi)$) is to recognize the letter and writer using only visual information.

$P(A|b \wedge c \wedge \lambda_{12} \wedge \lambda_{13} \wedge \pi)$ stands for recognizing the letter and writer using both perceptive and motor information.

$P(C|b \wedge \lambda_{12} \wedge \lambda_{13} \wedge \pi)$ is a letter copy taking into account the writer style. In a sense, the computation may be seen as, first, recognizing the read letter and, second, as generating the appropriate motor command to write the recognized letter.

Finally, $P(C|b \wedge \lambda_{23} \wedge \pi)$ is a trace copy, where the motor commands are generated without any recognition of the letter, but rather by trying to reproduce exactly the read trajectory even if it does not correspond to any known letter.

8.7 Cycles

8.7.1 Statement of the problem

Another common problem appears when you have several "tracks" of reasoning to draw the same conclusion.

In this case, you have cycles in your Bayesian graph which lead to problems expressing the model in the Bayesian programming formalism.

The most simple case may be expressing this with only three variables A, B, and C.

Let us suppose that we know, on the one hand, a dependency between A and C ($P(C|A)$) and, on the other hand, a dependency between A and B ($P(B|A)$) followed by a dependency between B and C ($P(C|B)$).

In this case, we cannot express the joint probability distribution as a product of these three elementary distributions, as C appears twice on the left.

An attractive solution is to write that $P(A \wedge B \wedge C) = P(A) \times P(B|A) \times P(C|A \wedge B)$, but then the known distributions $P(C|A)$ and $P(C|B)$ do not appear in the decomposition, instead in this decomposition the distribution $P(C|A \wedge B)$ appears, which is not known and may be very difficult to express.

Here again, the coherence variables offer an easy solution. C is the variable that may be deduced from A when a new variable C' may be deduced from the inference chain starting from A to infer B to finally infer C'. We then only need a coherence variable Λ to express that the distributions on C and C' should be "equal."

8.7.2 Bayesian program

This leads to the following Bayesian program:

$$
Pr\begin{cases}
Ds\begin{cases}
Sp(\pi)\begin{cases}
Va: \\
A, B, C, C', \Lambda \\
Dc: \\
\begin{cases}
 & P(A \wedge B \wedge C \wedge C' \wedge \Lambda|\pi) \\
= & P(A|\pi) \times P(C|A \wedge \pi) \\
 & \times P(B|A \wedge \pi) \times P(C'|B \wedge \pi) \\
 & \times P(\Lambda|C \wedge C' \wedge \pi)
\end{cases} \\
Fo: \\
\begin{cases} P(\Lambda|C \wedge C' \wedge \pi) = \delta_{C=C'} \end{cases}
\end{cases} \\
Id
\end{cases} \\
Qu: \\
P(C|a \wedge \lambda \wedge \pi)
\end{cases}
\tag{8.75}
$$

8.7.3 Instance and results

There are numerous such examples, however nice and simple instances may be extracted from robotic Computer Aided Design (CAD) systems taking into account uncertainty (see for instance the PhD work of Kamel Mekhnacha [Mekhnacha, 1999; Mekhnacha et al., 2001]).

Let us take a very simple example made of a robot in a one-dimensional environment (see Figure 8.4).

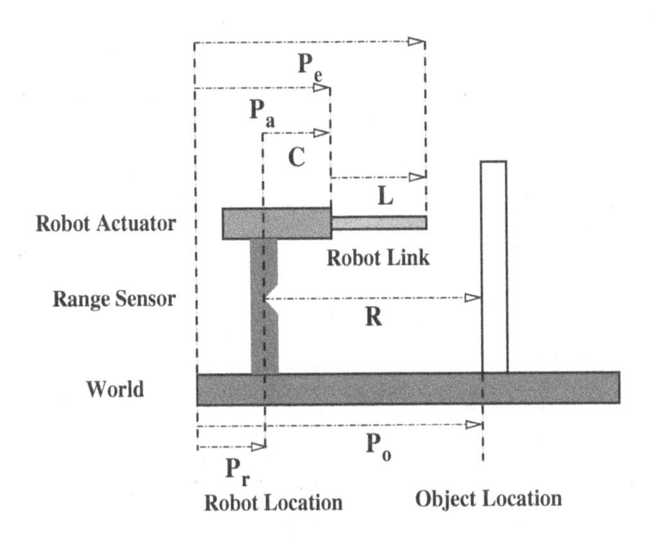

FIGURE 8.4: A simple planar robot with one degree of freedom.

The position of this robot's base in the world reference frame is stored in a variable P_r.

The robot has a range sensor that is able to measure the distance of an object (variable R). Knowing P_r and R you may infer the position of the object P_o as in a perfect world we would have: $P_o = P_r + R$.

The robot bears a prismatic joint. The command of this joint is the variable C. Knowing P_r and C you can infer the position of the link P_a as, yet in a perfect world we would have: $P_a = P_r + C$.

The length of the link is supposed to be L. Knowing P_a and L, we know the position of the extremity of the arm P_e as $P_e = P_a + L$.

However, the world is not perfect. We may have some uncertainty on the position of the robot $P(P_r)$, the precision of the sensor $P(P_o|P_r \wedge R)$, the command of the robot $P(P_a|P_r \wedge C)$, and even the length of the arm $P(P_e|P_a \wedge L)$.

The goal of the robot is to touch the object with its arm. When the contact is made, then we have $P_o < P_e$ and $P_e - P_o < \epsilon_t$. The two kinematic chains make a closed loop. This is modeled by a coherence variable Λ equal to one if and only if the contact is realized.

This finally leads to the following Bayesian program:

$$
Pr\begin{cases}
Ds\begin{cases}
Sp(\pi)\begin{cases}
Va: \\
P_r, R, P_o, C, P_a, L, P_e, \Lambda \\
Dc: \\
\begin{cases}
 & P(P_r \wedge \ldots \wedge \Lambda|\pi) \\
= & P(P_r|\pi) \times P(R|\pi) \times P(P_o|P_r \wedge R \wedge \pi) \\
 & \times P(C|\pi) \times P(P_a|P_r \wedge C \wedge \pi) \\
 & \times P(L|\pi) \times P(P_e|P_a \wedge L \wedge \pi) \\
 & \times P(\Lambda|P_o \wedge P_e \wedge \pi)
\end{cases} \\
Fo: \\
\begin{cases}
P(P_o|P_r \wedge R \wedge \pi) = Normal(P_r, \epsilon_r) \\
P(C|\pi) = Uniform \\
P(P_a|P_r \wedge C \wedge \pi) = Normal(p_r + c, \epsilon_c) \\
P(L|\pi) = Normal(L_0, \epsilon_L) \\
P(P_e|P_a \wedge L \wedge \pi) = \delta_{p_e = p_a + l} \\
P(\Lambda|P_o \wedge P_e \wedge \pi) = \delta_{0 \leq p_e - p_o \leq \epsilon_t}
\end{cases}
\end{cases} \\
Id
\end{cases} \\
Qu: \\
P(C|r \wedge \lambda \wedge \pi) \\
P(P_o|r \wedge c \wedge \lambda \wedge \pi) \\
P(L|r \wedge c \wedge \lambda \wedge \pi)
\end{cases}
\tag{8.76}
$$

This program assumes the following error models:

1. $Normal(P_r, \epsilon_r)$: error model for the sensor.
2. $Normal(p_r + c, \epsilon_c)$: error model for the control.
3. $Normal(L_0, \epsilon_L)$: error model for the manufacturing.
4. $\delta_{0 \leq p_e - p_o \leq \epsilon_t}$: error model for the task.

Numerous interesting questions may be asked about this model. Let us take three of them as examples:

$$P(C|r \wedge \lambda \wedge \pi) \tag{8.77}$$

where knowing the distance measured by the sensor we search the control that will drive the robot to the contact with the object (inverse kinematic).

$$(P_o|r \wedge c \wedge \lambda \wedge \pi) \tag{8.78}$$

where we look for the position of the object knowing both the reading of the sensor and the command that leads to contact (localization).

$$P(L|r \wedge c \wedge \lambda \wedge \pi) \tag{8.79}$$

where we derive the probability distribution on the length of the arm knowing the sensor's reading and the command (calibration).

The program "chapter8/inverse_k.py" is an implementation of Equation 8.76. It uses intervals and continuous variables. Uncertainties are modeled using conditional normals which use functions to compute the mean.

```
Pa =plSymbol("Pa",worldrange)
def actuator_model(Output_,Input_):
    Output_[0]=Input_[Pr]+Input_[C]

PPa=plCndNormal(Pa,Pr^C, \
                    plPythonExternalFunction(Pr^C, \
                    actuator_model),
                    2)
```

A functional Dirac is used to implement the distribution on the cohrence variable Λ:

```
Lambda = plSymbol("Lambda",plIntegerType(0,1))
def Coherence(Output_,Input_):
    r = Input_[Pe]- Input_[Po]
    if  r.to_float() > 0 and r.to_float() < 1 :
        Output_[Lambda]=1
    else :
        Output_[Lambda]=0

DiracLambda=plFunctionalDirac(Lambda,Pe^Po, \
               plPythonExternalFunction(Lambda,Pe^Po, \
               Coherence))

inverse_kinematic = model.ask_mc_sample(C,Lambda^R,500)
```

Approximate inference is made by controlling the number of samples used to approximate each integral of the inference process.

The inverse kinematic (Equation 8.77) is obtained with:

```
inverse_kinematic = model.ask_mc_sample(C,Lambda^R,500)
```

the location (Equation 8.78)

```
location= model.ask_mc_sample(C,Lambda^R,500)
```

and the calibration with (Equation 8.79)

```
calibration = model.ask_mc_sample(C,Lambda^R,500)
```

Chapter 9

Bayesian Programming Subroutines

Whatever the progress of human knowledge, there will always be room for ignorance, hence for chance and probability.[1]

Le Hasard
Emile Borel [1914]

The purpose of this chapter is to exhibit a first means to combine descriptions with one another in order to incrementally build more and more sophisticated probabilistic models. This is obtained by including in the decomposition calls to Bayesian subroutines. We show that, as in standard programming, it is possible to use existing probabilistic models to build more complex ones and to further structure the definition of complex descriptions as some reusability of previously defined models is possible.

[1]Quels que soient les progrès des connaissances humaines, il y aura toujours place pour l'ignorance et par suite pour le hasard et la probabilité.

9.1 The sprinkler model

9.1.1 Statement of the problem

Let's consider a simple toy model of a garden equipped with an automatic sprinkler designed not to operate on rainy days, which is classically used to present Bayesian nets. The model deals with three Boolean variables: *Rain*, *Sprinkler*, and *GrassWet*. These variables are short names for the corresponding predicates: "it rained," "the sprinkler was activated," and "the grass is wet." This Bayes net is hypothetically used to infer whether or not it rained based on the evidence given by the state of the grass. The corresponding Bayesian program is given in Equation 9.1.

$$
Pr\begin{cases}
Ds\begin{cases}
Sp(\pi)\begin{cases}
Va:\\
Rain, Sprinkler, GrassWet\\
Dc:\\
\begin{cases}
& P(Sprinkler \wedge Rain \wedge GrassWet|\pi_1)\\
= & P(Rain|\pi_1)\times P(Sprinkler|Rain\wedge\pi_1)\\
& \times P(GrassWet|Rain\wedge Sprinkler\wedge\pi_1)
\end{cases}\\
Fo:\\
P([Rain=1]|\pi_1)=\dfrac{171}{365}\\
P([Sprinkler=1]|[Rain=0]\wedge\pi_1)=0.40\\
P([Sprinkler=1]|[Rain=1]\wedge\pi_1)=0.01\\
P([GrassWet=1]|Rain\wedge Sprinkler\wedge\pi_1)\\
=\delta_{Rain\vee Sprinkler}
\end{cases}\\
Id
\end{cases}\\
Qu:\\
P(Rain|[GrassWet=1]\wedge\pi_1)
\end{cases}
\tag{9.1}
$$

where 171 is the number of rainy days in the considered area, 40% is the percentage of times the sprinkler triggers when the weather is dry, and 1% the percentage of times the sprinkler triggers when it should not as the rain already watered the vegetation.

The answer to the question may be computed by the following formula:

$$
\begin{aligned}
& P(Rain|[GrassWet=1]\wedge\pi_1)\\
\prec\; & P(Rain|\pi_1)\\
\times & \sum_{Sprinkler}\left[\begin{array}{c} P(Sprinkler|Rain\wedge\pi_1)\\ \times P([GrassWet=1]|Rain\wedge Sprinkler\wedge\pi_1)\end{array}\right]
\end{aligned}
\tag{9.2}
$$

Numerically it leads to:

$$P([Rain = 1] \mid [GrassWet = 1] \wedge \pi_1) = 69\% \tag{9.3}$$

Suppose now, that someone wants to take into account the status of another part of the house (say, the roof). One possibility would be to add one variable and to duplicate the previous code as in Equation 9.4.

$$Pr\begin{cases} Ds\begin{cases} Sp(\pi)\begin{cases} Va: \\ Rain, Sprinkler, GrassWet, RoofWet \\ Dc: \\ \begin{cases} & P(Sprinkler \wedge Rain \wedge GrassWet \wedge RoofWet|\pi_2) \\ = & P(Rain|\pi_2) \times P(Sprinkler|Rain \wedge \pi_2) \\ & \times P(GrassWet|Rain \wedge Sprinkler \wedge \pi_2) \\ & \times P(RoofWet|Rain \wedge \pi_2) \end{cases} \\ Fo: \\ P([Rain = 1]|\pi_2) = \frac{171}{365} \\ P([Sprinkler = 1] \mid [Rain = 0] \wedge \pi_2) = 0.40 \\ P([Sprinkler = 1] \mid [Rain = 1] \wedge \pi_2) = 0.01 \\ P([GrassWet = 1]|Rain \wedge Sprinkler \wedge \pi_2) \\ = \delta_{Rain \vee Sprinkler} \\ P([RoofWet = 1]|Rain \wedge \pi_2) = \delta_{[Rain=1]} \end{cases} \\ Id \end{cases} \\ Qu: \\ \left\{ P([RoofWet = 1] \mid [GrassWet = 1] \wedge \pi_2) \right. \end{cases} \tag{9.4}$$

The answer to the question may be computed by the following formula:

$$\begin{aligned} & P([RoofWet = 1] \mid [GrassWet = 1] \wedge \pi_2) \\ \propto & \sum_{Rain} \left[\begin{array}{l} P([RoofWet = 1]|Rain \wedge \pi_2) \\ P(Rain|\pi_2) \\ \sum_{Sprinkler} \left[\begin{array}{l} P(Sprinkler|Rain \wedge \pi_2) \\ P([GrassWet = 1]|Rain \wedge Sprinkler \wedge \pi_2) \end{array} \right] \end{array} \right] \end{aligned} \tag{9.5}$$

As $P([RoofWet = 1]|Rain \wedge \pi_2) = \delta_{[Rain=1]}$, we finally get:

$$\begin{aligned} & P([RoofWet = 1] \mid [GrassWet = 1] \wedge \pi_2) \\ = & P([Rain = 1] \mid [GrassWet = 1] \wedge \pi_2) \\ = & 69\% \end{aligned} \tag{9.6}$$

9.1.2 Bayesian program

Another possibility is to write a new Bayesian program (Equation 9.7) using directly the information on the coupling between $Rain$ and $GrassWet$ provided by the Bayesian program (Equation 9.1):

$$
Pr \begin{cases} Ds \begin{cases} Sp(\pi) \begin{cases} Va: \\ Rain, GrassWet, RoofWet \\ Dc: \\ \begin{cases} & P(Rain \wedge GrassWet \wedge RoofWet|\pi_3) \\ = & P(Rain \wedge GrassWet|\pi_3) \\ & \times P(RoofWet|Rain \wedge \pi_3) \end{cases} \\ Fo: \\ P(Rain \wedge GrassWet|\pi_3) = P(Rain \wedge GrassWet|\pi_1) \\ P(RoofWet|Rain \wedge \pi_3) = \delta_{Rain=1} \end{cases} \\ Id \end{cases} \\ Qu: \\ \left\{ P([RoofWet = 1] | [GrassWet = 1] \wedge \pi_3) \right. \end{cases} \tag{9.7}
$$

$P(Rain \wedge GrassWet|\pi_3) = P(Rain \wedge GrassWet|\pi_1)$ may be seen as calling the Bayesian program (9.1) as a probabilistic subroutine. We get directly:

$$
\begin{aligned} & P([RoofWet = 1] | [GrassWet = 1] \wedge \pi_3) \\ = & P([Rain = 1] | [GrassWet = 1] \wedge \pi_1) \\ = & 69\% \end{aligned} \tag{9.8}
$$

9.1.3 Instance and results

9.1.3.1 Advantages of Bayesian programming subroutines

There are several advantages to Bayesian programming subroutine calls:

- As in standard programming, subroutines make the code more compact and easier to read.
- As in standard programming the use of a submodel allows the hiding of the details regarding the definition of this model (here the existence and effect of the sprinklers and the detailed decomposition used in π_1).
- Finally, calling subroutines gives the ability to use Bayesian programs that have been specified and learned by others. For instance, the value 171/365 does not appear in π_3 and could have been fixed or learned in π_1.

However, it is probabilistic as, contrary to standard subroutine calls, it does not transmit to the calling code a single value but a whole probability distribution.

The program "chapter9/sprinkler.py" makes use of Bayesian subroutines:

- $P(GrassWet \wedge \pi_{Paris})$,
- $P(Rain \mid GrassWet \wedge \pi_{Paris})$.

A submodel is built: it uses *Sprinkler* as a variable.

```
submodel=plJointDistribution(Rain^Sprinkler^GrassWet,\
                             jointlist)
```

The following statements build a description for a new model by using questions made with the previously defined submodel.

```
#define the new decomposition
#using a question to another program
jointlist=plComputableObjectList()
jointlist.push_back(submodel.ask(GrassWet^Rain))
jointlist.push_back(plCndDistribution(Roof,Rain,[1,0,0,1]))

model=plJointDistribution(Rain^Roof^GrassWet,\
                          jointlist)
```

In this model the variable *Sprinkler* is not used, but it will produce the exact same result for all the questions with the variable *Rain*, *Roof*, and *GrassWet* as the extended model built for verification:

```
extendedmodel=plJointDistribution(Rain^Roof^GrassWet^Sprinkler,\
                                  jointlist)
```

9.1.3.2 Defining new subroutines with data

As Bayesian programs, subroutines depend on the data used to instantiate the parametric forms.

For instance, 171 in the Bayesian program (Equation 9.1) is the average number of rainy days in Paris. It is a parameter that has been identified using a set of climate data δ_{Paris}. To be more exact, Bayesian program (Equation 9.1) should have been written:

$$Pr\begin{cases}Ds\begin{cases}Sp(\pi)\begin{cases}Va:\\ Rain, Sprinkler, GrassWet\\ Dc:\\ \begin{cases} & P(Sprinkler \wedge Rain \wedge GrassWet|\pi_1)\\ = & P(Rain|\pi_1)\times P(Sprinkler|Rain\wedge\pi_1)\\ & \times P(GrassWet|Rain\wedge Sprinkler\wedge\pi_1)\end{cases}\\ Fo:\\ P([Rain=1]|\delta_{Paris}\wedge\pi_1)=\frac{n}{365}\\ P([Sprinkler=1]|[Rain=0]\wedge\pi_1)=0.40\\ P([Sprinkler=1]|[Rain=1]\wedge\pi_1)=0.01\\ P([GrassWet=1]|Rain\wedge Sprinkler\wedge\pi_1)\\ =\delta_{Rain\vee Sprinkler}\end{cases}\\ Id:\\ \text{learn } n \text{ as the average number of rainy days in the data set } \delta_{Paris}\end{cases}\\ Qu:\\ P(Rain|[GrassWet=1]\wedge\delta_{Paris}\wedge\pi_1)\end{cases} \tag{9.9}$$

Using another set of data as, for instance, δ_{nice}, would lead to another value of this parameter n, namely 88.

Of course, the questions:

$$P(Rain|[GrassWet=1]\wedge\delta_{Paris}\wedge\pi_1) \tag{9.10}$$

and

$$P(Rain|[GrassWet=1]\wedge\delta_{Nice}\wedge\pi_1) \tag{9.11}$$

lead to different results:

$$\begin{aligned}P([Rain=1]|[GrassWet=1]\wedge\delta_{Paris}\wedge\pi_1)=69\%\\ P([Rain=1]|[GrassWet=1]\wedge\delta_{Nice}\wedge\pi_1)=44\%\end{aligned} \tag{9.12}$$

The program "chapter9/sprinkler.py" contains examples to replace one distribution by another in an already defined model. A first possibility is to locally redefine the submodel by changing $P(Rain \wedge \pi_{Paris})$ by $P(Rain \wedge \pi_{Nice})$. This can be done with the "replace" method.

```
PRainNice=plProbTable(Rain,[0.9,0.1])
submodel.replace(Rain,PRainNice)
```

And then it is possible to completely redefine the model.

```
#define the new decomposition using question to another program
jointlist=plComputableObjectList()
jointlist.push_back(submodel.ask(GrassWet^Rain))
jointlist.push_back(plCndDistribution(Roof,Rain,[1,0,0,1]))
model=plJointDistribution(Rain^Roof^GrassWet,\
                          jointlist)
```

It is also possible to directly make the change in the calling model. For example, if we want to go back to the model working in Paris we could use the following instruction:

```
model.replace(Rain,PRainParis)
```

9.2 Calling subroutines conditioned by values

9.2.1 Statement of the problem

The next step would be to be able to call different subroutines conditionally to values of a given variable, for example, using either the "Paris" or the "Nice" model according to the place of interest.

9.2.2 Bayesian program

We can introduce a new variable *Location* with two values: *nice* and *paris* and use a conditional probability distribution to select the submodel we would like to use knowing our location.

The resulting Bayesian program is the following:

$$Pr\begin{cases} Ds\begin{cases} Sp(\pi)\begin{cases} Va: \\ Rain, GrassWet, RoofWet, Location \\ Dc: \\ \begin{cases} & P(Rain \wedge GrassWet \wedge RoofWet \wedge Location|\pi_4) \\ = & P(Location|\pi_4) \\ & \times P(Rain \wedge GrassWet|Location \wedge \pi_4) \\ & \times P(RoofWet|Rain \wedge \pi_4) \end{cases} \\ Fo: \\ P(Location|\pi_4) = any \\ P(Rain \wedge GrassWet|[Location = paris] \wedge \pi_4) \\ = P(Rain \wedge GrassWet|\delta_{paris} \wedge \pi_1) \\ P(Rain \wedge GrassWet|[Location = nice] \wedge \pi_4) \\ = P(Rain \wedge GrassWet|\delta_{nice} \wedge \pi_1) \\ P([RoofWet = 1]|Rain \wedge \pi_4) = \delta_{[Rain=1]} \end{cases} \\ Id \end{cases} \\ Qu: \\ \begin{cases} P([RoofWet = 1]|[GrassWet = 1] \wedge [Location = paris] \wedge \pi_4) \\ P([RoofWet = 1]|[GrassWet = 1] \wedge [Location = nice] \wedge \pi_4) \end{cases} \end{cases} \tag{9.13}$$

where we state that knowing the location, we call the Bayesian program specified by preliminary knowledge π_1 with learning done either on the data set δ_{Paris}

$$\begin{aligned} & P(Rain \wedge GrassWet|[Location = paris] \wedge \pi_4) \\ = \; & P(Rain \wedge GrassWet|\delta_{Paris} \wedge \pi_1) \end{aligned} \tag{9.14}$$

or on the data set δ_{Nice}

$$\begin{aligned} & P(Rain \wedge GrassWet|[Location = nice] \wedge \pi_4) \\ = \; & P(Rain \wedge GrassWet|\delta_{Nice} \wedge \pi_1) \end{aligned} \tag{9.15}$$

9.2.3 Instance and results

You can easily check that the results for the two questions are as expected:

$$\begin{aligned} & P([RoofWet = 1]|[GrassWet = 1] \wedge [Location = paris] \wedge \pi_4) \\ = \; & P([Rain = 1]|[GrassWet = 1] \wedge \delta_{Paris} \wedge \pi_1) = 69\% \\ & P([RoofWet = 1]|[GrassWet = 1] \wedge [Location = nice] \wedge \pi_4) \\ = \; & P([Rain = 1]|[GrassWet = 1] \wedge \delta_{Nice} \wedge \pi_1) = 44\% \end{aligned} \tag{9.16}$$

In "chapter9/sprinkler.py" we again use the "replace" method to change the preliminary knowledge of an existing description. This description is used according to the value of a key: *Location* to perform the same inferences but with different data to define the description.

```
#selecting subroutines
#defines a new variable
Location = plSymbol(''Location", plLabelType(['Paris','Nice']))
locval=plValues(Location)
jointlist=plComputableObjectList()
#
#push a uniform distribution for the location
jointlist.push_back(plUniform(Location))
#
#define the two distributions corresponding to Paris and Nice
PGrasswetkLocation=plDistributionTable(GrassWet,Location)
locval[Location]='Paris'
submodel.replace(Rain,PRainParis)
PGrasswetkLocation.push(submodel.ask(GrassWet),locval)
locval[Location]='Nice'
submodel.replace(Rain,PRainNice)
PGrasswetkLocation.push(submodel.ask(GrassWet),locval)
#and push it in the joint distribution list
jointlist.push_back(PGrasswetkLocation)
#
#idem for the conditional ditribution on Rain
PRainkGrasswetLocation=\
     plDistributionTable(Rain,GrassWet^Location,Location)
locval[Location]='Paris'
submodel.replace(Rain,PRainParis)
PRainkGrasswetLocation.push(submodel.ask(Rain,GrassWet),locval)
locval[Location]='Nice'
submodel.replace(Rain,PRainNice)
PRainkGrasswetLocation.push(submodel.ask(Rain,GrassWet),locval)
#and push it in the joint distribution list
jointlist.push_back(PRainkGrasswetLocation)

#dirac model: when it has rained,the roof is wet
jointlist.push_back(plCndDistribution(Roof,Rain,[1,0,0,1]))

model=plJointDistribution(Rain^Roof^GrassWet^Location,\
                                   jointlist)
```

9.3 Water treatment center revisited (final)

9.3.1 Statement of the problem

Despite its very modular structure, the water treatment center model was quite cumbersome to write and read (see Bayesian program in Section 5.2.3).

We would like to give a simpler final specification of this model using Bayesian programing subroutine calls and the model of a single unit as specified by the Bayesian program 4.29.

9.3.2 Bayesian program

The Bayesian program for the water treatment center using subroutine calls to the Bayesian program for water treatment units is very simple and compact:

$$
Pr\begin{cases}
Ds\begin{cases}
Sp(\pi)\begin{cases}
Va: \\
I_0, I_1, S_O, C_0, O_0, S_1, C_1, O_1, S_2, C_2, O_2, I_3, S_3, C_3, O_3 \\
Dc: \\
\begin{cases}
P\left(I_0 \wedge I_I \wedge \cdots \wedge O_3 | \pi_{center}\right) \\
= \begin{bmatrix}
P\left(I_0 \wedge I_1 \wedge I_3 | \pi_{center}\right) \\
P\left(S_0 \wedge C_0 \wedge O_0 | I_0 \wedge I_1 \wedge \pi_{unit1}\right) \\
P\left(S_1 \wedge C_1 \wedge O_1 | I_0 \wedge I_1 \wedge \pi_{unit2}\right) \\
P\left(S_2 \wedge C_2 \wedge O_2 | O_0 \wedge O_1 \wedge \pi_{unit3}\right) \\
P\left(S_3 \wedge C_3 \wedge O_3 | I_3 \wedge O_2 \wedge \pi_{unit4}\right)
\end{bmatrix}
\end{cases} \\
Fo:
\end{cases} \\
Id
\end{cases} \\
Qu:
\end{cases}
\tag{9.17}
$$

The obtained results for the different questions are evidently the same but for the diagnosis ones. Indeed, in Equation 9.17 we have chosen to hide the F_i variables and consequently we cannot anymore ask questions using them. Here again, it is similar to what occurs with the use of classical subroutine calls where you cannot, in the calling program, use internal variables of the subroutines. An alternative would have been to explicitly use the F_i variables in the above program to preserve the ability to ask diagnosis questions.

9.4 Fusion of subroutines

9.4.1 Statement of the problem

Another example where Bayesian subroutines may be used to hide the implementation details or specificities can be found in fusion models. For example, we may refine the model of a sensor by further modeling its behavior, in particular we may describe a fault tree leading to a probability of false alarm. No matter how complex this model is, we can use Bayesian subroutines to encapsulate it into a simple form which gives the probability distribution on the state of the system knowing the readings, hiding all the details concerning the variables used in the fault tree.

9.4.2 Bayesian program

We may once more revisit the sensor fusion example of Chapter 7 but this time taking into account two models of sensors with two different models of false alarms. Bayesian subroutines may be used to hide all the details about the sensor models and can directly be used in a standard fusion program:

$$
Pr\begin{cases}
Ds\begin{cases}
Sp(\pi)\begin{cases}
Va: \\
X, Y, D_2, D_3 \\
Dc: \\
\begin{cases}
 & P(X \wedge Y \wedge D_2 \wedge D_3 \mid \pi) \\
= & P(X \wedge Y \mid \pi) \times \\
 & P(D_2 \mid X \wedge Y \wedge \pi) \times \\
 & P(D_3 \mid X \wedge Y \wedge \pi)
\end{cases} \\
Fo: \\
\begin{cases}
P(X \wedge Y \mid \pi) = Uniform \\
P(D_2 \mid X \wedge Y \wedge \pi) = P(D_2 \mid X \wedge Y \wedge \pi_{S2}) \\
P(D_3 \mid X \wedge Y \wedge \pi) = P(D_3 \mid X \wedge Y \wedge \pi_{S3})
\end{cases}
\end{cases} \\
Id
\end{cases} \\
Qu: \\
P(X, Y \mid d_2 \wedge d_3 \wedge \pi)
\end{cases}
\tag{9.18}
$$

The first sensor becomes faulty if two conditions A_2 and B_2 are met while the other becomes faulty if one of the other conditions A_3 or B_3 is met. We obtain two descriptions π_{S1} and π_{S2} which relate the position X, Y to the sensor readings.

$$
Pr\begin{cases}
Ds\begin{cases}
Sp(\pi)\begin{cases}
Va: X, Y, F_2, D_2, A_2, B_2 \\
Dc: \\
\begin{cases}
& P(X \wedge Y \wedge F_2 \wedge D_2 \wedge A_2 \wedge B_2 \mid \pi_{S2}) \\
= & P(X \wedge Y \mid \pi_{S2}) P(A_2 \mid \pi_{S2}) P(B_2 \mid \pi_{S2}) \\
& P(F_2 \mid A_2 \wedge B_2 \wedge \pi_{S2}) P(D_2 \mid X \wedge Y \wedge F_2 \wedge \pi_{S2})
\end{cases} \\
Fo: \\
\begin{cases}
P(X \wedge Y \mid \pi_{S2}) = Uniform \\
P(A_2 = true \mid \pi_{S2}) = 0.2, P(B_2 = true \mid \pi_{S2}) = 0.1 \\
P(F_2 \mid A_2 \wedge B_2 \wedge \pi_{S2}) = \delta_{\mathbf{A_2} \wedge \mathbf{B_2}} \\
P(D_2 \mid X \wedge Y \wedge F_2 \wedge \pi_{S2}) \\
\begin{cases}
[F_2 = 0] : B\left(\left[\mu = f_d^2(X, Y)\right], \left[\sigma = g_d^2(X, Y)\right]\right) \\
\begin{cases}
f_d^2 = \sqrt{(X+50)^2 + Y^2} \\
g_d^2 = 1 + \dfrac{f_d^2(X, Y)}{10}
\end{cases} \\
[F_2 = 1] : Uniform
\end{cases}
\end{cases}
\end{cases} \\
Id
\end{cases} \\
Qu: \\
P(D_2 \mid X \wedge Y \wedge \pi_{S2})
\end{cases}
\tag{9.19}
$$

$$
Pr\begin{cases}
Ds\begin{cases}
Sp(\pi)\begin{cases}
Va: X, Y, F_3, D_3, A_3, B_3 \\
Dc: \\
\begin{cases}
& P(X \wedge Y \wedge F_3 \wedge D_3 \wedge A_3 \wedge B_3 \mid \pi_{S3}) \\
= & P(X \wedge Y \mid \pi_{s3}) P(A_3 \mid \pi_{s3}) P(B_3 \mid \pi_{s3}) \\
& P(F_3 \mid A_3 \wedge B_3 \wedge \pi_{s3}) P(D_3 \mid X \wedge Y \wedge F_3 \wedge \pi_{s3})
\end{cases} \\
Fo: \\
\begin{cases}
P(X \wedge Y \mid \pi_{S3}) = Uniform \\
P(A_3 = true \mid \pi_{S3}) = 0.01, P(B_3 = true \mid \pi_{S3}) = 0.03 \\
P(F_3 \mid A_3 \wedge B_3 \wedge \pi_{s3}) = \delta_{\mathbf{A_3} \vee \mathbf{B_3}} \\
P(D_3 \mid X \wedge Y \wedge F_3 \wedge \pi_{S3}) \\
\begin{cases}
[F_3 = 0] : B\left(\left[\mu = f_d^2(X, Y)\right], \left[\sigma = g_d^3(X, Y)\right]\right) \\
\begin{cases}
f_d^3 = \sqrt{X^2 + (Y+50)^2} \\
g_d^3 = 1 + \dfrac{f_d^3(X, Y)}{10}
\end{cases} \\
[F_3 = 1] : Uniform
\end{cases}
\end{cases}
\end{cases} \\
Id
\end{cases} \\
Qu: \\
P(D_3 \mid X \wedge Y \wedge \pi_{S3})
\end{cases}
\tag{9.20}
$$

These two descriptions differ by the type of fault model: $\delta_{A_2 \wedge B_2}$ versus $\delta_{A_3 \vee B_3}$ and also by the sensor models which have to take into account the location of the landmarks: $f_d^2 = \sqrt{(X+50)^2 + Y^2}$ versus $f_d^3 = \sqrt{X^2 + (Y+50)^2}$.

In "chapter9/subroutinefusion.py" we have defined two sensor models: sensor_model2 and sensor_model3, corresponding to the programs 9.19 and 9.20. These two models are used to build a simple "main" Bayesian program to perform the fusion of the two sensors.

```
JointDistributionList=plComputableObjectList()
JointDistributionList.push_back(plComputableObject\
(plUniform(X)*plUniform(Y))
JointDistributionList.push_back(sensor_model3.ask(D3,X^Y))
JointDistributionList.push_back(sensor_model2.ask(D2,X^Y))
main_model=plJointDistribution(X^Y^D3^D2,JointDistributionList)
question=main_model.ask(X^Y,D2^D3)
```

9.5 Superposition

9.5.1 Statement of the problem

In some occasions, it may be necessary to select the ranges of values for which the result of a model is or is not valid. For example, in the localization example we may consider the use of bearings is only valid if the result is in a given region, for example: $XB \geq 0, YB \geq 0$, where XB and YB are the localization parameters given by the bearings. If the result is not within the specified range we may decide to use a uniform distribution on our final estimation of the location X, Y. More generally, we can use different sensors in different places. For example, we may use the distances only in the region $XD \leq 0, YD \leq 0$ and the bearing in the region $XB \geq 0, YB \geq 0$.

9.5.2 Bayesian program

The standard fusion program for our localization problem (see Section 7.3) can be restated as follows (at first, dropping the distances for the sake of simplicity):

$$
Pr\begin{cases}
Ds\begin{cases}
Sp\,(\pi_B)\begin{cases}
Va: \\
X, Y, X_B, Y_B, B_1, B_2, B_3 \\
Dc: \\
\begin{cases}
P\left(X \wedge Y \wedge X_B \wedge Y_B \wedge B_1 \wedge B_2 \wedge B_3 | \pi_B\right) \\
= P\left(B1 \wedge B2 \wedge B3 \wedge \pi_B\right) \\
\times P\left(X_B \wedge Y_B \mid B_1 \wedge B_2 \wedge B_3 \wedge \pi_B\right) \\
\times P\left(X \wedge Y \mid X_B \wedge Y_B \wedge B_1 \wedge B_2 \wedge B_3 \wedge \pi_B\right)
\end{cases} \\
Fo: \\
\begin{cases}
P\left(B_1 \wedge B_2 \wedge B_3 | \pi_B |=\right) Uniform \\
P\left(X_B \wedge Y_B \mid B_1 \wedge B_2 \wedge B_3 \wedge \pi_B\right) = \\
P\left(X_B \wedge Y_B \mid B_1 \wedge B_2 \wedge B_3 \wedge \pi_k\right) \\
P\left(X \wedge Y \mid X_B \wedge Y_B \wedge B_1 \wedge B_2 \wedge B_3 \wedge \pi_B\right) = \\
\begin{cases}
if XB > 0, YB > 0 : P\left(X \wedge Y \mid B_1 \wedge B_2 \wedge B_3 \wedge \pi_k\right) \\
else : Uniform
\end{cases}
\end{cases}
\end{cases} \\
Id
\end{cases} \\
Qu: \\
P\left(X \wedge Y | b_1 \wedge b_2 \wedge b_3 \wedge \pi_B\right)
\end{cases}
\tag{9.21}
$$

The program 9.21 implements the following idea: if the position corresponding to the measurements is within the specified region the distribution of the location remains unchanged; otherwise it is unknown.

9.5.3 Instance and results

The Figure 9.1 represents the distribution for a given measurement of the bearings corresponding to the location $X = Y = 0$.

We may generalize the previous example, by associating other regions to other sensors. For example, if we assume $P\left(X_D \wedge Y_D \mid D_1 \wedge D_2 \wedge D_3 \wedge \pi_k\right)$ is the question allowing us to locate the boat with the distance measurements, we may allocate these sensors to the region $X_D \leq 0, Y_D \leq 0$ (see Figure 9.1b).

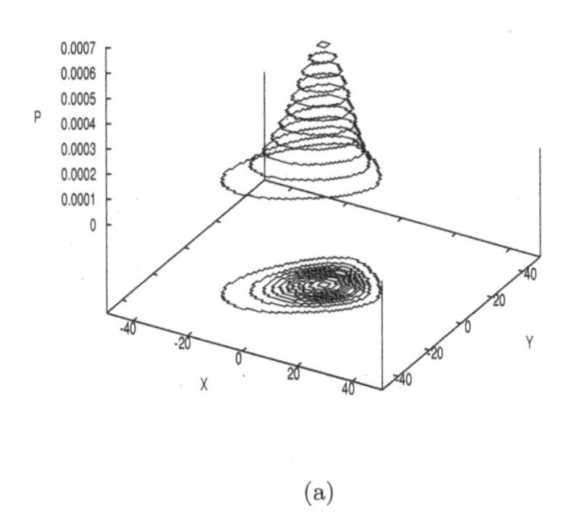

(a)

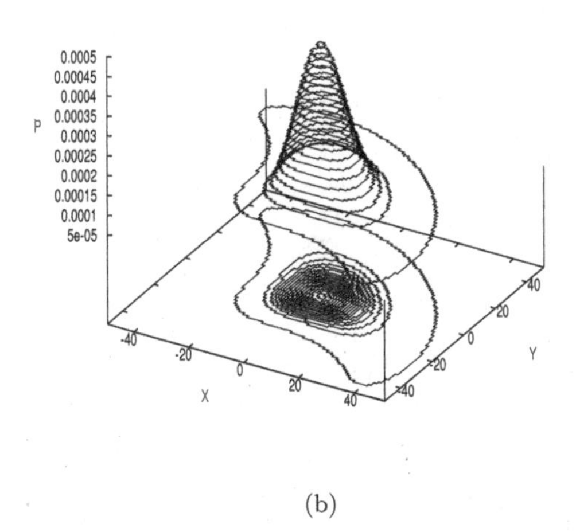

(b)

FIGURE 9.1: Superposition with a uniform distribution:
(a): $P(X \wedge Y | b_1 = 225 \wedge b_2 = 180 \wedge b_3 = 270 \wedge \pi)$
(b): $P(X \wedge Y | d_1 = 70 \wedge d_2 = 50 \wedge d_3 = 50 \wedge \pi)$

Since the two regions do not overlap we can can stitch our two localization procedures on the same space with the program in Equation 9.22. The result is presented Figure 9.2. The sensor superposition still gives good results on the boundaries of the regions.

Should the regions overlap, it is then possible to fuse all the sensors in that region. For example, if we use the program (Equation 7.6) and if we assume a new valid region $XB \geq -20, YB \geq -20$ for the localization with bearings we may introduce two new variables XF, YF and use the sensor fusion program in the region $-20 \leq XF \leq 0, -20 \leq YF \leq 0$.

$$
Pr \begin{cases} Ds \begin{cases} Sp\,(\pi_S) \begin{cases} Va: \\ X, Y, X_B, Y_B, X_D, Y_D, B_1, B_2, B_3, D_1, D_2, D_3, C_B, C_D \\ Dc: \\ \begin{cases} P(X \wedge Y \wedge X_B \wedge Y_B \wedge X_D \wedge Y_D \wedge B_1 \ldots \wedge D_3 \wedge \pi_S) \\ = P(B1 \wedge B2 \wedge B3 \wedge D1 \wedge D2 \wedge D3 \wedge \pi_S) \\ \times P(X_B \wedge Y_B \mid B_1 \wedge B_2 \wedge B_3 \wedge \pi_S) \\ \times P(C_B \mid X_B \wedge Y_B \wedge \pi_s) \\ \times P(X_D \wedge Y_D \mid D_1 \wedge D_2 \wedge D_3 \wedge \pi_S) \\ \times P(C_D \mid X_D \wedge Y_D \wedge \pi_s) \\ \times P(X \wedge Y \mid C_B \wedge C_D) \end{cases} \\ Fo: \\ \begin{cases} P(B_1 \ldots \wedge D_3 | \pi_S \mid =) \, Uniform \\ P(X_B \wedge Y_B \mid B_1 \wedge B_2 \wedge B_3 \wedge \pi_S) = \\ P(X_B \wedge Y_B \mid B_1 \wedge B_2 \wedge B_3 \wedge \pi_k) \\ P(X_D \wedge Y_D \mid D_1 \wedge D_2 \wedge D_3 \wedge \pi_S) = \\ P(X_D \wedge Y_D \mid D_1 \wedge D_2 \wedge D_3 \wedge \pi_k) \\ P(C_B \mid X_B \wedge Y_B \wedge \pi_s) = \\ \begin{cases} if X_B \geq 0, Y_B \geq 0 : P(C_B = 1) = 1 \\ else : P(C_B = 1) = 0 \end{cases} \\ P(C_D \mid X_D \wedge Y_D \wedge \pi_s) = \\ \begin{cases} if X_D < 0, Y_D < 0 : P(C_D = 1) = 1 \\ else : P(C_D = 1) = 0 \end{cases} \\ P(X \wedge Y \mid C_D \wedge C_B \wedge \pi_S) = \\ \begin{cases} if C_D = 1 : P(X_D \wedge Y_D \mid D_1 \wedge D_2 \wedge D_3 \wedge \pi_k) \\ else : if C_D = 1 : P(X_B \wedge Y_B \mid B_1 \wedge B_2 \wedge B_3 \wedge \pi_k) \\ else : Uniform \end{cases} \end{cases} \end{cases} \\ Id \end{cases} \\ Qu: \\ P(X \wedge Y | d_1 \wedge d_2 \wedge d_3 \wedge b_1 \wedge b_2 \wedge b_3 \wedge \pi_s) \end{cases}
\tag{9.22}
$$

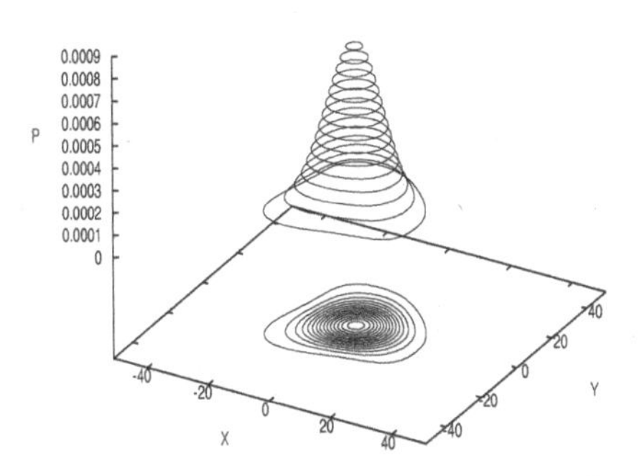

FIGURE 9.2: Probability of location at the boundary of two sensing regions.

In "chapter9/stitching.py" we use these questions as subroutines:

- ```
  PXBYB_K_B1B2B3=\
  localisation_model_with_bearings.ask(XB^YB,B1^B2^B3)
  ```
- ```
  PXDYD_K_D1D2D3=\
  localisation_model_with_bearings.ask(XD^YD,D1^D2^D3)
  ```

Since these conditional distributions will be used as a model to the conditional distribution on X, Y it is necessary to make a copy of them and to rename the variables using:

```
PXBYB_K_B1B2B3bis=plCndDistribution(PXBYB_K_B1B2B3)
PXY_K_B1B2B3=PXBYB_K_B1B2B3bis.rename(X^Y^B1^B2^B3)
```

Dirac distributions on constraints may be defined as follows:

```
def cdf(out, XdYd):
    out[0] = XdYd[0]
    out[1] = XdYd[1]
CD = plSymbol(''CD", PL_BINARY_TYPE)
PCD__XdYd = plIneqConstraint(CD,\
plPythonExternalFunction(XD^YD, cdf), 2)
JointDistributionList.push_back(PCD__XdYd)
```

Finally, the conditional distribution on X and Y is defined as:

```
PXY = plDistributionTable(X^Y, D1^D2^D3^B1^B2^B3^CD^CB, CD^CB)
vcdcb = plValues(CD^CB)
vcdcb[CD] = 1
vcdcb[CB] = 0
PXY.push(PXY_K_D1D2D3, vcdcb)
vcdcb[CD] = 0
vcdcb[CB] = 1
PXY.push(PXY_K_B1B2B3, vcdcb)
PXY.push_default( plUniform(X^Y) )
```

Chapter 10

Bayesian Programming Conditional Statement

> The probability of ten consecutive heads is 0.1 percent; thus, when you have millions of coin tossers, or investors, in the end there will be thousands of very successful practitioners of coin tossing, or stock picking.
>
> *The Age of Turbulence*
> Alan Greenspan [2007]

The purpose of this chapter is to introduce probabilistic branching statements. We will start by describing the probabilistic "if-then-else" statement which, as in standard programming, can naturally be extended to a probabilistic "case" statement. From an inference point of view, the probabilistic if-then-else statement is simply the integration over the probability distribution on a binary variable representing the truth value of the condition used in the classical "if" statement. The main difference with the classical approach is that the Bayesian program will explore both branches when the truth value of the condition is given by a probability distribution. This allows us to mix behaviors and to recognize models.

10.1 Bayesian if-then-else

10.1.1 Statement of the problem

Let's recall the Khepera robot (see Chapter 4) and formalize the programs to push a part or to follow its contour. The probabilistic variables Dir, $Prox$, and Rot, respectively, denote the direction of the nearest obstacle, its proximity, and the rotational speed of the robot (the robot is assumed to move at constant speed).

$$
Pr\begin{cases}
Ds\begin{cases}
Sp(\pi_b \wedge \delta_b):\\
\begin{cases}
Va: Dir, Prox, Rot\\
Dc:\\
\begin{cases}
P\left(Dir \wedge Prox \wedge Rot \mid \pi_b \wedge \delta_b\right)\\
= P\left(Dir \wedge Prox \mid \pi_b \wedge \delta_b\right)\\
\times P\left(Rot \mid Dir \wedge Prox \wedge \pi_b \wedge \delta_b\right)
\end{cases}\\
Fo:\\
\begin{cases}
P\left(Dir \wedge Prox \mid \pi_b \wedge \delta_b\right) = Uniform\\
P\left(Rot \mid Dir \wedge Prox \wedge \pi_b \wedge \delta_b\right) = Normal(\sigma_b, \mu_b)
\end{cases}
\end{cases}\\
Id:\\
\begin{cases}
\sigma_b(Dir, Prox) \leftarrow \delta_b\\
\mu_b(Dir, Prox) \leftarrow \delta_b
\end{cases}
\end{cases}\\
Qu: P\left(Rot \mid Dir \wedge Prox \wedge \pi_b \wedge \delta_b\right)
\end{cases}
\tag{10.1}
$$

As explained in Chapter 4, the desired program is obtained by using the data set δ_b recorded during the learning of the desired behavior. For example, it is possible to obtain a program to avoid obstacles by controlling the robot to do so during the learning phase and by building the probability table $P\left(Rot \mid Dir \wedge Prox \wedge \pi_{avoidance} \wedge \delta_{avoidance}\right)$ based on the data $\delta_{avoidance}$ recorded during learning.

The light sensors of the robot may also be used to build a new variable θ_l indicating the direction of a light beam in the robot reference frame. This new variable may be used to move toward the light (phototaxis) using the program in Equation 10.2.

$$
Pr\begin{cases}
Ds\begin{cases}
Sp(\pi_{phototaxis}):\\
\begin{cases}
Va:\Theta_l, Rot\\
Dc:\\
\begin{cases}
P(\Theta_l \wedge Rot \mid \pi_{phototaxis})\\
= P(\Theta_l \mid \pi_{phototaxis})\\
\times P(Rot \mid \Theta_l \wedge \pi_{phototaxis})
\end{cases}\\
Fo:\\
\begin{cases}
P(\Theta_l \mid \pi_{phototaxis}) = Uniform\\
P(Rot \mid \Theta_l \wedge \pi_{phototaxis})\\
= Normal(\mu = \Theta_l, \sigma = 2)
\end{cases}
\end{cases}\\
Id:
\end{cases}\\
Qu: P(Rot \mid \Theta_l \wedge \pi_{phototaxis})
\end{cases}
\tag{10.2}
$$

Contrary to the program in Equation 10.1, the description for "phototaxy behavior" does not require any learning.

The probabilistic conditional statement will be used to combine the two behaviors: "phototaxy" and "avoidance" into a more complex behavior ("home") leading the robot to reach its base (where the light is located) while avoiding the obstacles.

To do so, we define a binary variable H to switch from one behavior to another. The probability distribution over this variable will be conditioned by the distance ($Prox$) to the obstacles: if we are close to one obstacle, the robot is asked to perform the "avoidance" behavior or it is asked to perform the "phototaxy" behavior.

10.1.2 Bayesian program

The program in Equation 10.3 will achieve the "home" behavior.

$$Pr\left\{\begin{array}{l} Ds\left\{\begin{array}{l} Sp(\pi_{home}): \\ \left\{\begin{array}{l} Va: Dir, Prox, \Theta_l, H, Rot \\ Dc:\left\{\begin{array}{l} P\left(Dir \wedge Prox \wedge \Theta_l \wedge H \wedge Rot \mid \pi_{home}\right) \\ = P\left(Dir \wedge Prox \wedge \Theta_l \mid \pi_{home}\right) \\ \times P\left(H \mid Prox \wedge \pi_{home}\right) \\ \times P\left(Rot \mid Dir \wedge Prox \wedge H \wedge \Theta_l \wedge \pi_{home}\right) \end{array}\right. \\ Fo:\left\{\begin{array}{l} P\left(Dir \wedge Prox \wedge \Theta_l \mid \pi_{home}\right) = Uniform \\ P\left(H = avoidance \mid Prox \wedge pi_{home}\right) = S\_Shape(Prox) \\ P\left(Rot \mid Dir \wedge Prox \wedge H \wedge \Theta_l \wedge \pi_{home}\right) = \\ \left\{\begin{array}{l} H = avoidance: \\ P\left(Rot \mid Dir \wedge Prox \wedge \pi_{avoidance} \wedge \delta_{avoidance}\right) \\ H = phototaxy: \\ P\left(Rot \mid \Theta_l \wedge \pi_{phototaxy}\right) \end{array}\right. \end{array}\right. \end{array}\right. \\ Id: \end{array}\right. \\ Qu: P\left(Rot \mid Dir \wedge Prox \wedge \Theta_l \wedge \pi_{home}\right) \end{array}\right. \tag{10.3}$$

The S_Shape function is used to tune the level of mixing according to the distance to the obstacles. For example, Figure 10.1 represents a discretized version of the function used to compute the probability distribution $P\left(H \mid Prox \wedge \pi_{home}\right)$.

$$P\left(H \mid Prox \wedge \pi_{home}\right) = \frac{1}{1 + e^{\beta(\alpha - prox)}} \tag{10.4}$$

The parameter α of Equation 10.4 is the threshold at which the probability for doing the "phototaxy" behavior becomes greater than the probability for doing the "avoidance" behavior. The β ($\beta \geq 0$) parameter tunes how close from a standard choice based on a threshold the final behavior will be. When $\beta \rightarrow \infty$ then the program 10.3 will lead to the same behavior as the one programmed with a classical program (see 10.5). On the contrary, if $\beta \rightarrow 0$ the result will lead to a random choice between the two behaviors.

$$\textbf{IF } Prox > \alpha \text{ "phototaxy" } \textbf{ELSE} \text{ "avoidance"} \tag{10.5}$$

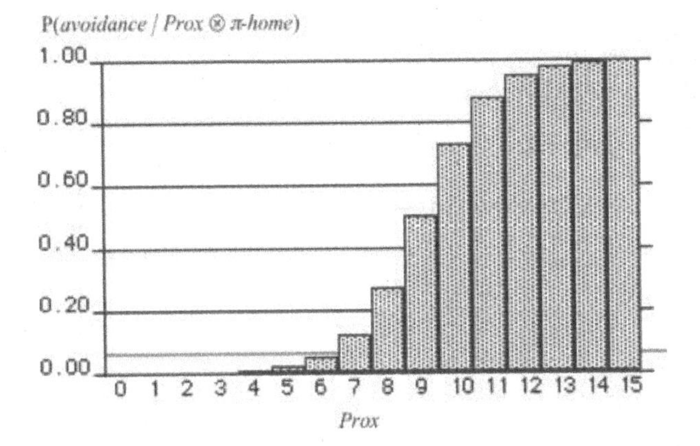

FIGURE 10.1: The shape of the sigmoid defines the way to mix behaviors. Far from the obstacle ($Prox = 0$) the probability of $[H = 1]$ is 0 meaning that you want to do phototaxy, on the contrary, close to the obstacle ($Prox = 15$) the probability of $[H = 1]$ is 1 meaning that you only care about avoiding the obstacle. In between the behavior will be a combination of phototaxy and avoidance behaviors (see below).

A program similar to the one used to learn the "pushing/following" behavior, "chapter4/kepera.py" may be used with the appropriate data ($\delta_{avoidance}$) to create the Bayesian program described in 10.1 to learn the "avoidance behavior." The conditional probability distribution on *Rot* for the "avoidance" behavior is obtained by:

```
# define the description
render_description= \
        plJointDistribution(Vrot^Dir^Prox,Decomposition)
# define the associated question
render_question= \
        render_description.ask(Vrot,Dir^Prox)
```

The conditional probability distribution on *Rot* for the "phototaxy" behavior is obtained with the following program ("chapter10/Bif.py")

```
Theta = plSymbol('Theta',plIntegerType(-10,10))
PTheta=plUniform(Theta)
Decomposition=plComputableObjectList()
Decomposition.push_back(PTheta)
Decomposition.push_back(plCndNormal(Vrot,Theta,2))
phototaxy_description=plJointDistribution(Decomposition)
phototaxy_question=phototaxy_description.ask(Vrot,Theta)
```

The Bayesian If-Then-Else is defined by a conditional distribution on H based on an S_Shape function:

```
def generic_H_prob_function(alpha,beta):
    def H_prob_function(Input_) :
        v = 1.0/(1.0+exp(beta*(alpha-Input_[1].to_float())))
        if Input_[0].to_int() == 0 : #H=0 means Avoidance
            return v
        else:
            return 1-v
    return H_prob_function
```

The conditional probability distribution on H is then defined as:

```
PH = plCndAnonymousDistribution(\
     H,Prox,plPythonExternalProbFunction( \
     H^Prox,generic_H_prob_function(9,0.25 )))
```

To combine the two behaviors, the following program may be used:

```
#H is used as a key to select the proper behavior
PVrot=plDistributionTable(Vrot,Dir^Prox^Theta^H,H)
PVrot.push(phototaxy_question,1) #phototaxy
PVrot.push(render_question,0) #avoidance
#define the decomposition
JointDistributionList=plComputableObjectList()
JointDistributionList.push_back(plUniform(Prox))
JointDistributionList.push_back(plUniform(Theta))
JointDistributionList.push_back(plUniform(Dir))
JointDistributionList.push_back(PH)
JointDistributionList.push_back(PVrot)

#define the specification and the question
home_specification=plJointDistribution(JointDistributionList)
home_question=home_specification.ask(Vrot,Dir^Prox^Theta)
```

10.1.3 Instance and results

Figures 10.2 and 10.3 present the probability distributions obtained when a robot must avoid an obstacle on the left with a light source also on the left.[1] When the object is on the left, the robot needs to turn right to avoid it. This is what happens when the robot is close to the objects (see Figure 10.2). However, when the robot is further from the object, the presence of the light source on the left influences the way the robot will avoid obstacles. In this case, the robot may turn left despite the presence of the obstacle (see Figure 10.3).

[1]These results are given in Lebeltel et al. [2004].

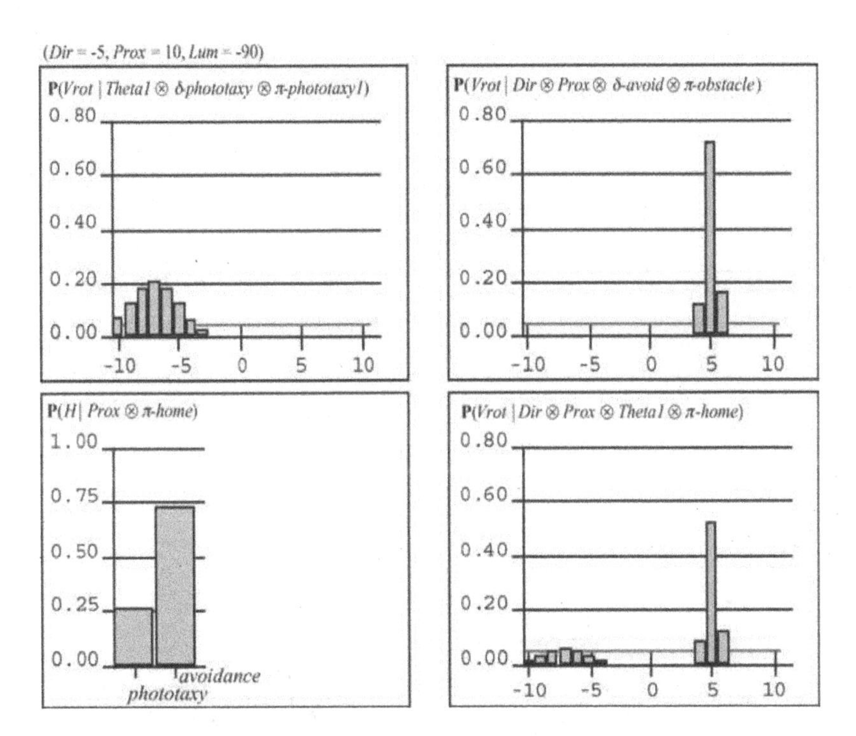

FIGURE 10.2: The top left distribution shows the knowledge on *Rot* given by the phototaxy description; the top right is the probability on *Rot* given by the "avoidance" description; the bottom left shows the knowledge of the "command variable" H; finally, the bottom right shows the probability distribution on *Rot* resulting from the marginalization (weighted sum) of variable H, and the robot will most probably turn right.

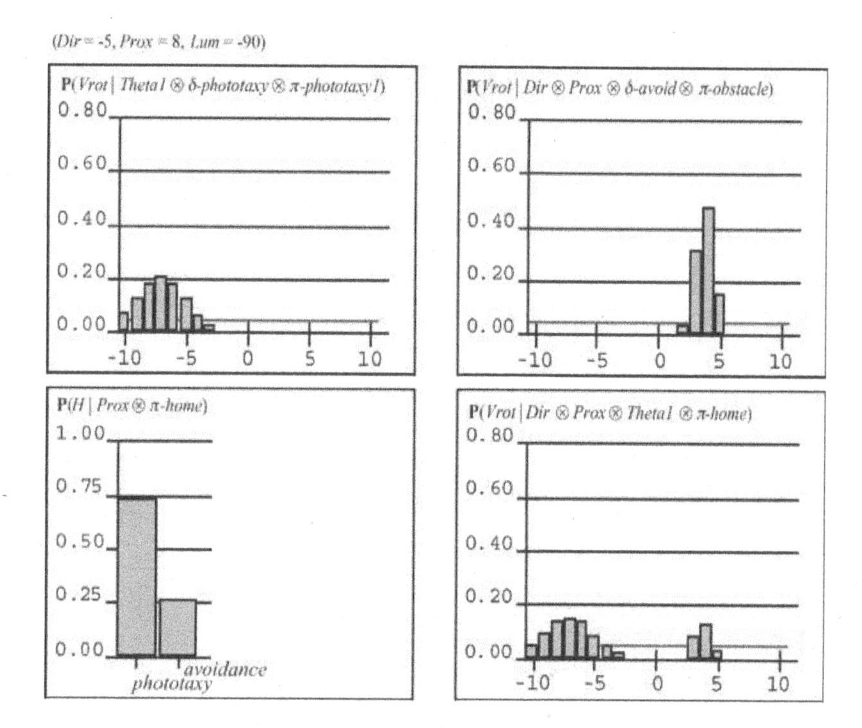

FIGURE 10.3: We are now far from the obstacle; the probability of $[H = \textit{phototaxy}]$ is higher than the probability of $[H = \textit{avoidance}]$ (bottom left). Consequently, the result of the combination is completely different than in the previous case and the robot will most probably turn left.

10.2 Behavior recognition

10.2.1 Statement of the problem

In the previous example, the variable H is used to select a particular behavior: "avoidance" or "phototaxy." This makes it possible to combine existing models. The same specification may be used to infer the behavior from the sensors and the motor readings. In other words, if someone is maneuvering the robot it is possible to infer the behavior that is currently being used by the operator. This method provides a first approach to behavior recognition and model selection. It can be easily extended to several behaviors or models by simply extending the cardinality of the variable H.

10.2.2 Bayesian program

The specification is identical to the program in 10.3 but the question now concerns the variable H:

$$
Pr: \begin{cases} Ds: \begin{cases} Sp(\pi_{home}): \\ \begin{cases} Va: Dir, Prox, \Theta_l, H, Rot \\ Dc: \text{identical to 10.3} \end{cases} \end{cases} \\ Qu: P\left(H \mid Dir \wedge Prox \wedge Theta_l \wedge \Theta_l \wedge \pi_{home}\right) \end{cases} \tag{10.6}
$$

10.2.3 Instance and results

The question in Equation 10.6 may be used at any time to infer the current behavior of the robot from its current sensory-motor status. This is a way to abstract a state within a large dimension ($Rot \times Dir \times Prox \times \Theta_l$) into a much smaller dimension space H.

This basic idea has been considerably extended to build "Bayesian maps" in the PhD thesis of Julien Diard [Diard, 2003; Diard et al., 2004; Diard and Bessière, 2008; Diard et al., 2010].

Using the description in file "chapter10/Bif.py" the questionon H is expressed as:

```
#define a new question on H
behavior_question=home_specification.ask(H,Vrot^Dir^Prox^Theta)
```

10.3 Mixture of models and model recognition

In the previous section, the probability distribution on the variable H is conditioned by a sensor value. The probability distribution on H may also be given as a prior or may be learned. The program in Equation 10.7 is the general form of a mixture of models.

$$Pr:\begin{cases} Ds:\begin{cases} Sp(\pi_M):\begin{cases} Va: H\in[1,\ldots,n], I, S_1,\ldots,S_n \\ Dc:\begin{cases} P(H\wedge I\wedge S_1\ldots S_n\mid\pi_M) \\ = P(I\mid\pi_M) \\ \times P(H\mid I\wedge\pi_M) \\ \times P(S_1\mid H\wedge I\wedge\pi_M) \\ \ldots \\ \times P(S_n\mid H\wedge I\wedge\pi_M) \end{cases} \\ Fo:\begin{cases} P(I\mid\pi_M), P(H\mid I\wedge\pi_M) = \\ \begin{cases} H=1: \\ P(S\mid I\wedge\pi_1) \\ \ldots \\ H=n: \\ P(S\mid I\wedge\pi_n) \end{cases} \end{cases} \end{cases} \\ Id: \end{cases} \\ Qu: P(S\mid I\wedge\pi_M) \end{cases} \tag{10.7}$$

As the variable H is used to select the appropriate model π_i in the parametric forms, the answer to the question in Equation 10.7 is a weighted sum of the submodels given by:

$$P(S\mid I\wedge\pi_M) = \frac{1}{Z}\times\sum_{h=1,\ldots,n}\left(P(H=h\mid I\wedge\pi_M)\,P(S\mid I\wedge\pi_h)\right) \tag{10.8}$$

Many variations exist around this general framework which depend on the parametric forms and on the choices made for the identification. For example, by learning $P(H\mid I\wedge\pi_{home})$ it is possible to register several behaviors during a learning phase and to identify them when they take place in another experiment. In the program 10.9 we assume the two behaviors "Avoidance" and "Phototaxy" have been previously defined using learning and explicit programming. The program uses the data δ_{home} to learn $P(H\mid Prox\wedge\pi_{home}\wedge\delta_{home})$ with the Expectation-Maximization (EM) algorithm (see 15.2) during a "Homing" behavior. Instead of having to set this conditional distribution with an algebraic formula, the learning approach allows the program to adapt itself to what is considered by the operator as "to be closed" from the obstacle.

$$
Pr \begin{cases} Ds \begin{cases} Sp(\pi_{home} \wedge \delta_{home}): \\ \begin{cases} Va: Dir, Prox, \Theta_l, H, Rot \\ Dc: \begin{cases} P(Dir \wedge Prox \wedge \Theta_l \wedge H \wedge Rot \mid \pi_{home}) \\ = P(Dir \wedge Prox \wedge \Theta_l \mid \pi_{home}) \\ \times P(H \mid Prox \wedge \pi_{home}) \\ \times P(Rot \mid Dir \wedge Prox \wedge H \wedge \Theta_l \wedge \pi_{home}) \end{cases} \\ Fo: \begin{cases} P(Dir \wedge Prox \wedge \Theta_l \mid \pi_{home}) = Uniform \\ P(H = avoidance \mid Prox \wedge pi_{home}) = Histogram \\ P(Rot \mid Dir \wedge Prox \wedge H \wedge \Theta_l \wedge \pi_{home}) = \\ \begin{cases} H = avoidance: \\ P(Rot \mid Dir \wedge Prox \wedge \pi_{avoidance} \wedge \delta_{avoidance}) \\ H = phototaxy: \\ P(Rot \mid \Theta_l \wedge \pi_{phototaxy}) \end{cases} \end{cases} \end{cases} \\ Id: P(H = avoidance \mid Prox \wedge \pi_{home} \wedge \delta_{home}) = EM \end{cases} \\ Qu: P(H \mid Rot \wedge Dir \wedge Prox \wedge \Theta_l \wedge \pi_{home} \wedge \delta_{home}) \end{cases}
\tag{10.9}
$$

The file "chapter10/EMbehavior.py" implements the program in Equation 10.9. It makes use of a generic EM learner. First the probability distribution $P(Rot \mid Dir \wedge Prox \wedge H \wedge \Theta_l \wedge \pi_{home})$ is defined using H as a switch:

```
PVrot=plDistributionTable(Vrot,Dir^Prox^Theta^H,H)
PVrot.push(phototaxy_question,1) #phototaxy
PVrot.push(render_question,0) #avoidance
```

Since the EM learning algorithm is used to learn

$$P(H \mid Prox \wedge \delta_{home} \wedge \pi_{home})$$

we specify an initial distribution: for each value of $Prox$ the binomial law on H is chosen at random.

```
#init the distribution
PH_init= plDistributionTable(H,Prox,Prox)
random=1
for i in plValues(Prox):
    PH_init.push(plProbTable(H,random),i)
```

The user specifies which kind of distribution on H must be learned.

```
PH_learned = plCndLearnHistogram(H,Prox)
```

We start defining an "plEMLearner." The first argument is the initial value for the joint distribution to be learned. The second argument is the list of learners used to infer the joint distribution from data:

The *plLearnFrozenDistribution* object is used to tell the algorithm that the distribution passed as an argument that should not be learned.

The learner object *PH_learned* is part of this list to tell the algorithm which learner to use for H.

```
#define the distribution which needed to be learned
HLearner =plEMLearner(\
               plUniform(Prox)* \
               plUniform(Dir)* \
               plUniform(Theta)* \
               PH_init * \
               PVrot\
               ,\
               [plLearnFrozenDistribution( plUniform(Prox)),\
                plLearnFrozenDistribution( plUniform(Dir)),\
                plLearnFrozenDistribution( plUniform(Theta)),\
                PH_learned,\
                plLearnFrozenDistribution(PVrot)])
```

The following instruction runs the learning algorithm on data previously recorded during a homing experiment. The second argument tells the precision at which the EM algorithm should stop.

```
HLearner.run(datahoming,0.01)
```

Finally, the joint distribution is obtained and can be used as in the previous examples.

```
#Extract a  model from the learner
learned_model = HLearner.get_joint_distribution()
#use this question at 0.1 Hz to drive the robot
home_question=learned_model.ask(Vrot,Dir^Prox^Theta)
#use this question to classify the behavior
behavior_question=home_specification.ask(^H,Vrot,Dir^Prox^Theta)
```

Chapter 11

Bayesian Programming Iteration

Probability does pervade the universe, and in this sense, the old chestnut about baseball imitating life really has validity. The statistics of streaks and slumps, properly understood, do teach an important lesson about epistemology, and life in general. The history of a species, or any natural phenomenon, that requires unbroken continuity in a world of trouble, works like a batting streak. All are games of a gambler playing with a limited stake against a house with infinite resources. The gambler must eventually go bust. His aim can only be to stick around as long as possible, to have some fun while he's at it, and, if he happens to be a moral agent as well, to worry about staying the course with honor!

The Streak of Streaks Stephen Jay Gould [1988]

In this chapter we propose to define a description with distributions indexed by integers. By setting the range of indexes, we define the number of distributions used in the description. This way we define generic descriptions only depending on the range of the indexes, just as we fixed the number of iterations in a "for" loop. In pursuing this idea, we can redefine the notion of filter where each new evidence is incorporated into the result of a previous inference. If the index represents successive time intervals, we can then use these techniques to study time sequences and use the Markov assumption to simplify the description. The approach is useful to implement dynamic Bayesian networks within the Bayesian programming formalism.

11.1 Generic iteration

11.1.1 Statement of the problem

To define an iteration we may index by i the variables and the distributions found in a description. For example, considering we are throwing dice N times we may be interested in summing the values which turn up each time. By modeling this process with the proper description we can answer many questions related to the sum obtained after N throws. We call S_i the sum after the step i and O_i the value obtained at throw i. With this notation we can define a generic program only depending on the number N of throws.

11.1.2 Bayesian program

The program 11.1 is designed to model a game with N fair dices. It used N probabilistic variables $S_i \in [i, i \times 6]$ and $O_i \in [1, \dots 6]$ indexed by $i \in [1..N]$ to model the outcomes of the throws and the sum after each throw. *Uniform* distributions are used to model the outcomes and the conditional probability distributions on the sums are modeled with Dirac distribution $P(S_i \mid O_i \wedge S_{i-1}) = \delta_{o_i + s_{i-1}}$ as they are deterministic.

$$
Pr \begin{cases}
Ds \begin{cases}
Sp \begin{cases}
Va: \begin{cases}
O_1 \in [1,6], S_1 \in [1,6] \\
\dots \\
O_N \in [i,6], S_N \in [N, N \times 6]
\end{cases} \\
Dc: \begin{cases}
P(O_1 \wedge S_1, \dots O_i \wedge S_i \dots O_N \wedge S_N) \\
= P(O_1) \times P(S_1 \mid O_1) \times \\
P(O_2) P(S_2 \mid S_1 \wedge O_2) \times \\
\dots \\
P(O_N) P(S_N \mid S_{N-1} \wedge O_N)
\end{cases} \\
Fo: \begin{cases}
P(O_1) = P(O_2) = \dots = P(O_N) = Uniform \\
P(S_1 \mid O_1) = \delta_{o_1} \\
P(S_2 \mid S_1 \wedge O_2) = \delta_{s_1 + o_2} \\
\dots \\
P(S_N \mid S_{N-1} \wedge O_N) = \delta_{s_{N-1} + o_N}
\end{cases}
\end{cases} \\
Id:
\end{cases} \\
Qu: P(S_N)
\end{cases}
\tag{11.1}
$$

11.1.3 Instance and results

The Bayesian program in Equation 11.1 allows for several questions:

- $P(S_N)$: to obtain probability of the sum when throwing N fair dices.
- $P(S_N \mid o_m \wedge \ldots \wedge o_k)$: the probability of the sum knowing the outcomes of throws $m \ldots k$.
- $P(O_m \wedge \ldots \wedge O_k \mid s_N)$: the probability of the outcomes given the sum

For example, for a game with three dices, Figure 11.1 represents the probability of the sum after the game knowing the outcome of the first roll $P(S_3 \mid [O_1 = 4])$ and the probability distribution for the last roll knowing the final sum $P(O_3 \mid [S_3 = 14])$.

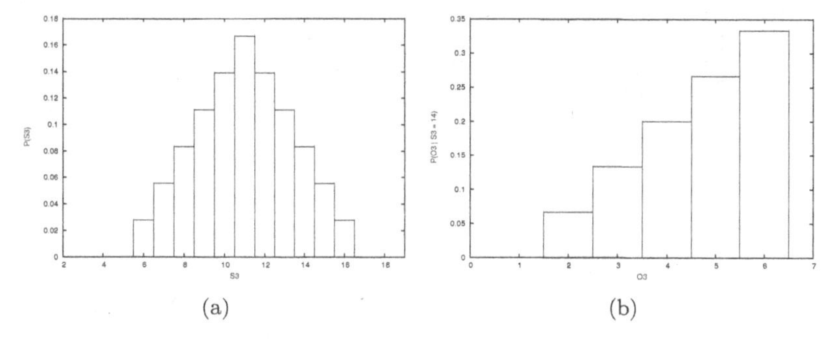

FIGURE 11.1: Distributions are obtained with the description in 11.1:
(a): $P(S_3 \mid [O_1 = 4])$
(b): $P(O_3 \mid [S_3 = 14])$

The file "chapter11/sumdices.py" contains a function to create the description of program in Equation 11.1 with N dices with two arrays of probabilistic variables O and S of dimension $N+1$.

```
D3=dices_game(O,S)
```

Figure 11.1a,b is obtained with the following instructions:

```
D3=dices_game(O,S)
PS3kO1=D3.ask(S[3],O[1])
PS3=PS3kO1.instantiate(4).compile()
#to draw the distribution used in the book
PS3.plot('fileS3'  )
PO3kS3=D3.ask(O[3],S[3])
PO3=PO3kS3.instantiate(14).compile()
PO3.plot(ExDir+'chapter11/figures/PO3kS3'  )
```

11.2 Generic Bayesian filters

11.2.1 Statement of the problem

Often a sequence of measurements helps to better characterize the state of a system. Bayesian filters serve this purpose. They may be seen as special cases of dynamic Bayesian networks and are often used to process time series of sensor readings.

11.2.2 Bayesian program

The following program (Equation 11.2) defines a generic Bayesian filter.

$$
Pr\begin{cases}
Ds\begin{cases}
Sp\begin{cases}
Va:\begin{cases} S^t, \forall t \in [0,\ldots,T] : S^t \in D_S \\ O^t, \forall t \in [1,\ldots,T] : O^t \in D_O \end{cases} \\
Dc:\begin{cases} P\left(S^0 \wedge O^1, \ldots S^t \wedge O^t \ldots S^T \wedge O^T\right) = \\ P\left(S^0\right) \prod\limits_{t\in[1\ldots T]} \left(P\left(S^t \mid S^{t-1}\right) P\left(O^t \mid S^t\right)\right) \end{cases} \\
Fo:\begin{cases} P\left(S^0\right) = \text{Initial condition} \\ P\left(S^t \mid S^{t-1}\right) = \text{Transition Model} \\ P\left(O^t \mid S^t\right) = \text{Sensor Model} \end{cases}
\end{cases} \\
Id:
\end{cases} \\
Qu: P\left(S^T \mid o^1 \ldots o^T\right)
\end{cases}
\tag{11.2}
$$

The variables S^t have the same definition domain and denote the states of the system at time t. For example, a sequence $s^0, s^1, ..., s^T$ represents a possible evolution for the system. The variable O^t denotes the observation of the system at time t. These variables share the same definition domain. The Bayesian program 11.2 encodes the two main hypotheses used in classical Bayesian filters.

- The probability distribution on the current state only depends on the previous state $P\left(S^t \mid S^{t-1}\right)$ (order 1 Markov hypothesis).
- The observation only depends on the current state: $P\left(O^t \mid S^t\right)$.

The distribution $P\left(S^0\right)$ describes the knowledge about the initial conditions. The parametric form $P\left(S^t \mid S^{t-1}\right)$ does not depend on t and defines the priors relative to the evolution of the system. If D_S is discrete, $P\left(S^t \mid S^{t-1}\right)$ is usually represented as a probabilistic transition matrix. $P\left(O^t \mid S^t\right)$ is the sensor model.

As such, the Bayesian program in Equation 11.2 permits answering several interesting questions:

- Filtering: $P\left(S^T \mid o^1 \dots o^T\right)$: the purpose of this question is to infer the current state according to the past sequence of measurements.
- Smoothing: $P\left(S^k \mid o^1 \dots o^k \dots o^T\right)$ $(k < T)$ estimates a past state of the system by taking into account more recent measurements.
- Forecasting: $P\left(S^T \mid o^1 \dots o^k\right)$ estimates the state of the system in the future $(T > k)$ based on the current measurements.

The filtering question $P\left(S^T \mid o^1 \dots o^T\right)$ has the very appealing property to lead to a recursive computation. Indeed, the answer to this question is obtained by marginalizing the missing variables:

$$P\left(S^T|o^1 \wedge \cdots \wedge o^T\right) \prec \sum_{S^1 \dots S^{T-1}} \left[\prod_{t=1}^{T} \left[P\left(o^t|S^t\right) P\left(S^t|S^{t-1}\right)\right] P(S^0)\right] \quad (11.3)$$

The term $P\left(o^T|S^T\right) P\left(S^T|S^{T-1}\right)$ may be factored out of the sum of variables $S^1 \cdots S^{T-2}$:

$$\begin{aligned} & P\left(S^T|o^1 \wedge \cdots \wedge o^T\right) \\ \prec \; & P\left(o^T|S^T\right) \sum_{S^{T-1}} \left[\begin{array}{c} P\left(S^T|S^{T-1}\right) \\ \sum_{S^1 \dots S^{T-2}} \left[\prod_{t=1}^{T-1} \left[P\left(o^t|S^t\right) P\left(S^t|S^{t-1}\right)\right] P(S^0)\right] \end{array} \right] \end{aligned} \quad (11.4)$$

In the term $\sum_{S^1 \dots S^{T-2}} \left[\prod_{t=1}^{T-1} \left[P\left(o^t|S^t\right) P\left(S^t|S^{t-1}\right)\right] P(S^0)\right]$ we recognize the same filtering question at the preceding instant:

$$P\left(S^{T-1}|o^1 \wedge \cdots \wedge o^{T-1}\right) \prec \sum_{S^1 \dots S^{T-2}} \left[\prod_{t=1}^{T-1} \left[P\left(o^t|S^t\right) P\left(S^t|S^{t-1}\right)\right] P(S^0)\right] \quad (11.5)$$

And we finally get the recursive expression:

$$\begin{aligned} & P\left(S^T|o^1 \wedge \cdots \wedge o^T\right) \\ \prec \; & P\left(o^T|S^T\right) \sum_{S^{T-1}} \left[\begin{array}{c} P\left(S^T|S^{T-1}\right) \\ P\left(S^{T-1}|o^1 \wedge \cdots \wedge o^{T-1}\right) \end{array} \right] \end{aligned} \quad (11.6)$$

This recursive expression is a fundamental characteristic of Bayesian filtering as it ensures that the corresponding computations are kept tractable.[1]

[1]Note that this recursive property is only true for the filtering question and not for the smoothing or forecasting ones. These two last questions lead to much cumbersome computation as some supplementary marginalizations are required.

The computation of $\sum_{S^{T-1}} \left[P\left(o^T|S^{T-1}\right) P\left(S^{T-1}|o^1 \wedge \cdots \wedge o^{T-1}\right)\right]$ is often called the prediction step where we try to predict the state at time T knowing the observation until time $T-1$:

$$\sum_{S^{T-1}} \left[P\left(o^T|S^{T-1}\right) P\left(S^{T-1}|o^1 \wedge \cdots \wedge o^{T-1}\right)\right] = P\left(S^T|o^1 \wedge \cdots \wedge o^{T-1}\right) \quad (11.7)$$

The product $P\left(o^T|S^T\right) P\left(S^T|o^1 \wedge \cdots \wedge o^{T-1}\right)$ is called the estimation step as we update the prediction using the new observation o^T.

11.2.3 Instance and results

Again we use the localization problem presented in Section 7.1.3. The state variable is the location of the boat $S = X \wedge Y$. The observation variables $O = B_1 \wedge B_2 \wedge B_3$ are the bearing angles as in the program (Equation 9.21). This time, the bearing measurements will be taken T times $t \in 1, \ldots, T : B^t_{i=1,2,3}$. The probability on the initial state $X^0 \wedge Y^0$ is supposed to be uniform.

The sensor model and a transition model are defined by:

- The sensor model uses a fixed precision:

$$P\left(B^t_i \mid X^t \wedge Y^t \wedge \pi\right) = B\left(\left[\mu = f^i_b\left(X^t, Y^t\right)\right], [\sigma = 10]\right) \quad (11.8)$$

- The transition model assumes the boat to be almost stationary during the sequence of measurements:

$$P\left(X^t \mid X^{t-1} \wedge \pi\right) = B\left(\left[\mu = X^{t-1}\right], [\sigma = 5]\right) \quad (11.9)$$

$$P\left(Y^t \mid Y^{t-1} \wedge \pi\right) = B\left(\left[\mu = Y^{t-1}\right], [\sigma = 5]\right) \quad (11.10)$$

This can be summarized by the Bayesian program in Equation 11.11.

Figure 11.2 shows how the uncertainty is reduced as more readings are added into the model. The program in Equation 11.11 has a strong similarity with a Kalman filter with the noticeable difference that it uses a nonlinear observational model.

$$
Pr\begin{cases}
Ds\begin{cases}
Sp(\pi)\begin{cases}
Va: X^0, Y^0, \cdots, X^T, Y^T, B_1^1, \cdots, B_3^T \\
Dc: \begin{cases}
P\left(X^0 \wedge \cdots \wedge B_3^T\right) \\
= \prod_{t=1}^{T}\left[\begin{array}{c} P\left(X^t|X^{t-1}\right)P\left(Y^t|Y^{t-1}\right) \\ \prod_{i=1}^{3}\left[P\left(B_i^t|X^t \wedge Y^t\right)\right] \end{array}\right] \times P\left(X^0 \wedge Y^0\right)
\end{cases} \\
Fo: \begin{cases}
P\left(X^0 \wedge Y^0\right) = Uniform \\
P\left(X^t|X^{t-1}\right) = B\left(\left[\mu = X^{t-1}\right], [\sigma = 5]\right) \\
P\left(Y^t|Y^{t-1}\right) = B\left(\left[\mu = Y^{t-1}\right], [\sigma = 5]\right) \\
P\left(B_i^t|X^t \wedge Y^t\right) = B\left(\left[\mu = f_b^i\left(X^t, Y^t\right)\right], [\sigma = 10]\right)
\end{cases}
\end{cases} \\
Id
\end{cases} \\
Qu: P\left(X^T \wedge Y^T | B_1^0 \wedge \cdots \wedge B_3^T\right)
\end{cases}
\tag{11.11}
$$

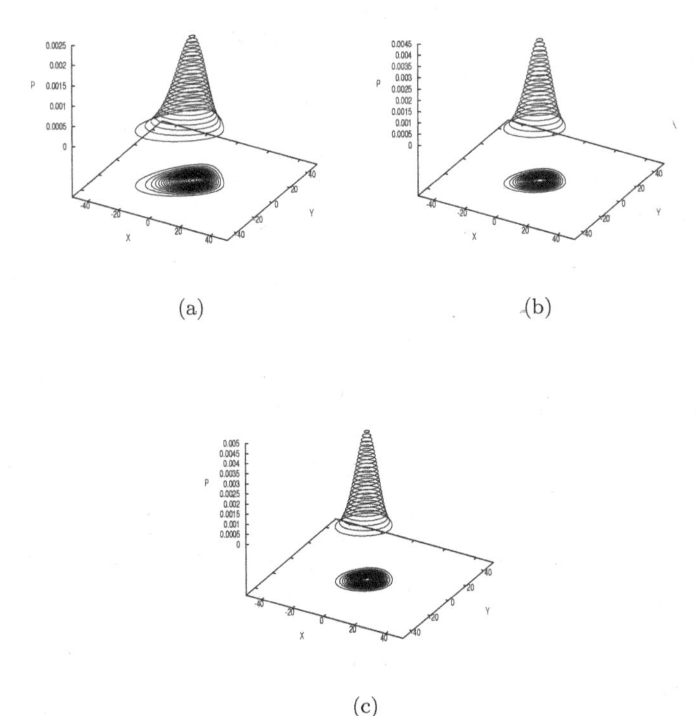

FIGURE 11.2: The precision of the location is improved by feeding the model with several measurements: (a) 1, (b) 2, and (c) 3 measurements.

The file "chapter11/pseudokalman.py" has been used to create the drawings in Figure 11.2. There are two parts in the file:

1. The first part defines the recursive Bayesian program.
2. The second part shows how to use this program in a "read the sensor" and "evaluate the probability distribution on states" loop.

To define a recursive program we have to create a special kind of distribution denoted as "Mutable" on $X_{t-1} \wedge Y_{t-1}$ and to initialize it.

```
PXt_1Yt_1=plMutableDistribution(plUniform(Xt_1^Yt_1))
```

Mutable distributions may be considered as global distributions. When their values are changed, the new values are used in all the previously defined definitions. For example, if the value of $P(X_{t-1} \wedge Y_{t-1})$ is changed to $P(X_t \wedge Y_t \mid b_1, b_2, b_3)$, then the question at time $t+1$ will take $P(X_t \wedge Y_t \mid b_1, b_2, b_3)$ as prior on the location of the boat and effectively compute $P(X_{t+1} \wedge Y_{t+1} \mid B_1 \wedge B_2 \wedge B_3)$.

```
JointDistributionList=plComputableObjectList()
#use the mutable distribution as the state distribution
JointDistributionList.push_back(PXt_1Yt_1)
#define the transition model
JointDistributionList.push_back(plCndNormal(Xt,Xt_1,5))
JointDistributionList.push_back(plCndNormal(Yt,Yt_1,5))
#define the sensor model
JointDistributionList.push_back(plCndNormal(B1,Xt^Yt, \
                        plPythonExternalFunction(Xt^Yt,f_b_1), \
                        10.0))
JointDistributionList.push_back(plCndNormal(B2,Xt^Yt, \
                        plPythonExternalFunction(Xt^Yt,f_b_2), \
                        10.0))
JointDistributionList.push_back(plCndNormal(B3,Xt^Yt, \
                        plPythonExternalFunction(Xt^Yt,f_b_3), \
                        10.0))
#define the joint distribution
filtered_localisation_model= \
   plJointDistribution(JointDistributionList)

#define the question to get  the new state
PXY_K_B1B2B3=filtered_localisation_model.ask(Xt^Yt,B1^B2^B3)
```

Finally, a standard filter may be written as a loop which sequentially performs the following tasks:

1. Reads the observations.
2. Estimates the current state.
3. Sets the prior knowledge for the next iteration with the "mutate" method.

This sequence is implemented as follows:

```
for val in V:
    #read the observations
    sensor_reading_values[B1]= val[0]
    sensor_reading_values[B2]= val[1]
    sensor_reading_values[B3]= val[2]
    #estimate the state
    PXY=PXY_K_B1B2B3.instantiate(sensor_reading_values)
    compiled_PXY=PXY.compile()
    outputfile = 'C:PXY_{0}'.format(i)
    compiled_PXY.plot(outputfile)
    #Prepare the next iteration
    compiled_PXY.rename(Xt_1^Yt_1)
    PXt_1Yt_1.mutate(compiled_PXY)
    i=i+1
```

11.3 Markov localization

11.3.1 Statement of the problem

The generic Bayesian filter as it is described by the program in Equation 11.2 offers a large variety of refinements. For example, we can change the structure as in the autoregressive hidden Markov model (HMM) where the observation at time t depends on the state at time t and $t-1$, leading to the observational model $P(O_t \mid S_t \wedge S_{t-1})$. Another direction is to add "control" variables to the state variables. In this case the transition model could reflect the future state of the system knowing the current state and the control. This approach is useful to keep track of a moving vehicle with external sensors knowing its current speed.

11.3.2 Bayesian program

The resulting Bayesian program is the following:

$$
Pr\begin{cases}
Ds\begin{cases}
Sp\begin{cases}
Va:\begin{cases}
S^0\ldots S^t,\forall t\in[0,\ldots,T]:S^t\in D_S\\
M^0\ldots M^t,\forall t\in[0,\ldots,T]:M^t\in D_M\\
O^1\ldots O^t,\forall t\in[0,\ldots,T]:O^t\in D_O
\end{cases}\\
Dc:\begin{cases}
P\left(S^0\wedge M^0,\ldots S^t\wedge O^t\ldots S^T\wedge O^T\right)=\\
P\left(S^0\right)P\left(M^0\right)\\
\times\prod_{t\in[1\ldots T]}\left[P\left(M^t\right)P\left(S^t\mid S^{t-1}\wedge M^{t-1}\right)P\left(O^t\mid S^t\right)\right]
\end{cases}\\
Fo:\begin{cases}
P\left(S^0\right)=\text{Initial condition}\\
P\left(M^t\right)=\text{Priors on commands}\\
P\left(S^t\mid S^{t-1}\wedge M^{t-1}\right)=\text{Transition Model}\\
P\left(O^t\mid S^t\right)=\text{Sensor Model}
\end{cases}
\end{cases}\\
Id:
\end{cases}\\
Qu:P\left(S^T\mid o^0\wedge m^0\wedge o^1\wedge m^1\ldots o^T\right)
\end{cases}
\tag{11.12}
$$

The specification of Equation 11.2 permits adding several interesting questions to the questions already attached to a Bayesian filter, for instance:

- Filtering: $P\left(S^T\mid m^0\wedge o^1\wedge m^1\ldots\wedge m^{T-1}\wedge o^T\right)$, the purpose of this question is to infer the current state according to the past sequence of measurements and commands.

- Forecasting: $P\left(S^k\mid m^0\wedge o^1\wedge m^1\ldots o^T\wedge m^T\wedge m^{T+1}\ldots m^k\right)$, to estimate the state of the system in the future ($k>T$) with the past and present measurements and the past and future commands.

- Control: $P\left(M^T\mid m^0\wedge o^1\wedge m^1\ldots o^T\wedge s^k\right)$ $(k>T)$, to search the current control to reach a given state in the future (s^k) knowing the past and present measurements and the past commands.

- Interpreting: $P\left(S^0\wedge S^1\ldots S^T\mid o^0\wedge o^1\wedge m^1\ldots o^T\right)$, to infer the set of states from the past observations and commands.

Note that the recursive property is only valid for the filtering question. The other question, even if completely valid from a mathematical point of view, may be intractable in practice due to some cumbersome computations.

11.3.3 Instance and results

Using our previous example, we replace the initial assumption about the stationarity of the boat, assuming it is now moving. At each time step t the boat is instructed to move with constant speed Vx, Vy during the next time interval. This hypothesis leads to the following transition model:

$$P\left(X^t|X^{t-1} \wedge V_x^{t-1}\right) = B\left(\left[\mu = X^{t-1} + Vx^{t-1} \times \delta t\right], [\sigma = 5]\right) \tag{11.13}$$

$$P\left(Y^t|Y^{t-1} \wedge V_y^{t-1}\right) = B\left(\left[\mu = Y^{t-1} + Vy^{t-1} \times \delta t\right], [\sigma = 5]\right) \tag{11.14}$$

This transition model is used to modify the program in Equation 11.11 into the program in Equation 11.15, which implements an elementary version of the Markov localization.

$$Pr\begin{cases} Ds\begin{cases} Sp(\pi)\begin{cases} Va: \\ X^0, Y^0, V_x^0, V_y^0 \cdots, V_y^T, B_1^1, \cdots, B_3^T \\ Dc: \\ \begin{cases} P\left(X^0 \wedge \cdots \wedge B_3^T\right) \\ = \prod_{t=1}^{T}\left[\begin{array}{l} P\left(V_x^t\right)P\left(V_y^t\right) \\ P\left(X^t|X^{t-1} \wedge V_x^{t-1}\right)P\left(Y^t|Y^{t-1} \wedge V_y^{t-1}\right) \\ \prod_{i=1}^{3}\left[P\left(B_i^t|X^t \wedge Y^t\right)\right] \end{array}\right] \\ \times P\left(X^0 \wedge Y^0\right)P\left(V_x^0\right)P\left(V_y^0\right) \end{cases} \\ Fo: \\ P\left(X^0 \wedge Y^0\right) = Uniform \\ P\left(V_x^t\right) = Constant \\ P\left(V_y^t\right) = Constant \\ P\left(X^t|X^{t-1} \wedge V_x^{t-1}\right) = B\left(\left[\mu = X^{t-1} + Vx^{t-1} \times \delta t\right], [\sigma = 5]\right) \\ P\left(Y^t|Y^{t-1} \wedge V_y^{t-1}\right) = B\left(\left[\mu = Y^{t-1} + Vy^{t-1} \times \delta t\right], [\sigma = 5]\right) \\ P\left(B_i^t|X^t \wedge Y^t\right) \\ = B\left(\left[\mu = f_b^i\left(X^t, Y^t\right)\right], [\sigma = 10]\right) \end{cases} \\ Id \end{cases} \\ Qu: \\ P\left(X^T \wedge Y^T|v_x^0 \wedge \cdots \wedge b_3^T\right) \end{cases} \tag{11.15}$$

This program may be, for instance, used to estimate the position of the boat knowing successive observations of the bearings and assuming a constant velocity toward the upper right corner as in Figure 11.3.

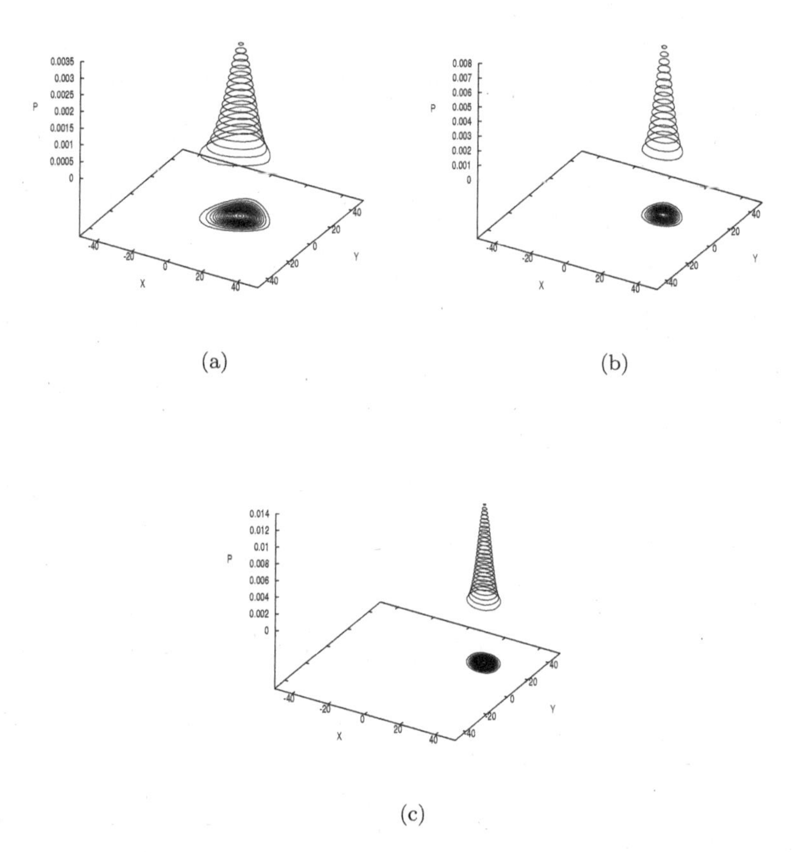

FIGURE 11.3: The precision of the location is increasing even with the boat moving at constant speed toward the upper right corner. Such a model is very similar to what is used in a GPS (global positioning system).

The file "chapter11/markovloc.py" has been used to create the drawings of Figure 11.3. The structure of the program is identical to the stationary case. The only difference lies in the specification which now includes a " prediction" term: $P(X_t \mid X_{t-1} \wedge Mx_{t-1})$ where $Mx_{t-1}+X_{t-1}$ is used as the mean for X_t. An external function f_vx is used to compute $Mx_{t-1} + X_{t-1}$.

```
#recursive specification:
#define the initial distribution  X_0 Y_0
PXt_1Yt_1=plMutableDistribution(plUniform(Xt_1^Yt_1))

JointDistributionList=plComputableObjectList()
#use the mutable distribution as the prior
#on the state distribution
JointDistributionList.push_back(PXt_1Yt_1)
#use avialable knowledge on M_t:
JointDistributionList.push_back(plUniform(Mxt_1^Myt_1))
#define the prediction term
JointDistributionList.push_back(plCndNormal(Xt,Xt_1^Mxt_1,\
                  plPythonExternalFunction(Xt_1^Mxt_1,f_vx),2))
JointDistributionList.push_back(plCndNormal(Yt,Yt_1^Myt_1, \
                  plPythonExternalFunction(Yt_1^Myt_1,f_vy),2))
#define the sensor model
JointDistributionList.push_back(plCndNormal(B1,Xt^Yt, \
                       plPythonExternalFunction(Xt^Yt,f_b_1), \
                                                    10.0))
JointDistributionList.push_back(plCndNormal(B2,Xt^Yt, \
                       plPythonExternalFunction(Xt^Yt,f_b_2), \
                                                    10.0))
JointDistributionList.push_back(plCndNormal(B3,Xt^Yt, \
                       plPythonExternalFunction(Xt^Yt,f_b_3), \
                                                    10.0))
#define the joint distribution
filtered_localisation_model = \
                    plJointDistribution(JointDistributionList)
```

Part III

Bayesian Programming Formalism and Algorithms

Chapter 12

Bayesian Programming Formalism

It is a truth very certain that when it is not in our power to determine what is true we ought to follow what is most probable.

Discours de la Méthode

Descartes [1637]

The purpose of this chapter is to present Bayesian Programming formally and to demonstrate that it is very simple and very clear but, nevertheless, very powerful and subtle. Probability is an extension of logic, as mathematically sane and simple as logic, but with more expressive power than logic.

It may seem unusual to present the formalism toward the middle of the book. We have done this to help comprehension and to assist intuition without sacrificing rigor. After reading this chapter, anyone will be able to check that all the examples and programs presented earlier comply with the formalism.

12.1 Logical propositions

The first concept we use is the usual notion of logical propositions. Propositions are denoted by lowercase names. Propositions may be composed to obtain new propositions using the usual logical operators: $a \wedge b$, denoting the conjunction of propositions a and b, $a \vee b$ their disjunction, and $\neg a$, the negation of proposition a.

12.2 Probability of a proposition

To be able to deal with uncertainty, we attach probabilities to propositions.

We consider that, to assign a probability to a proposition a, it is necessary to have at least some *preliminary knowledge*, summed up by a proposition π. Consequently, the probability of a proposition a is always conditioned, at least, by π. For each different π, $P(\bullet|\pi)$ is an application that assigns to each proposition a a unique real value $P(a|\pi)$ in the interval $[0,1]$.

Of course, we are interested in reasoning about the probabilities of conjunctions, disjunctions, and negations of propositions, denoted, respectively, by $P(a \wedge b|\pi)$, $P(a \vee b|\pi)$, and $P(\neg a|\pi)$.

We are also interested in the probability of proposition a conditioned by both the preliminary knowledge π and some other proposition b. This is denoted $P(a|b \wedge \pi)$.

12.3 Normalization and conjunction postulates

Probabilistic reasoning requires only two basic rules:

1. The *conjunction rule*, which gives the probability of a conjunction of propositions.

$$\begin{aligned} P(a \wedge b|\pi) &= P(a|\pi) \times P(b|a \wedge \pi) \\ &= P(b|\pi) \times P(a|b \wedge \pi) \end{aligned} \tag{12.1}$$

2. The *normalization rule*, which states that the sum of the probabilities of a and $\neg a$ is one.

$$P(a|\pi) + P(\neg a|\pi) = 1 \tag{12.2}$$

In this book, we take these two rules as postulates.[1]

As in logic, where the resolution principle (Robinson [1965], Robinson [1979]) is sufficient to solve any inference problem, in discrete probabilities, these two rules (Equations 12.1 and 12.2) are sufficient for any probabilistic computation. Indeed, we may derive all the other necessary inference rules from these two.[2]

12.4 Disjunction rule for propositions

For instance, the rule concerning the disjunction of propositions

$$P(a \vee b|\pi) = P(a|\pi) + P(b|\pi) - P(a \wedge b|\pi) \tag{12.3}$$

may be derived as follows:

$$\begin{aligned}
P(a \vee b|\pi) &= 1 - P(\neg a \wedge \neg b|\pi) \\
&= 1 - P(\neg a|\pi) \times P(\neg b|\neg a \wedge \pi) \\
&= 1 - P(\neg a|\pi) \times (1 - P(b|\neg a \wedge \pi)) \\
&= 1 - P(\neg a|\pi) + P(\neg a|\pi) \times P(b|\neg a \wedge \pi) \\
&= P(a|\pi) + P(\neg a \wedge b|\pi) \\
&= P(a|\pi) + P(b|\pi) \times P(\neg a|b \wedge \pi) \\
&= P(a|\pi) + P(b|\pi) \times (1 - P(a|b \wedge \pi)) \\
&= P(a|\pi) + P(b|\pi) - P(a \wedge b|\pi)
\end{aligned} \tag{12.4}$$

12.5 Discrete variables

The notion of *discrete variable* is the second concept we require. Variables are denoted by names starting with one uppercase letter.

By definition, a discrete variable X is a set of logical propositions x_i, such that these propositions are mutually exclusive (for all i, j with $i \neq j$, $x_i \wedge x_j$ is false) and exhaustive (at least one of the propositions x_i is true). x_i means that variable X takes its ith value. $Card(X)$ denotes the cardinality of the set X (the number of propositions x_i).

---

[1] For sources giving justifications of these two rules, see Chapter 16, Section 16.8 on the Cox theorem.

[2] These two rules are sufficient as long as we work with discrete variables. To use continuous variables much more elaborated math is required. See Chapter 16, "Discrete versus continuous variables" (Section 16.9) for a discussion on this matter.

12.6 Variable conjunction

The conjunction of two variables X and Y, denoted $X \wedge Y$, is defined as the set of $Card(X) \times Card(Y)$ propositions $x_i \wedge x_j$. $X \wedge Y$ is a set of mutually exclusive and exhaustive logical propositions. As such, it is a new variable.[3]

Of course, the conjunction of n variables is also a variable and, as such, it may be renamed at any time and considered as a unique variable.

12.7 Probability on variables

For simplicity and clarity, we also use probabilistic formulas with variables appearing instead of propositions.

By convention, each time a variable X appears in a probabilistic formula $\Phi(X)$, it should be understood as: $\forall x_i \in X, \Phi(x_i)$.

For instance, given three variables X, Y, and Z:

$$P(X \wedge Y|Z \wedge \pi) = P(X|Z \wedge \pi) \tag{12.5}$$

stands for:

$$\begin{array}{c} \forall x_i \in X, \forall y_j \in Y, \forall z_k \in Z, \\ P(x_i \wedge y_j|z_k \wedge \pi) = P(x_i|z_k \wedge \pi) \end{array} \tag{12.6}$$

12.8 Conjunction rule for variables

$$\begin{array}{rcl} P(X \wedge Y|\pi) & = & P(X|\pi) \times P(Y|X \wedge \pi) \\ & = & P(Y|\pi) \times P(X|Y \wedge \pi) \end{array} \tag{12.7}$$

According to our convention for probabilistic formulas including variables, this may be restated as:

$$\begin{array}{c} \forall x_i \in X, \forall y_j \in Y, \\ P(x_i \wedge y_j|\pi) = P(x_i|\pi) \times P(y_j|x_i \wedge \pi) = P(y_j|\pi) \times P(x_i|y_j \wedge \pi) \end{array} \tag{12.8}$$

which may be directly deduced from the conjunction rule for propositions (Equation 12.1).

[3] In contrast, the disjunction of two variables, defined as the set of propositions $x_i \vee x_j$, is not a variable. These propositions are not mutually exclusive.

12.9 Normalization rule for variables

$$\sum_{X} [P(X|\pi)] = 1 \tag{12.9}$$

The normalization rule may obviously be derived as follows:

$$\begin{aligned} 1 &= P(x_1|\pi) + P(\neg x_1|\pi) \\ &= P(x_1|\pi) + P\left(x_2 \vee \cdots \vee x_{Card(X)}|\pi\right) \\ &= P(x_1|\pi) + P(x_2|\pi) + \cdots + P\left(x_{Card(X)}|\pi\right) \\ &= \sum_{x_i \in X} [P(x_i|\pi)] \end{aligned} \tag{12.10}$$

where the first equality derives from the normalization rule for propositions (Equation 12.2), the second from the exhaustiveness of propositions x_i, and the third from both the application of Equation 12.3 and the mutual exclusivity of propositions x_i.

12.10 Marginalization rule

$$\sum_{X} [P(X \wedge Y|\pi)] = P(Y|\pi) \tag{12.11}$$

The marginalization rule is derived by the successive application of the conjunction rule (Equation 12.7) and the normalization rule (Equation 12.9):

$$\begin{aligned} \sum_{X} [P(X \wedge Y|\pi)] &= \sum_{X} [P(Y|\pi) \times P(X|Y \wedge \pi)] \\ &= P(Y|\pi) \times \sum_{X} [P(X|Y \wedge \pi)] \\ &= P(Y|\pi) \end{aligned} \tag{12.12}$$

12.11 Bayesian program

We define a *Bayesian program* as a means of specifying a family of probability distributions.

The constituent elements of a Bayesian program are presented below:

$$Program \begin{cases} Description \begin{cases} Specification\ (\pi) \begin{cases} Variables \\ Decomposition \\ Forms\ (Parametric\ or\ Program) \end{cases} \\ Identification\ (based\ on\ \delta) \end{cases} \\ Question \end{cases}$$

1. A program is constructed from a description and a question.

2. A description is constructed using some specification (π) as given by the programmer and an identification or learning process for the parameters not completely specified by the specification, using a data set (δ).

3. A specification is constructed from a set of pertinent variables, a decomposition, and a set of forms.

4. Forms are either parametric forms or questions to other Bayesian programs.

12.12 Description

The purpose of a description is to specify an effective method of computing a joint distribution on a set of variables $\{X_1, X_2, \cdots, X_N\}$ given a set of experimental data δ and some specification π. This joint distribution is denoted as: $P(X_1 \wedge X_2 \wedge \cdots \wedge X_N | \delta \wedge \pi)$.

12.13 Specification

To specify preliminary knowledge, the programmer must undertake the following:

1. Define the set of relevant variables $\{X_1, X_2, \cdots, X_N\}$ on which the joint distribution is defined.

2. Decompose the joint distribution:

Given a partition $\{X_1, X_2, \cdots, X_N\}$ into K subsets, we define K variables $L_1, \cdots, L_K$, each corresponding to one of these subsets.

Each variable L_k is obtained as the conjunction of the variables $\{X_{k_1}, X_{k_2}, \cdots\}$ belonging to the kth subset. The conjunction rule (Equation 12.7) leads to:

$$\begin{aligned} & P(X_1 \wedge X_2 \wedge \cdots \wedge X_N | \delta \wedge \pi) \\ = & P(L_1 \wedge \cdots \wedge L_K | \delta \wedge \pi) \\ = & P(L_1 | \delta \wedge \pi) \times P(L_2 | L_1 \wedge \delta \wedge \pi) \\ & \times \cdots \times P(L_K | L_{K-1} \wedge \cdots \wedge L_1 \wedge \delta \wedge \pi) \end{aligned} \tag{12.13}$$

Conditional independence hypotheses then allow further simplifications. A conditional independence hypothesis for variable L_k is defined by choosing some variable X_n among the variables appearing in conjunction $L_{k-1} \wedge \cdots \wedge L_2 \wedge L_1$, calling R_k the conjunction of these chosen variables and setting:

$$P(L_k | L_{k-1} \wedge \cdots \wedge L_1 \wedge \delta \wedge \pi) = P(L_k | R_k \wedge \delta \wedge \pi) \tag{12.14}$$

We then obtain:

$$\begin{aligned} & P(X_1 \wedge X_2 \wedge \cdots \wedge X_N | \delta \wedge \pi) \\ = & P(L_1 | \delta \wedge \pi) \times P(L_2 | R_2 \wedge \delta \wedge \pi) \times \cdots \times P(L_K | R_K \wedge \delta \wedge \pi) \end{aligned} \tag{12.15}$$

Such a simplification of the joint distribution as a product of simpler distributions is called a decomposition.

This ensures that each variable appears at the most once on the left of a conditioning bar, which is the necessary and sufficient condition to write mathematically valid decompositions.

3. Define the forms:

   Each distribution $P(L_k | R_k \wedge \delta \wedge \pi)$ appearing in the product is then associated with either a parametric form (i.e., a function $f_\mu(L_k)$) or a question to another Bayesian program. In general, μ is a vector of parameters that may depend on R_k or δ or both. Learning takes place when some of these parameters are computed using the data set δ.

12.14 Questions

Given a description (i.e., $P(X_1 \wedge X_2 \wedge \cdots \wedge X_N|\delta \wedge \pi)$), a question is obtained by partitioning $\{X_1, X_2, \cdots, X_N\}$ into three sets: the searched variables, the known variables, and the free variables.

We define the variables *Searched*, *Known*, and *Free* as the conjunction of the variables belonging to these sets. We define a question as the set of distributions:

$$P(Searched|Known \wedge \delta \wedge \pi) \tag{12.16}$$

made of as many "instantiated questions" as the cardinal of *Known*, each instantiated question being the distribution:

$$P(Searched|known \wedge \delta \wedge \pi) \tag{12.17}$$

12.15 Inference

Given the joint distribution $P(X_1 \wedge X_2 \wedge \cdots \wedge X_N|\delta \wedge \pi)$, it is always possible to compute any possible question using the following general inference:

$$\begin{aligned}
& P(Searched|known \wedge \delta \wedge \pi) \\
= & \sum_{Free} [P(Searched \wedge Free|known \wedge \delta \wedge \pi)] \\
= & \frac{\sum_{Free} [P(Searched \wedge Free \wedge known|\delta \wedge \pi)]}{P(known|\delta \wedge \pi)} \\
= & \frac{\sum_{Free} [P(Searched \wedge Free \wedge known|\delta \wedge \pi)]}{\sum_{Free \wedge Searched} [P(Searched \wedge Free \wedge known|\delta \wedge \pi)]} \\
= & \frac{1}{Z} \times \sum_{Free} [P(Searched \wedge Free \wedge known|\delta \wedge \pi)]
\end{aligned} \tag{12.18}$$

where the first equality results from the marginalization rule (Equation 12.11), the second results from the conjunction rule (Equation 12.7), and the third corresponds to a second application of the marginalization rule. The denominator appears to be a normalization term. Consequently, by convention, we will replace it by Z.

Theoretically, this allows us to solve any Bayesian inference problem. In practice, however, the cost of computing exhaustively and exactly

$P\left(Searched | known \wedge \delta \wedge \pi\right)$ is too great in most cases. Chapter 14 reviews and explains the main techniques and algorithms that can be used to deal with this inference problem.

Before that, Chapter 13 revisits the main Bayesian models using the present formalism.

Chapter 13

Bayesian Models Revisited

Remember that all models are wrong; the practical question is how wrong do they have to be to not be useful

Empirical Model-Building and Response Surfaces
Box and Draper [1987]

The goal of this chapter is to review the main probabilistic models currently used.

We systematically use the Bayesian Programming formalism to present these models, because it is precise and concise, and it simplifies their comparison.

We mainly concentrate on the definition of these models. Discussions about inference and computation are postponed to Chapter 14 and discussions about learning and identification are postponed to Chapter 15.

We chose to divide the different probabilistic models into three categories: the general purpose probabilistic models, the engineering oriented probabilistic models, and the cognitive oriented probabilistic models.

In the first category, the modeling choices are made independently of any specific knowledge about the modeled phenomenon. Most of the time, these choices are essentially made to keep the inference tractable. However, the technical simplifications of these models may be compatible with large classes of problems and consequently may have numerous applications.

In the second category, on the contrary, the modeling choices and simplifications are decided according to some specific knowledge about the modeled phenomenon. These choices could eventually lead to very poor models from a computational viewpoint. However, most of the time, problem-dependent knowledge, such as conditional independence between variables, leads to very significant and effective simplifications and computational improvements.

Finally, in the cognitive oriented probabilistic models category, the different models are presented according to a cognitive classification where common cognitive problems are links to common Bayesian solutions.

Several of these models were already presented with more detail in previous chapters. Certain models will appear several times in different categories but are presented with a different point of view for each presentation. We think that these repetitions are useful as our goal in this chapter is to give a synthetic overview of all these models.

13.1 General purpose probabilistic models

13.1.1 Graphical models and Bayesian networks

13.1.1.1 Bayesian networks

Bayesian networks (BNs), first introduced by Pearl [1988], have emerged as a primary method for dealing with probabilistic and uncertain information. They are the result of the marriage between the theory of probabilities and the theory of graphs.

BNs are defined by the following Bayesian program:

$$Pr \begin{cases} Ds \begin{cases} Sp(\pi) \begin{cases} Va: \\ X_1, \cdots, X_N \\ Dc: \\ \begin{cases} & P(X_1 \wedge \cdots \wedge X_N | \pi) \\ = & \prod_{n=1}^{N} [P(X_n | R_n \wedge \pi)] \end{cases} \\ Fo: \\ any \end{cases} \\ Id \end{cases} \\ Qu: \\ P(X_n | known) \end{cases} \tag{13.1}$$

- The pertinent variables are not constrained and have no specific semantics.

- The decomposition, on the contrary, is specific: it is a product of distributions with one and only one variable X_n conditioned by a conjunction of other variables R_n, called its parents. An obvious bijection exists between joint probability distributions defined by such decompositions and *directed acyclic graphs* (DAG): nodes are associated with variables, and oriented edges are associated with conditional dependencies. Using graphs in probabilistic models leads to an efficient way to define hypotheses over a set of variables, an economic representation of joint probability distribution, and, most importantly, an easy and efficient way to perform probabilistic inference (see Chapter 14).

- The parametric forms are not constrained but they are very often restricted to probability tables.

- Very efficient inference techniques have been developed to answer questions $P(X_n | known)$, however, some difficulties appear with more general questions (see Chapter 14).

Readings on Bayesian networks and graphical models should start with the following introductory textbooks: *Probabilistic Reasoning in Intelligent Systems: Networks of Plausible Inference* [Pearl, 1988], *Graphical Models* [Lauritzen, 1996], *Learning in Graphical Models* [Jordan, 1999], and *Graphical Models for Machine Learning and Digital Communication* [Frey, 1998].

See FAQ-FAM, Chapter 16, Section 16.3 for a summary of the differences between Bayesian programming and Bayesian networks.

13.1.1.2 Dynamical Bayesian networks

To deal with time and to model stochastic processes, the framework of BNs has been extended to dynamic Bayesian networks (DBNs) (see [Dean and Kanazawa, 1988]). Given a graph representing the structural knowledge at time t, supposing this structure to be time invariant and time to be discrete, the resulting DBN is the repetition of the first structure from a start time to a final time. Each part at time t in the final graph is called a time slice.

They are defined by the following Bayesian program:

$$Pr\begin{cases} Ds\begin{cases} Sp(\pi)\begin{cases} Va: \\ X_1^0,\cdots,X_N^0,\cdots,X_1^T,\cdots,X_N^T \\ Dc: \\ \begin{cases} & P\left(X_1^0\wedge\cdots\wedge X_N^T|\pi\right) \\ = & \prod_{t=0}^{T}\left[\prod_{n=1}^{N}\left[P\left(X_n^t|R_n^t\wedge\pi\right)\right]\right] \end{cases} \\ Fo: \\ any \end{cases} \\ Id \end{cases} \\ Qu: \\ P\left(X_n^t|known\right) \end{cases} \tag{13.2}$$

- R_n^t is a conjunction of variables taken in the set $\left\{X_1^t,\cdots,X_{n-1}^t\right\}\cup\left\{X_1^{t-1},\cdots,X_N^{t-1}\right\}$. This means that X_n^t depends only on its parents at time t ($\left\{X_1^t,\cdots,X_{n-1}^t\right\}$), as in a regular BN and on some variables from the previous time slice ($\left\{X_1^{t-1},\cdots,X_N^{t-1}\right\}$).

- $\prod_{n=1}^{N}\left[P\left(X_n^t|R_n^t\wedge\pi\right)\right]$ defines a graph for a time slice and all time slices are identical when the time index t changes.[1]

- A DBN, as a whole, "unrolled" over time, may be considered as a large regular BN. Consequently, the usual inference techniques applicable to BNs are still valid for such "unrolled" DBNs.

The best introduction, survey, and starting point on DBNs is the PhD thesis of K. Murphy, *Dynamic Bayesian Networks: Representation, Inference and Learning* [Murphy, 2002].

---

[1]The first time slice may be different as it expresses initial conditions.

13.1.2 Recursive Bayesian estimation

13.1.2.1 Bayesian filtering, prediction, and smoothing

Recursive Bayesian estimation is the generic name for a very large applied class of probabilistic models of time series.

They are defined by the following Bayesian program:

$$
Pr\begin{cases}
Ds\begin{cases}
Sp(\pi)\begin{cases}
Va: \\
S^0,\cdots,S^T,O^0,\cdots,O^T \\
Dc: \\
\begin{cases}
& P\left(S^0\wedge\cdots\wedge S^T\wedge O^0\wedge\cdots\wedge O^T|\pi\right) \\
= & P\left(S^0\wedge O^0\right)\times\prod_{t=1}^{T}\left[P\left(S^t|S^{t-1}\right)\times P\left(O^t|S^t\right)\right]
\end{cases} \\
Fo: \\
\begin{cases}
P\left(S^0\wedge O^0\right) \\
P\left(S^t|S^{t-1}\right) \\
P\left(O^t|S^t\right)
\end{cases}
\end{cases} \\
Id
\end{cases} \\
Qu: \\
\begin{cases}
P\left(S^{t+k}|O^0\wedge\cdots\wedge O^t\right) \\
(k=0)\equiv Filtering \\
(k>0)\equiv Prediction \\
(k<0)\equiv Smoothing
\end{cases}
\end{cases}
\tag{13.3}
$$

- Variables $S^0,\cdots,S^T$ are a time series of state variables considered to be on a time horizon ranging from 0 to T. Variables $O^0,\cdots,O^T$ are a time series of observation variables on the same horizon.

- The decomposition is based:

  - on $P\left(S^t|S^{t-1}\right)$, called the system model, transition model, or dynamic model, which formalizes the transition from the state at time $t-1$ to the state at time t;
  - on $P\left(O^t|S^t\right)$, called the observation model, which expresses what can be observed at time t when the system is in state S^t;
  - on an initial state at time 0: $P\left(S^0\wedge O^0\right)$.

- The question usually asked of these models is $P\left(S^{t+k}|O^0\wedge\cdots\wedge O^t\right)$: what is the probability distribution for the state at time $t+k$ knowing the observations from instant 0 to t? The most common case is Bayesian filtering where $k=0$, which means that one searches for the present state, knowing the past observations. However it is also possible to do prediction ($k > 0$), where one tries to extrapolate a future state from past observations, or to do smoothing ($k < 0$), where one tries to recover a past state from observations made either before or after that instant. However, some more complicated questions may also be asked (see the later section on hidden Markov models).

Bayesian filters ($k = 0$) have a very interesting recursive property, which contributes greatly to their attractiveness. $P\left(S^t|O^0 \wedge \cdots \wedge O^t\right)$ may be computed simply from $P\left(S^{t-1}|O^0 \wedge \cdots \wedge O^{t-1}\right)$ with the following formula:

$$\begin{aligned} & P\left(S^t|O^0 \wedge \cdots \wedge O^t\right) \\ = & P\left(O^t|S^t\right) \times \sum_{S^{t-1}} \left[P\left(S^t|S^{t-1}\right) \times P\left(S^{t-1}|O^0 \wedge \cdots \wedge O^{t-1}\right)\right] \end{aligned} \tag{13.4}$$

Another interesting point of view for this equation is to consider that there are two phases, a prediction phase and an estimation phase:

- During the prediction phase, the state is predicted using the dynamic model and the estimation of the state at the previous moment:

$$\begin{aligned} & P\left(S^t|O^0 \wedge \cdots \wedge O^{t-1}\right) \\ = & \sum_{S^{t-1}} \left[P\left(S^t|S^{t-1}\right) \times P\left(S^{t-1}|O^0 \wedge \cdots \wedge O^{t-1}\right)\right] \end{aligned} \tag{13.5}$$

- During the estimation phase, the prediction is either confirmed or invalidated using the last observation:

$$\begin{aligned} & P\left(S^t|O^0 \wedge \cdots \wedge O^t\right) \\ = & P\left(O^t|S^t\right) \times P\left(S^t|O^0 \wedge \cdots \wedge O^{t-1}\right) \end{aligned} \tag{13.6}$$

13.1.2.2 Hidden Markov models

Hidden Markov models (HMMs) are a very popular specialization of Bayesian filters.

They are defined by the following Bayesian program:

$$
Pr\begin{cases}
Ds\begin{cases}
Sp(\pi)\begin{cases}
Va: \\
S^0,\cdots,S^T,O^0,\cdots,O^T \\
Dc: \\
\begin{cases}
P\left(S^0\wedge\cdots\wedge O^T|\pi\right) \\
= \left[\begin{array}{c} P\left(S^0\wedge O^0|\pi\right) \\ \prod_{t=1}^{T}\left[P\left(S^t|S^{t-1}\wedge\pi\right)\times P\left(O^t|S^t\wedge\pi\right)\right]\end{array}\right]
\end{cases} \\
Fo: \\
\begin{cases}
P\left(S^0\wedge O^0|\pi\right)\equiv Matrix \\
P\left(S^t|S^{t-1}\wedge\pi\right)\equiv Matrix \\
P\left(O^t|S^t\wedge\pi\right)\equiv Matrix
\end{cases}
\end{cases} \\
Id
\end{cases} \\
Qu: \\
Max_{S^1\wedge\cdots\wedge S^{T-1}}\left[P\left(S^1\wedge\cdots\wedge S^{T-1}|S^T\wedge O^0\wedge\cdots\wedge O^T\wedge\pi\right)\right]
\end{cases}
\tag{13.7}
$$

- Variables are treated as being discrete.
- The transition model $P\left(S^t|S^{t-1}\wedge\pi\right)$ and the observation model $P\left(O^t|S^t\wedge\pi\right)$ are both specified using probability matrices.
- The question most frequently asked of HMMs is:

$$Max_{S^1\wedge\cdots\wedge S^{T-1}}\left[P\left(S^1\wedge\cdots\wedge S^{T-1}|S^T\wedge O^0\wedge\cdots\wedge O^T\wedge\pi\right)\right] \tag{13.8}$$

  What is the most probable series of states that leads to the present state, knowing the past observations?[2]

This particular question may be answered with a specific and very efficient algorithm called the *Viterbi algorithm*, which is presented in Chapter 14.

A specific learning algorithm called the *Baum–Welch algorithm* has also been developed for HMMs (see Chapter 15).

A good introduction to HMMs is Rabiner's tutorial [Rabiner, 1989].

13.1.2.3 Kalman filters

The very well-known *Kalman filters* [Kalman, 1960] are another specialization of Bayesian filters.

They are defined by the following Bayesian program:

[2] A common example of the application of HMM is automatic speech recognition. The states are either words or phonemes and one wants to recognize the most probable sequence of states (the sentence) corresponding to the observations (the heard frequencies).

$$Pr\left\{\begin{array}{l} Ds\left\{\begin{array}{l} Sp(\pi)\left\{\begin{array}{l} Va: \\ S^0,\cdots,S^T,O^0,\cdots,O^T \\ Dc: \\ \left\{\begin{array}{l} P\left(S^0\wedge\cdots\wedge O^T|\pi\right) \\ = \left[\begin{array}{c} P\left(S^0\wedge O^0|\pi\right) \\ \prod_{t=1}^{T}\left[P\left(S^t|S^{t-1}\wedge\pi\right)\times P\left(O^t|S^t\wedge\pi\right)\right] \end{array}\right] \end{array}\right. \\ Fo: \\ \left\{\begin{array}{l} P\left(S^t|S^{t-1}\wedge\pi\right)\equiv G\left(S^t,A\bullet S^{t-1},Q\right) \\ P\left(O^t|S^t\wedge\pi\right)\equiv G\left(O^t,H\bullet S^t,R\right) \end{array}\right. \end{array}\right. \\ Id \end{array}\right. \\ Qu: \\ P\left(S^T|O^0\wedge\cdots\wedge O^T\wedge\pi\right) \end{array}\right. \tag{13.9}$$

- Variables are continuous.

- The transition model $P\left(S^t|S^{t-1}\wedge\pi\right)$ and the observation model $P\left(O^t|S^t\wedge\pi\right)$ are both specified using Gaussian laws with means that are linear functions of the conditioning variables.

With these hypotheses, and using the recursive formula in Equation 13.4, it is possible to solve the inference problem analytically to answer the usual $P\left(S^T|O^0\wedge\cdots\wedge O^T\wedge\pi\right)$ question. This leads to an extremely efficient algorithm, which explains the popularity of Kalman filters and the number of their everyday applications.[3]

When there are no obvious linear transition and observation models, it is still often possible, using a first-order Taylor's expansion, to treat these models as locally linear. This generalization is commonly called *extended Kalman filters.*

A good tutorial by Welch and Bishop may be found on the Web (http://www.cs.unc.edu/ welch/kalman/). For a more complete mathematical presentation, one should refer to a report by Barker et al. [1994], but these are only two sources from the vast literature on this subject.

---

[3]A very popular application of Kalman filters is the GPS (global positioning system). The recursive evaluation of the position explains why the precision is poor when you turn on your GPS and improves rapidly after a while. The dynamic model takes into account the previous position and speed to predict the future position (Equation 13.5), when the observation model confirms (or invalidates) this position knowing the signal coming from the satellites (Equation 13.6).

13.1.2.4 Particle filters

The fashionable particle filters may also be seen as a specific implementation of Bayesian filters.

The distribution $P\left(S^{t-1}|O^0 \wedge \cdots \wedge O^{t-1} \wedge \pi\right)$ is approximated by a set of N particles having weights proportional to their probabilities. The recursive Equation 13.4 is then used to inspire a dynamic process that produces an approximation of $P\left(S^t|O^0 \wedge \cdots \wedge O^t \wedge \pi\right)$. The principle of this dynamic process is that the particles are first moved according to the transition model $P\left(S^t|S^{t-1} \wedge \pi\right)$, then their weights are updated according to the observation model $P\left(O^t|S^t \wedge \pi\right)$.

Arulampalam's tutorial gives a good overview of this [Arulampalam et al., 2002].

13.1.3 Mixture models

Mixture models try to approximate a distribution on a set of variables $\{X_1, \cdots, X_N\}$ by adding up (mixing) a set of simple distributions.

The most popular mixture models are Gaussian mixtures in which the component distributions are Gaussian. However, the component distributions may be of any nature, for instance, logistic or Poisson distributions.

Such a mixture is usually defined as follows:

$$Pr\begin{cases} Ds\begin{cases} Sp(\pi)\begin{cases} Va: \\ X_1, \cdots, X_N \\ Dc: \\ \begin{cases} & P\left(X_1 \wedge \cdots \wedge X_N|\pi\right) \\ = & \sum_{m=1}^{M}\left[\alpha_m P\left(X_1 \wedge \cdots \wedge X_N|\pi_m\right)\right] \end{cases} \\ Fo: \\ \begin{cases} For-instance: \\ P\left(X_1 \wedge \cdots \wedge X_N|\pi_m\right) \equiv G\left(X_1 \wedge \cdots \wedge X_N, \mu_m, \sigma_m\right) \end{cases} \end{cases} \\ Id: \end{cases} \\ Qu: \end{cases} \tag{13.10}$$

It should be noted that this is not a valid Bayesian program. In particular, the decomposition does not have the right form:

$$P\left(X_1 \wedge \cdots \wedge X_N|\pi\right) = P\left(L_1|\pi\right) \times \prod_{k=2}^{K}\left[P\left(L_k|R_k \wedge \pi\right)\right] \tag{13.11}$$

It is, however, a very popular and convenient way to specify distributions $P\left(X_1 \wedge \cdots \wedge X_N|\pi\right)$, especially when the types of the component distributions

$P(X_1 \wedge \cdots \wedge X_N | \pi_m)$ are chosen to ensure efficient analytical solutions to some of the inference problems.

Furthermore, it is possible to specify such a mixture as a correct Bayesian program by adding one model selection variable H to the previous definition:

$$
Pr \begin{cases} Ds \begin{cases} Sp(\pi) \begin{cases} Va: \\ X_1, \cdots, X_N, H \\ Dc: \\ \begin{cases} & P(X_1 \wedge \cdots \wedge X_N \wedge H | \pi) \\ = & P(H|\pi) \times P(X_1 \wedge \cdots \wedge X_N | H \wedge \pi) \end{cases} \\ Fo: \\ \begin{cases} P(H|\pi) \equiv Table \\ P(X_1 \wedge \cdots \wedge X_N | [H = m] \wedge \pi) \\ \equiv P(X_1 \wedge \cdots \wedge X_N | \pi_m) \end{cases} \end{cases} \\ Id \end{cases} \\ Qu: \\ \begin{cases} & P(X_1 \wedge \cdots \wedge X_N | \pi) \\ = & \sum_{m=1}^{M} [P([H = m] | \pi) \times P(X_1 \wedge \cdots \wedge X_N | \pi_m)] \end{cases} \end{cases} \tag{13.12}
$$

- H is a discrete variable taking M values. H is used as a selection variable. Knowing the value of H, we suppose that the joint distribution is reduced to one of its component distributions:

$$
P(X_1 \wedge \cdots \wedge X_N | [H = m] \wedge \pi) = P(X_1 \wedge \cdots \wedge X_N | \pi_m) \tag{13.13}
$$

- In these simple and common mixture models, H, the selection variable, is assumed to be independent of the other variables. We saw in Chapter 10 a discussion of more elaborate mixing models in which H depends on some of the other variables. This is also the case in expert mixture models, as first described by Jordan and Jacobs [1994].

- Identification is a crucial step for these models, when the values of $P(H|\pi)$ and the parameters of the component distributions $P(X_1 \wedge \cdots \wedge X_N | \pi_m)$ are searched to find the best possible fit between the observed data and the joint distribution. This is usually done using the EM algorithm or some of its variants (see Chapter 15).

- The questions asked of the joint distribution are of the form $P(X_1 \wedge \cdots \wedge X_N | \pi)$ where the selection variable H is unknown. Consequently, solving the question assumes a summation for the possible value of H, and we finally retrieve the usual mixture form:

$$P(X_1 \wedge \cdots \wedge X_N|\pi) = \sum_{m=1}^{M} [P([H=m]|\pi) \times P(X_1 \wedge \cdots \wedge X_N|\pi_m)] \tag{13.14}$$

A reference on mixture models is provided in McLachlan and Peel's book *Finite Mixture Model* [2000].

13.1.4 Maximum entropy approaches

Maximum entropy approaches play a very important role in physical applications. The late E.T. Jaynes, in his regrettably unfinished book [Jaynes, 2003], gives a wonderful presentation of these approaches as well as a fascinating apologia for the subjectivist epistemology of probabilities.

The maximum entropy models may be described by the following Bayesian program:

$$Pr\begin{cases} Ds\begin{cases} Sp(\pi)\begin{cases} Va: \\ X_1, \cdots, X_N \\ Dc: \\ \begin{cases} & P(X_1 \wedge \cdots \wedge X_N|\pi) \\ = & \prod_{m=0}^{M}\left[e^{-[\lambda_m \times f_m(X_1 \wedge \cdots \wedge X_N)]}\right] \\ = & e^{-\sum_{m=0}^{M}[\lambda_m \times f_m(X_1 \wedge \cdots \wedge X_N)]} \end{cases} \\ Fo: \\ \begin{cases} f_0(X_1 \wedge \cdots \wedge X_N) = 1 \\ f_1, \cdots, f_M \end{cases} \end{cases} \\ Id: \end{cases} \\ Qu: \end{cases} \tag{13.15}$$

- The variables $X_1, \cdots, X_N$ are not constrained.

- The decomposition is made of a product of exponential distributions $e^{-[\lambda_m \times f_m(X_1 \wedge \cdots \wedge X_N)]}$ where each f_m is called an observable function. An observable function may be any real function on the space defined by $X_1, \cdots, X_N$, such that its expectation may be computed:

$$\begin{aligned}&\langle f_m (X_1 \wedge \cdots \wedge X_N) \rangle \\ = &\sum_{X_1 \wedge \cdots \wedge X_N} [P(X_1 \wedge \cdots \wedge X_N | \pi) \times f_m (X_1 \wedge \cdots \wedge X_N)]\end{aligned} \tag{13.16}$$

- The constraints on the problem are usually expressed by M real values F_m called levels of constraint, which impose the condition $\langle f_m (X_1 \wedge \cdots \wedge X_N) \rangle = F_m$. These levels of constraint may either be arbitrary values fixed a priori by the programmer, or the results of experimental observation. In this latter case, the levels of constraint are equal to the observed mean values of the observable functions on the data set. The constraint imposed on the distribution is that the expectations of the observable functions according to the distribution should be equal to the means of these observable functions.

- The identification problem is, then, knowing the level of constraint F_m, to find the *Lagrange multipliers* λ_m that maximize the entropy of the distribution $P(X_1 \wedge \cdots \wedge X_N | \pi)$.

The maximum entropy approach is a very general and powerful way to represent probabilistic models and to explain what is going on when one wants to identify the parameters of a distribution, choose its form, or even compare models. Unfortunately, finding the values of the Lagrange multipliers λ_m can be very difficult.

A sound introduction is of course Jaynes' book *Probability Theory — The Logic of Science* [Jaynes, 2003]. Other references are the edited proceedings of the yearly MaxEnt conferences, which cover both theory and applications since 1979 (see Mohammad-Djafari et al. [2010], for the most recent one).

13.2 Engineering oriented probabilistic models

13.2.1 Sensor fusion

Sensor fusion is a very common and crucial problem for both living systems and artifacts. The problem is as follows: given a phenomenon and some sensors, how can we derive information on the phenomenon by combining the information from the different sensors?

The most common and simple Bayesian modeling for sensor fusion is the following:

$$
Pr\begin{cases} Ds\begin{cases} Sp(\pi)\begin{cases} Va: \\ \Phi, R_1, \cdots, R_N \\ Dc: \\ \begin{cases} & P(\Phi \wedge R_1 \wedge \cdots \wedge R_N|\pi) \\ = & P(\Phi|\pi) \times \prod_{n=1}^{N}\left[P(R_n|\Phi \wedge \pi)\right] \end{cases} \\ Fo: \\ any \end{cases} \\ Id: \end{cases} \\ Qu: \\ P(\Phi|r_1 \wedge \cdots \wedge r_N \wedge \pi) \end{cases} \tag{13.17}
$$

- Φ is the variable used to describe the phenomenon, when $R_1, \cdots, R_N$ are the variables encoding the readings of the sensors.
- The decomposition:

$$
P(\Phi \wedge R_1 \wedge \cdots \wedge R_N|\pi) = P(\Phi|\pi) \times \prod_{n=1}^{N}\left[P(R_n|\Phi \wedge \pi)\right] \tag{13.18}
$$

may seem peculiar, as the readings of the different sensors are obviously not independent from one another. The exact meaning of this equation is that the phenomenon Φ is considered to be the main reason for the contingency of the readings. Consequently, it is stated that knowing Φ, the readings R_n are independent. Φ is the cause of the readings and, knowing the cause, the consequences are independent. Indeed, this is a very strong hypothesis, far from always being satisfied. However, it very often gives satisfactory results and has the main advantage of considerably reducing the complexity of the computation.

- The distributions $P(R_n|\Phi \wedge \pi)$ are called sensor models. Indeed, these distributions encode the way a given sensor responds to the observed phenomenon. When dealing with industrial sensors, this is the kind of information directly provided by the device manufacturer. However, these distributions may also be identified very easily by experiment.
- The most common question asked of this fusion model is:

$$
P(\Phi|r_1 \wedge \cdots \wedge r_N \wedge \pi) \tag{13.19}
$$

It should be noted that this is an inverse question as the model has been specified the other way around by giving the distributions $P(R_n|\Phi \wedge \pi)$. The capacity to answer such inverse questions easily is one of the main advantages of probabilistic modeling, thanks to Bayes' rule.

13.2.2 Classification

The classification problem may be seen as the same as the sensor fusion problem just described. Usually, the problem is called a classification problem when the possible value for Φ is limited to a small number of classes and it is called a sensor fusion problem when Φ can be interpreted as a "measure."

A slightly more subtle definition of classification uses one more variable. In this model, not only is there the variable Φ, used to merge the information, but there is C, used to classify the situation. C has far fewer values than Φ and it is possible to specify $P(\Phi|C)$, which, for each class, makes the possible values of Φ explicit. Answering the classification question $P(C|r_1 \wedge \cdots \wedge r_N)$ supposes a summation over the different values of Φ.

The Bayesian program then obtained is as follows:

$$
Pr \begin{cases} Ds \begin{cases} Sp(\pi) \begin{cases} Va: \\ C, \Phi, R_1, \cdots, R_N \\ Dc: \\ \begin{cases} & P(C \wedge \Phi \wedge R_1 \wedge \cdots \wedge R_N|\pi) \\ = & P(C|\pi) \times P(\Phi|C \wedge \pi) \times \prod_{n=1}^{N} [P(R_n|\Phi \wedge \pi)] \end{cases} \\ Fo: \\ any \end{cases} \\ Id: \end{cases} \\ Qu: \\ P(C|r_1 \wedge \cdots \wedge r_N \wedge \pi) \end{cases} \tag{13.20}
$$

13.2.3 Pattern recognition

Pattern recognition is another form of the same problem as the two preceding ones. However, it is called recognition because the emphasis is on deciding a given value for C rather than finding the distribution $P(C|r_1 \wedge \cdots \wedge r_N \wedge \pi)$.

Consequently, the pattern recognition community usually does not make a clear separation between the probabilistic inference part of the reasoning and the decision part, using a utility function. Both are considered as a single and integrated decision process.

13.2.4 Sequence recognition

The problem is to recognize a sequence of states knowing a sequence of observations and, possibly, a final state.

In Section 13.1.2.2 we presented hidden Markov models (HMMs) as a special case of Bayesian filters. These HMMs have been specially designed for sequence recognition, which is why the most common question asked of these models is $P\left(S^1 \wedge \cdots \wedge S^{T-1} | S^T \wedge O^0 \wedge \cdots \wedge O^T \wedge \pi\right)$ (see Equation 13.8).

13.2.5 Markov localization

Another possible variation of the Bayesian filter formalism is to add a control variable to the system. This extension is sometimes called an input–output HMM [Bengio and Frasconi, 1995; Ghahramani, 2002]. However, in the field of robotics, it has received more attention under the name of Markov localization [Burgard et al., 1996; Thrun et al., 2005]. In this field, such an extension is natural, as the robot can observe its state by sensors, but can also influence its state via motor commands.

Starting from a Bayesian filter structure, the control variable is used to refine the transition model $P\left(S^t | S^{t-1} \wedge \pi\right)$ of the Bayesian filter into $P\left(S^t | S^{t-1} \wedge A^{t-1} \wedge \pi\right)$, which is then called the action model. The rest of the Bayesian filter is unchanged. The Bayesian program then obtained is as follows:

$$
Pr \begin{cases} Ds \begin{cases} Sp(\pi) \begin{cases} Va: \\ S^0, \cdots, S^T, A^0, \cdots, A^T, O^0, \cdots, O^T \\ Dc: \\ \begin{cases} P\left(S^0 \wedge \cdots \wedge O^T | \pi\right) \\ = \begin{bmatrix} P\left(S^0 \wedge O^0 | \pi\right) \times \prod_{t=0}^{T} \left[P\left(A^t | \pi\right)\right] \\ \prod_{t=1}^{T} \left[P\left(S^t | S^{t-1} \wedge A^{t-1} \wedge \pi\right) \times P\left(O^t | S^t \wedge \pi\right)\right] \end{bmatrix} \end{cases} \\ Fo: \end{cases} \\ Id \end{cases} \\ Qu: \\ P\left(S^T | a^0 \wedge \cdots \wedge a^{T-1} \wedge o^0 \wedge \cdots \wedge o^{T-1} \wedge \pi\right) \end{cases} \tag{13.21}
$$

The resulting model is used to answer the question

$$
P\left(S^T | a^0 \wedge \cdots \wedge a^{T-1} \wedge o^0 \wedge \cdots \wedge o^{T-1} \wedge \pi\right) \tag{13.22}
$$

which estimates the state of the robot, given past actions and observations. When this state represents the position of the robot in its environment, this amounts to localization.

A reference for Markov localization and its use in robotics is Thrun's book titled *Probabilistic Robotics* [Thrun et al., 2005].

13.2.6 Markov decision processes

13.2.6.1 Partially observable Markov decision processes

Partially observable Markov decision processes (POMDPs) are used to model a robot that must plan and execute a sequence of actions.

Formally, POMDPs use the same probabilistic model as Markov localization except that they are enriched by the definition of a reward (and/or cost) function.

This reward function R models those states that are good for the robot, and which actions are costly. In the most general notation, it is therefore a function that associates, for each state–action couple, a real-valued number. The reward function also helps to drive the planning process. Indeed, the aim of this process is to find an optimal plan in the sense that it maximizes a certain measure based on the reward function. This measure is most frequently the expected discounted cumulative reward:

$$\left\langle \sum_{t=0}^{T} \left[\gamma^t \times R^t\right] \right\rangle \tag{13.23}$$

where γ is a discount factor (less than one), $R(t)$ is the reward obtained at time t, and $\langle\rangle$ is the mathematical expectation. Given this measure, the goal of the planning process is to find an optimal mapping from probability distributions over states to actions (a policy). This planning process, which leads to intractable computation, is sometimes approximated using iterative algorithms called policy iteration or value iteration. These algorithms start with random policies, and improve them at each step until some numerical convergence criterion is met.

An introduction to POMDPs is provided by Kaelbling et al. [1998].

13.2.6.2 Markov decision process

Another approach for tackling the intractability of the planning problem in POMDPs is to suppose that the robot knows what state it is in. The state becomes observable, therefore the observation variable and model are no longer needed; the resulting formalism is called a (fully observable) MDP, and is summarized by the following Bayesian program:

$$Pr \begin{cases} Ds \begin{cases} Sp(\pi) \begin{cases} Va: \\ S^0, \cdots, S^T, A^0, \cdots, A^T \\ Dc: \\ \begin{cases} P\left(S^0 \wedge \cdots \wedge A^T | \pi\right) \\ = \begin{bmatrix} P\left(S^0|\pi\right) \times \prod_{t=0}^{T} \left[P\left(A^t|\pi\right)\right] \\ \prod_{t=1}^{T} \left[P\left(S^t|S^{t-1} \wedge A^{t-1} \wedge \pi\right)\right] \end{bmatrix} \end{cases} \\ Fo: \end{cases} \\ Id \end{cases} \\ Qu: \\ P\left(A^0 \wedge \cdots \wedge A^T | s^0 \wedge s^T \wedge \pi\right) \end{cases} \tag{13.24}$$

- The variables are: $S^0, \cdots, S^T$, a temporal sequence of states, and $A^0, \cdots, A^T$, a temporal sequence of actions.
- The decomposition makes a first-order Markov assumption by specifying that the state at time t depends on the state at time $t-1$ and also on the action taken at time $t-1$.
- $P\left(S^t|S^{t-1} \wedge A^{t-1} \wedge \pi\right)$ is usually represented by a matrix and is called the "transition matrix" of the model.
- The question addressed to the MDP is $P\left(A^0 \wedge \cdots \wedge A^T|s^0 \wedge s^T \wedge \pi\right)$: What is the sequence of actions required to go from state s^0 to state s^T?

A introductory review of POMDPs and MDPs is proposed by Boutilier et al. [1999]. MDPs can cope with planning in state spaces bigger than POMDPs, but are still limited to some hundreds of states. Therefore, many research efforts have been aimed toward hierarchical decomposition of the planning and modeling problems in MDPs, especially in robotics, where the full observability hypothesis makes their practical use difficult [Hauskrecht et al., 1998; Lane and Kaelbling, 2001; Pineau and Thrun, 2002; Diard et al., 2004].

13.3 Cognitive oriented probabilistic models

It is remarkable that a wide variety of common cognitive issues (in the sense that they appear frequently) can be tackled by a small set of common models (in the sense that these models are shared by these issues). In other words, a few template mathematical constructs, based only on probabilities and Bayes' rule, can be applied to a large assortment of problems that have to be addressed by cognitive systems.

Our purpose, in this section, is to demonstrate these assertions by proposing a step by step inspection of these cognitive problems and, for each of them, by describing a candidate Bayesian model.[4] The book titled *Probabilistic Reasoning and Decision Making in Sensory-Motor Systems* [Bessière et al., 2008] proposed much more detail on several of these models. A recent paper in *Science* by Tenenbaum et al. [2011] offers an interesting general overview of this matter.

13.3.1 Ambiguities

Natural cognitive systems are immersed in rich and widely variable environments. It would be difficult to assume that such systems apprehend their environments in all their details, all the time, if only because of limited sensory or memory capacities. As a consequence, relations between the characteristics of external phenomena and internal states cannot always be bijections. In other words, internal states will sometimes be ambiguous with respect to external situations.

13.3.1.1 Inverse problem

A problem is said to be *inverse* when we know a direct (or forward) relation and we seek the reverse relation. The inversion of a deterministic function, which often does not have a closed-form solution, can be very difficult.

The Bayesian program corresponding to an inverse problem is the following:

$$Pr\begin{cases} Ds\begin{cases} Sp(\pi)\begin{cases} Va: \\ \Phi, S \\ Dc: \\ \begin{cases} & P(\Phi \wedge S|\pi) \\ = & P(\Phi|\pi)\times P(S|\Phi\wedge\pi) \end{cases} \\ Fo: \\ any \end{cases} \\ Id \end{cases} \\ Qu: \\ P(\Phi|s\wedge\pi) = \dfrac{P(\Phi|\pi)\times P(s|\Phi\wedge\pi)}{\sum\limits_{\Phi}[P(\Phi|\pi)\times P(s|\Phi\wedge\pi)]} \end{cases} \tag{13.25}$$

[4]A large part of this section was originally published in *Acta Biotheoretica* under the title "Common bayesian models for common cognitive issues" by Colas et al. [2010]

- In the Bayesian framework, an inverse problem is addressed using the symmetry of Bayes' rule. In a generic example of perception, let Φ be a variable representing some characteristics of the phenomenon and let S be a variable representing the sensation. The joint probability distribution is typically factored as:

$$P(\Phi\ S) = P(\Phi)P(S \mid \Phi). \tag{13.26}$$

In this expression, $P(\Phi)$ is a *prior* on the phenomenon; that is, the expectation about the phenomenon before any observation has occurred. $P(S \mid \Phi)$ is the probability distribution over sensations, given the phenomenon, which is also known as the *likelihood* of the phenomenon (when considered not as a probability distribution but as a function of Φ); it is the direct model.

- The probabilistic question of perception is $P(\Phi \mid S)$, the probability distribution on the phenomenon, based on a given sensation. This question, which is the *posterior* distribution on the phenomenon after some observation, is solved by Bayesian inference:

$$P(\Phi|s \wedge \pi) = \frac{P(\Phi|\pi) \times P(s|\Phi \wedge \pi)}{\sum_{\Phi} [P(\Phi|\pi) \times P(s|\Phi \wedge \pi)]} \tag{13.27}$$

Sensation is commonly defined as the effect of some phenomenon on the senses. Perception involves recovering information about the phenomenon, given the sensation. Perception is an inverse problem [Poggio, 1984; Yuille and Bülthoff, 1996; Pizlo, 2001]. Indeed, it is often easy to predict the sensations corresponding to a particular phenomenon (see Section 13.3.2). In this case, the direct function yields the sensation given the phenomenon, whereas perception is the inverse problem of extracting the phenomenon given the sensation.

For example, when we know the shape of an object, basic geometry allows us to derive its projection on the retina. Conversely, it is difficult to reconstruct the shape of an object given only its projection on the retina [Colas, 2006; Colas et al., 2008b].

13.3.1.2 Ill-posed problem

A problem is said to be *well posed* when it admits a unique solution. Conversely, an *ill-posed* problem admits either many solutions or no solution at all. In most of the nontrivial inverse problems, the direct functions are not injective. Therefore, the inverse relation is not properly a function. When the direct function is not injective, the inverse problem is ill-posed.

Perception is often an ill-posed problem. One illustrative example (see Figure 13.1) is the well-known Necker's cube, in which a wire-frame 2-D drawing of a cube is often perceived as a cube in one of two possible positions (even if it can actually correspond to the projection of an infinite number of 3-D structures).

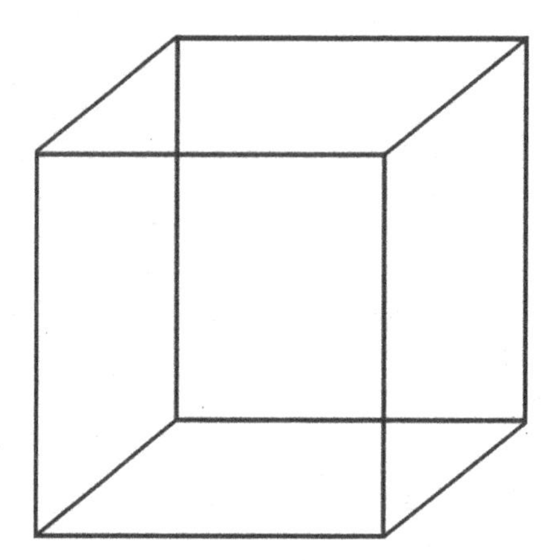

FIGURE 13.1: Necker's cube: bistable percept.

There are many other examples of ill-posed perceptions, including recovering the shape of an object from lighting or motion. Such examples have led to the development of Bayesian models of perception, which are reviewed in the book by Knill and Richards [1996] and in the more recent article by Kersten et al. [2004].

It has also been argued that illusions arise when ill-posed problems are solved by choosing a solution that does not correspond to the reality of the percept (an example can be found in Geisler and Kersten [2002]).

In robotics, a common instance of an ill-posed problem is *perceptual aliasing*, which occurs in robot localization when two different locations in an environment produce identical sensor readings [Kuipers, 2000; Thrun, 2000b].

In speech recognition or generation, recovering the seven-dimensional shape of the vocal tract [Maeda, 1990] from the first four dimensions (formants) of the acoustic signal alone would be another example of an ill-posed problem. More generally, the control of any redundant motor system (i.e., one with more degrees of freedom than the dimension of the space to be accessed) is an ill-posed problem.

In the Bayesian framework, the focus is not on finding a solution matching all the constraints of the problem. Instead, a probability distribution is computed. The analogy of a well-posed problem is a distribution with exactly one mode. An ill-posed problem will typically yield a multimodal distribution, or a distribution with a plateau.

The theoretical issue with inverse problems is that a direct function does not always admit an inverse function. In the Bayesian framework, the probability distribution can always be reversed, even if it is difficult to compute. Notice that the property of an ill-posed problem is defined by the result of the inference. On the other hand, a problem is inverse by virtue of its structure. Therefore, an ill-posed problem is not necessarily an inverse problem.

13.3.1.3 Discussion

An ambiguity is a difficulty in finding a unique interpretation for a given stimulus. This often leads to multistable percepts, in which case, different interpretations can be actually perceived in the same conditions. Perceptual reversals can be involved when a transition occurs from one percept to another. Ambiguity often arises as a consequence of an ill-posed, inverse problem.

However, an ambiguous likelihood for a given stimulus does not always induce multiple percepts. Indeed, as Equation 13.27 shows, the likelihood is multiplied by the prior, which can filter out part of the solutions. An illusion can occur precisely when the likelihood produces multiple interpretations and the prior rules out the correct one.

13.3.2 Fusion, multimodality, conflicts

Natural cognitive systems are equipped with a variety of rich sensors: rich, as they continuously provide several measurements about a given phenomenon (e.g., multiple cells in the retina), and various, as they can measure different manifestations of the same phenomenon (e.g., both hearing and seeing a falling object). The difficulty arises when several of these measurements are to be used together to recover characteristics of the phenomenon. We argue that the concepts of fusion, multimodality, and conflict can generally be cast into the same model structure. This model is structured around the conditional independency assumption. We also present extensions of this model that alleviate this assumption, thus trading model simplicity for expressiveness.

13.3.2.1 Fusion

Often, there are multiple sources of information that can be exploited for any given phenomenon. *Fusion* is the process of forming a single percept starting from multiple sources of information. The classical model of *naive Bayesian fusion* (sometimes called *weak fusion* in the psychology literature) is defined by the following Bayesian program which has already been presented several times:

$$
Pr\begin{cases}
Ds\begin{cases}
Sp(\pi)\begin{cases}
Va:\\
\Phi, S_1, S_2, \cdots, S_N\\
Dc:\\
\begin{cases}
& P\left(\Phi \wedge S_1 \wedge \cdots S_N|\pi\right)\\
= & P\left(\Phi|\pi\right) \times \prod_{n=1}^{N}\left[P\left(S_n|\Phi \wedge \pi\right)\right]
\end{cases}\\
Fo:\\
any
\end{cases}\\
Id
\end{cases}\\
Qu:\\
P\left(\Phi|s_1 \wedge \cdots s_N \wedge \pi\right) \propto P\left(\Phi|\pi\right) \times \prod_{n=1}^{N}\left[P\left(s_n|\Phi \wedge \pi\right)\right]
\end{cases}
\tag{13.28}
$$

- This model relies on the assumption that each piece of information is independent from the others, conditioned on knowledge of the phenomenon. Each piece of information is related to the phenomenon using an inverse model similar to that discussed in Section 13.3.1.1. For instance, once the cause is known, the consequences follow independently. This very strong hypothesis is far from always being satisfied. However, it often gives satisfactory results.

  With $S_1, \ldots, S_N$ representing the variables corresponding to the N pieces of information that are to be part of the fusion, and Φ being the variable corresponding to characteristics of interest in the phenomenon, the naive fusion model assumes the following decomposition of the joint probability distribution:

$$
P(\Phi\ S_1 \cdots S_N) = P(\Phi) \prod_{i=1}^{N} P(S_i \mid \Phi). \tag{13.29}
$$

- This model can be used to compute the probability distribution on the phenomenon, Φ, given all the sensations, $S_1, \ldots, S_N$:

$$
P(\Phi \mid s_1\ \cdots\ s_N) \propto P(\Phi) \prod_{i=1}^{N} P(s_i \mid \Phi).
$$

- This model offers the advantage of being much simpler than a complete model without any independence, as the joint is written as a product of low dimensional factors. The size of a complete joint probability distribution is exponential in the number of variables, whereas it is only linear when assuming naive fusion.

- Furthermore, the use of many information sources generally allows for an increase in the signal. For example, in the case of Gaussian uncertainty for the sensor models, $P(S_i \mid \Phi)$, and the prior, $P(\Phi)$, the uncertainty of the posterior distribution, $P(\Phi \mid s_1 \cdots s_N)$, can be proven to be smaller with such a model than if the pieces of information are not fused at all.

  Note that some models rely on *maximum likelihood estimation* (MLE). That is, they do not compute the posterior probability distribution, $P(\Phi \mid s_1 \cdots s_N)$, but instead they are concerned with the likelihood of the phenomenon, $P(s_1 \cdots s_N \mid \Phi)$. For a given set of sensations, this is a function of Φ that reflects how well the phenomenon explains the sensations. Thus, the MLE selects the phenomenon that best matches the data.

  Moreover, Bayes' rule is written as:

$$P(\Phi \mid S_1 \cdots S_N) \propto P(\Phi)P(S_1 \cdots S_N \mid \Phi) \tag{13.30}$$

  With a uniform prior, this states that the posterior is proportional to the likelihood. Therefore, the MLE is tantamount to assuming a uniform prior or dismissing it altogether.

A common example of fusion in the study of perception is *cue combination* (or intramodal fusion).

For instance, Jacobs [1999] studied the perception of the depth of a simulated rotated cylinder. The depth of the cylinder can be perceived by the motion of the texture, the deformation of the texture, or both.

Weiss et al. [2002] studied the perception of the motion of a rhombus, and proposed a model of fusion of the velocity estimates obtained from the edges that, in particular, can account for an illusion in the perception of the motion of a flat rhombus.

Hillis et al. [2004] proposed a model for combining texture and disparity cues for perceiving the slant of a plane.

One last example is given by Drewing and Ernst [2006], who modeled the perception of curvature from haptic feedback, using both position and force information.

13.3.2.2 Multimodality

Fusion is often considered within a sensory modality, such as vision or touch. However, it is often beneficial to consider multiple sources of information from various sensory modalities when forming a percept in what is known as *multimodality*.

The model for multimodality match the assumptions of the naive Bayesian fusion model presented above. The main assumption is that the probability distribution over each sensation is independent of the others given the phenomenon. Some of these models use MLE, while others compute the complete posterior distribution.

Bayesian modeling has often been used to study multimodal fusion.

For example, Anastasio et al. [2000] have proposed a model of multisensory enhancement in the superior colliculus. Enhancement occurs when the neural response to a stimulus in one modality is augmented by a stimulus in another modality. In their work, they proposed a model of the probability of a target in the colliculus with respect to the presence or absence of visual and auditory input.

Zupan et al. [2002] have proposed a model for visuo-vestibular interaction in which they use a sensory weighing model, interpreted as an MLE.

Gepshtein and Banks [2003], who studied visuo-haptic integration, used a maximum likelihood model in which the reliability of the visual cues is related to the projection.

Körding and Wolpert [2004] performed experiments on sensorimotor learning. Using a Bayesian model of integration between visual feedback and proprioception, they inferred and manipulated the prior distributions assumed for the participants in their experiments.

Jürgens and Becker [2006] used Bayesian fusion for the combination of vestibular, optokinetic, podokinesthetic, and cognitive information for the perception of rotation.

Finally, Haith et al. [2008] looks into sensorimotor adaptation in a visually guided reaching experiment and proposes a model that can also account for perceptual after-effects. This model predicts a visual shift after adaptation of reaching in a force field, which has been confirmed by an experiment.

Although the models are the same as those for intramodal fusion, they differ in robustness. The conditional independence hypothesis is never totally valid. Part of the uncertainty that exists between sensations may not be explained by Φ. These additional correlations are more likely smaller between different modalities than within the same modality, as different physical processes are involved. For example, while fog or lack of illumination can jointly degrade different types of visual information, such as color or motion, hearing performance is not affected.

13.3.2.3 Conflicts

When various sources of information are involved, each can sometimes lead to significantly different individual percepts. In experiments, a *conflict* arises when sensations from different modalities induce different behavior compared with when each modality is observed individually.

The models accounting for conflicts are still naive fusion and maximum likelihood (with a naive fusion decomposition). Conflicts arise when the probability distributions corresponding to each sensation are significantly different.

The concept of a conflict is of a similar nature to that of an ill-posed problem. Both are defined with respect to characteristics of the result. Conversely, inversion and fusion are both structural properties of a problem and its associated model.

Introducing conflicts offers good experimental leverage for providing a lot of information on the fusion process. For instance, the relative importance of different cues can be assessed by observing the frequencies of behavior corresponding to each of the cues. Sometimes, the response to conflicting situations is mixed behavior that is also interesting to study when compared with that occurring in the original situations.

To study the integration of visual and haptic information, Ernst and Banks [2002] designed experiments in which the height of a bar had to be evaluated either by simulated visual input, simulated tactile input, or both. In this latter condition when different heights for both visual and haptic stimuli were simulated, a maximum-likelihood model provided a good match to their data.

Battaglia et al. [2003] designed a localization task with conflicting visual and auditory input to compare the classical visual capture model with a maximum likelihood model. They argued for taking into account the observers' perceptual biases. One special case of this conflict is the ventriloquist effect [Alais and Burr, 2004; Banks, 2004].

13.3.2.4 Less naive fusion

Many models assume that if the phenomenon is known then the sensations may be considered as conditionally independent of one another. This is often too strong an assumption. A model could assume no conditional independence between the cues. This is sometimes called *strong fusion*, and leads to arbitrarily complex models that are not tested easily. Less naive fusion models lie between strong fusion and naive (or weak) fusion.

The corresponding Bayesian model has already been presented in Chapter 7, Equation 7.19:

$$
Pr\begin{cases}
Ds\begin{cases}
Sp(\pi)\begin{cases}
Va: \\
\Phi, A, S_1, \ldots, S_N \\
Dc: \\
\begin{cases}
& P(\Phi \wedge A \wedge S_1 \wedge \ldots \wedge S_N|\pi) \\
= & P(A \wedge \Phi|\pi) \times \prod_{n=1}^{N}\left[P(S_n|\Phi \wedge A \wedge \pi)\right]
\end{cases} \\
Fo: \\
any
\end{cases} \\
Id
\end{cases} \\
Qu: \\
\begin{cases}
P(\Phi|s_1 \wedge \ldots \wedge s_N \wedge \pi) \\
P(\Phi|a \wedge s_1 \wedge \ldots \wedge s_N \wedge \pi)
\end{cases}
\end{cases}
\tag{13.31}
$$

This model is the same as a naive fusion model but with a phenomenon augmented by the ancillary cues, $\Phi' = \Phi \wedge A$. The only difference is in the questions asked, this new model is only concerned with the part Φ of the global phenomenon Φ', either knowing the ancillary cues, A, or not.

For instance, Landy et al. [1995] introduced a *modified weak fusion* framework. They tackled the cue combination problem in depth estimation and argued for the addition of so-called *ancillary cues.*[5] These cues do not provide direct information on depth but help to assess the reliability of the various cues that appear in the fusion process. Ancillary cues are the basis for a *dynamic reweighing* of the depth cues.

Yuille and Bülthoff [1996] propose the term *strong coupling* for nonnaive Bayesian fusion. They consider the examples of shape recovery from shading and texture, and the coupling of binocular depth cues with monocular cues obtained from motion parallax.

13.3.2.5 Discussion

Fusion is often seen as a product of models. It can be used when the underlying models are defined independently so that they can be combined to form a shared variable. Each model links this variable to distinct properties, and the fusion operates on the conjunction (logical product) of these properties.

This idea of fusion as a product of independent models is also mirrored by the inference process. Indeed, most of the time, the result of the fusion process is proportional to the product of each individual result obtained by the underlying models, $P(\Phi \mid s_1 \cdots s_n) \propto \prod_{n=1}^{N} P(\Phi \mid s_n)$. However, this may not be the case, depending on the exact specification of the underlying models. There are some more complex fusion models that ensure that this inference product holds (see [Pradalier et al., 2003] and Section 8.4 on fusion with coherence variables).

It is also interesting to note that, when all the probability distributions in the product are Gaussian, the resulting probability distribution is also Gaussian. Its mean is a weighted sum of the means of each Gaussian weighted according to a function of their variance. Therefore, many weighted models proposed in the literature can be interpreted as Bayesian fusion models with Gaussian uncertainty. The weights also acquire the meaning of representing uncertainty, which can sometimes be manipulated.

Finally, a conflict can only occur when it is assumed that a unique object is to be perceived. When the discrepancy is too large, segmentation occurs, leading to the perception of separate objects (e.g., perception of transparency). Likewise, fusion is a process of combining information from different sources for one object or feature. Therefore, to account for segmentation, there is a need for more complex models. Such models can deal with either one or multiple objects, as well as provide a mechanism for deciding whether there is one or more objects. This theoretical issue is called *binding*, *unity assumption*, or

---

[5] See Section 7.4.

pairing [Sato et al., 2007], and has received recent attention using hierarchical Bayesian models.

13.3.3 Modularity, hierarchies

It would be difficult to assume that natural cognitive systems process their complex sensory inputs in a single layer of computation. Therefore, computations can probably be apprehended as processes that communicate intermediate results. This is similar to the notion of modularity, which is central to structured computer programming. A flow of computation can be broken down into structured sequences. First, we recall Bayesian models that amount to probabilistic versions of subroutine calls and conditional switches that have already been presented before but are presented here in the cognitive context. More complex constructs occur when parts of the computations are based on observing other computation processes. By so doing, we also recall Bayesian model recognition and Bayesian model abstraction.

13.3.3.1 Modularity

Hierarchies rely on the notion of modularity, the simplest instance of which is the subroutine; that is, the use of part of a model (submodel) within another model. A model can be seen as a resource that can be exploited by other models.

The corresponding Bayesian model is the following (see Chapter 9):

$$Pr\begin{cases} Ds\begin{cases} Sp(\pi)\begin{cases} Va: \\ A, B \\ Dc: \\ \begin{cases} & P(A \wedge B|\pi_1) \\ = & P(A|\pi_1) \times P(B|A \wedge \pi_1) \end{cases} \\ Fo: \\ P(B|A \wedge \pi_1) = P(B|A \wedge \pi_2) \end{cases} \\ Id \end{cases} \\ Qu: \\ any \end{cases} \tag{13.32}$$

- A call to a subroutine can be performed simply by specifying that a distribution for this model is the same as that for another model:

$$P(B|A \wedge \pi_1) = P(B|A \wedge \pi_2) \tag{13.33}$$

where π_2 is another model concerned with variables A and B. Equation 13.33 is an assumption that the distribution over B given A in model π_1 is the same as in π_2.

- Naturally, this requires model π_2 to be specified. Most of the time, $P(B|A \wedge \pi_2)$ is the result of an inference in model π_2. As a Bayesian Program, π_2 can be asked any probabilistic question related to its variables and can be arbitrarily complex. Moreover, many different models can question π_2. In this sense, π_2 can be considered as a resource.

For example, Laskey and Mahoney [1997] propose the *network fragments* framework, inspired by *object oriented* software analysis and design to define submodules of a global probabilistic model. They apply this tool to military situation assessment, where a military analyst has to reason on various levels (basic units, regiment, overall situation, etc.).

Koller and Pfeffer [1997] also propose *object oriented Bayesian networks* as a modeling language for Bayesian models based on an object oriented framework.[6]

13.3.3.2 Weighing and switching

A common idea in behavior modeling is to express a behavior as a weighted sum of more elementary ones. This is often referred to as "*weighing*" in the cognitive literature where it appears in various forms.

The generic Bayesian program for these approaches is the following (see Chapter 10):

$$
Pr\begin{cases}
Ds\begin{cases}
Sp(\pi)\begin{cases}
Va: \\
Y, H, X \\
Dc: \\
\begin{cases}
 & P(Y \wedge H \wedge X|\pi) \\
= & P(Y|\pi)\, P(H|Y \wedge \pi)\, P(X|H \wedge Y \wedge \pi)
\end{cases} \\
Fo: \\
\begin{cases}
P(Y|\pi) \equiv Any \\
P(H|Y \wedge \pi) \equiv Any \\
P(X|[H=m] \wedge Y \wedge \pi) \equiv P(X|Y \wedge \pi_m)
\end{cases}
\end{cases} \\
Id
\end{cases} \\
Qu: \\
\begin{cases}
 & P(X|y \wedge \pi) \\
\prec & \sum_{m=1}^{M} \left[P([H=m]|y \wedge \pi) \times P(X|y \wedge \pi_m) \right]
\end{cases}
\end{cases}
\tag{13.34}
$$

- The most simple form is obtained when there is no variable Y and when

[6]See more on this subject in the FAQ-FAM, Section 16.3 "Bayesian programming versus Bayesian networks."

$P(H|\pi)$ is defined as a table of constant values c_m, one for each of the M possible values of H.

In that case, we get:

$$P(X|\pi) \prec \sum_{m=1}^{M} [c_m \times P(X|\pi_m)] \tag{13.35}$$

which is the standard expression of a mixture.

- A more elaborate one is when Y conditions H but not X.

  In this case we get:

$$P(X|y \wedge \pi) \prec \sum_{m=1}^{M} [P([H=m]|y \wedge \pi) \times P(X|\pi_m)] \tag{13.36}$$

  where the weighting between the different behaviors is decided according to the value of Y.

- Finally, Y may conditions both H and X and we get:

$$P(X|y \wedge \pi) \prec \sum_{m=1}^{M} [P([H=m]|y \wedge \pi) \times P(X|y \wedge \pi_m)] \tag{13.37}$$

If Y gives all the necessary information without any uncertainty to choose between the different behaviors $P(X|y \wedge \pi_m)$, then for a given y, $P([H=m]|y \wedge \pi)$ is a Dirac with a value of 0 for all m but one. In that case the sum on m simplifies to a single term, and we get:

$$P(X|y \wedge \pi) = P(X|y \wedge \pi_m) \tag{13.38}$$

The value of Y is sufficient to switch from one behavior to another.

Weighing and *switching* may be seen as the same probabilistic model. Reducing the uncertainty on the choice variable H progressively transforms *weighting* into *switching*. This leads us to reconsider all the debate about "weighting versus switching" (see for instance [Stocker and Simoncelli, 2008] and [Aly and Yonelinas, 2012]) as low uncertainty makes them indistinguishable.

13.3.3.3 Learning as a probabilistic inference process

The weighting model allows for a combination of different models into one. Another issue that can be addressed using Bayesian modeling is model recognition. In particular, when the class of models is a parameter space and recognition is based on experimental data, this recognition amounts to a *machine learning* problem.

These models can be fit using the general framework of Bayesian model recognition. Let $\Delta = \{\Delta_i\}$ be the variables corresponding to the data used for learning (Δ_i, a variable for each datum), and let Π be the variable corresponding to the model. Generally, model recognition is performed using the following decomposition:

$$P(\Pi\ \Delta) = P(\Pi)P(\Delta \mid \Pi) \tag{13.39}$$

where $P(\Pi)$ is a prior on the various models and $P(\Delta \mid \Pi)$ is the probability of observations given the model (i.e., the likelihood of models).

Typically, the data are assumed to be independently and identically distributed (the i.i.d. assumption); that is, $P(\Delta \mid \Pi) = \prod_{i=1}^{N} P(\Delta_i \mid \Pi)$, where $P(\Delta_i \mid \Pi)$ does not depend on index i of each datum. For each possible model, π_j, the distribution, $P(\Delta_i \mid [\Pi = \pi_j])$, is a call to the submodel, π_j, as defined in Section 13.3.3.1.

The probabilistic question for model recognition is:

$$P(\Pi \mid \delta) \propto P(\Pi) \prod_{i=1}^{N} P(\delta_i \mid \Pi). \tag{13.40}$$

The expression 13.40 enables the computation of a probability distribution on the various models based on some of the data. This mechanism is hierarchical in the sense that we build a model for reasoning about an underlying set of models.

A common instance of model recognition is parameter learning. In this case, the models, Π, share a common parametric form, π', and the recognition occurs on the parameters, Θ: $\Pi = \Theta \wedge \pi'$. We can modify the decomposition presented above by including knowledge of the parametric form:

$$P(\Theta\ \Delta \mid \pi') = P(\Theta \mid \pi')P(\Delta \mid \Theta\ \pi') = P(\Theta \mid \pi') \prod_{i=1}^{N} P(\Delta_i \mid \Theta\ \pi') \tag{13.41}$$

where $P(\Theta \mid \pi')$ is a prior distribution on the parameters and $P(\Delta \mid \Theta\ \pi')$ (or $\prod_{i=1}^{N} P(\Delta_i \mid \Theta\ \pi')$ with the i.i.d. assumption) is the likelihood of parameters.

So, learning can be accomplished by using the following question:

$$P(\Theta \mid \delta\ \pi') \propto P(\Theta \mid \pi')P(\delta \mid \Theta\ \pi') \propto P(\Theta \mid \pi') \prod_{i=1}^{N} P(\delta_i \mid \Theta\ \pi') \tag{13.42}$$

The likelihood functions are usually completely specified by the parametric form, π', and the parameters, Θ. However, the prior on the parameters, $P(\Theta \mid \pi')$, may need some additional parameters called *hyper-parameters*. The

Bayesian formalism allows for these hyper-parameters in the same way. Let Λ be the variable representing these hyper-parameters. We write the joint probability distribution as:

$$P(\Lambda\ \Theta\ \Delta \mid \pi'') = P(\Lambda \mid \pi'')P(\Theta \mid \Lambda\ \pi'')P(\Delta \mid \Theta\ \pi'') \qquad (13.43)$$

where $P(\Lambda \mid \pi'')$ is a prior on hyper-parameters, $P(\Theta \mid \Lambda\ \pi'')$ is the distribution on parameters Θ according to hyper-parameters and $P(\Delta \mid \Theta\ \pi'')$ is the likelihood function, as above. As a result, inference on the parameters is modified slightly:

$$P(\Theta \mid \delta\ \pi'') \propto \sum_{\Lambda} [P(\Lambda \mid \pi'')P(\Theta \mid \Lambda\ \pi'')]P(\delta \mid \Theta\ \pi'') \qquad (13.44)$$

The prior on the hyper-parameters could also be parametric, and it is possible to add another set of parameters. It all amounts to what knowledge the modeler wants to include in the model. Moreover, it can be shown that deep layers of priors have much less influence on parameter estimation [MacKay, 2003].

All the entropy principles (maximum entropy principle, minimum relative entropy principle, Kullback–Leibler divergence) and their numerous applications are closely and simply related to the above models.

Indeed, to take the simplest case of Equation 13.40, if we consider that there are K different possible observations (i.e., the variable Δ_i can take K different values), by gathering the same observations, we can restate this equation as:

$$P(\Pi \mid \delta) \propto P(\Pi) \prod_{k=1}^{K} P([\Delta_i = k] \mid \Pi)^{n_k} \qquad (13.45)$$

where n_k is the number of times that the observation, $[\Delta_i \doteq k]$, has been made. To a first approximation, this number is proportional to the probability $P([\Delta_i = k] \mid \Pi)$ itself and we obtain, with N the total number of observations:

$$P(\Pi \mid \delta) \propto P(\Pi) \prod_{k=1}^{K} P([\Delta_i = k] \mid \Pi)^{NP([\Delta_i=k] \mid \Pi)} \qquad (13.46)$$

Finally, if we assume a uniform prior on the different models, and if we take the logarithm of Equation 13.46, we obtain the maximum entropy principle:

$$\log\left(P(\Pi \mid \delta)\right) = N \sum_{k=1}^{K} [P([\Delta_i = k] \mid \Pi) \log\left(P([\Delta_i = k] \mid \Pi)\right)] + C \qquad (13.47)$$

For example, Gopnik and Schulz [2004] studied the learning of causal dependencies by young children. The experiments included trying to decide which objects are "blickets" (imaginary objects that are supposed to illuminate a given machine). Some objects are put on a machine that lights up depending on which objects are placed on it. The patterns of response were predicted well by a causal Bayesian network, even after adding some prior knowledge ("blickets are rare"). The learning phase involved selecting the causal dependency structure that matches the observations among all the possibilities.

Xu and Garcia [2008] made this impressive experiment with 8 month old children demonstrating that they are already able to do model recognition. They are, indeed, able to estimate the proportion of blue and red balls in an urn from few samples.

Another example is the application of *embedded hidden Markov models*. These are models in which a top-level Bayesian model reasons on nodes that are themselves Bayesian models. Nefian and Hayes [1999] proposed such a formalism in the context of face recognition. Each submodel is responsible for the recognition of a particular feature of the face (forehead, eyes, nose, mouth, and chin). The global model ensures perception of the facial structure by recognizing each feature in the correct order. Neal et al. [2003] applied embedded HMMs to the tracking of 3-D human motion from 2-D tracker images. The high dimensionality of the problem proved to be less an issue for their algorithm than it was for the previous approach.

13.3.3.4 Abstraction

Usually, a modeler uses learning in order to select a unique model or set of parameter values, which is then applied to the problem at hand. That is, the learning process computes a probability distribution over models (or their parameters) to be applied, and a decision, based on this probability distribution, is used to select only one model or parameter set.

Another way to use model recognition is to include it as part of a higher-level program in order to maintain the uncertainty on the models at the time of their application. This is called model *abstraction*.

Let Π be the variable representing the submodels, let Δ be the variable representing the data, and let X be the sought-after variable that depends on the model. The joint probability distribution can be decomposed in the following way:

$$P(X \; \Delta \; \Pi) = P(\Pi) P(\Delta \mid \Pi) P(X \mid \Pi) \tag{13.48}$$

where $P(\Pi)$ and $P(\Delta \mid \Pi)$ are the priors and likelihood functions defined as before and $P(X \mid \Pi)$ describes the influence of the model, Π, on the variable of interest, X.

The question is concerned with the distribution over X given the data, Δ:

$$P(X \mid \delta) \propto \sum_{\Pi} \left[P(\Pi) P(\delta \mid \Pi) P(X \mid \Pi) \right]. \tag{13.49}$$

This inference is similar to model recognition except for the factor $P(X \mid \Pi)$.

With respect to the question, $P(X \mid \delta)$, the details of the models can be abstracted.

When applied to classes of models and their parameters (i.e., when X is Θ), this abstraction model yields the *Bayesian model selection* (BMS) method. It can also be used to jointly compute the distribution over joint models and parameters, using $P(\Pi\ \Theta \mid \Delta)$ [Kemp and Tenenbaum, 2008].

Diard and Bessière [2008] used abstraction for robot localization. They defined several *Bayesian maps* corresponding to various locations in the environment. Each map is a model of sensorimotor interactions with a part of the environment. Then, they built an *abstracted map* based on these models. In this new map, the location of the robot is defined in terms of the submap that best fits the observations obtained from the robot's sensors. The aim of their abstracted map was to navigate in the environment. Therefore, they were more interested in the action to be taken than in the actual location. However, the choice of an action was made with respect to the uncertainty of the location.

A similar model was also recently applied to the domain of multimodal perception, under the name of causal inference [Körding et al., 2007; Sato et al., 2007]. When sensory cues are close, they could originate from a single source, and the small spatial discrepancies could be explained away by noise; on the other hand, when cues are largely separated, they more probably originate from distinct sources, instead, and their spatial positions are not correlated. The optimal strategy, when estimating the positions of these cues, is then to have both alternative models coexist and to integrate over the number of sources during the final estimation.

13.3.4 Loops

It would be hard to assume that natural cognitive systems process their complex sensory systems in a single direction, uniquely from sensory input toward motor outputs. Indeed, neurophysiology highlights a variety of ways that neural system activity can be fed back in loops. Models that include loops are mainly temporal in nature and deal with memory systems. Examples are mostly taken from models of artificial systems, and especially from the robotics community.

13.3.4.1 Temporal series of observations

In order to model the temporal behavior of a phenomenon, it is usually assumed that a sequential series of observations is available. These are usually used for *state estimation*: recovering the evolution of some internal state that is not observed.

The corresponding Bayesian program (13.50) has been already presented a number of times (see Chapter 11 and Section 13.1.2).

Kalman filters are the most common examples of models in this category, probably because of strong assumptions that lead to a closed-form solution for state estimation [Kalman, 1960; Ghahramani et al., 1997]. They are widely used in robotics [Thrun et al., 2005] and in the life sciences [van der Kooij et al., 1999; Kiemel et al., 2002]. When the state space can be assumed to be discrete, hidden Markov models can be applied [Rabiner, 1989; Rabiner and Juang, 1993]. A common technique for approximating the inference required for state estimation is to model the state probability distribution using particle filters; this can also be seen as the application of a mixture model to a loop model [Arulampalam et al., 2002]. More generally, these models are instances of Bayesian filters [Leonard et al., 1992; Bessière et al., 2003]. When the independence assumptions do not vary over time, this class of models is called dynamic Bayesian networks [Dean and Kanazawa, 1989; Murphy, 2002]. These models have also been extensively covered in the statistics literature, sometimes using different vocabularies [Harvey, 1992; Brockwell and Davis, 2000]. Interesting extensions have been proposed to occupancy grids for application in the automotive industry [Coué, 2003; Coué et al., 2006; Tay et al., 2007, 2008].

$$
Pr\begin{cases}
Ds\begin{cases}
Sp(\pi)\begin{cases}
Va: \\
S^0,\cdots,S^T,O^0,\cdots,O^T \\
Dc: \\
\begin{cases}
 & P\left(S^0\wedge\cdots\wedge S^T\wedge O^0\wedge\cdots\wedge O^T|\pi\right) \\
= & P\left(S^0\wedge O^0\right)\times\prod_{t=1}^{T}\left[P\left(S^t|S^{t-1}\right)\times P\left(O^t|S^t\right)\right]
\end{cases} \\
Fo: \\
\begin{cases}
P\left(S^0\wedge O^0\right) \\
P\left(S^t|S^{t-1}\right) \\
P\left(O^t|S^t\right)
\end{cases}
\end{cases} \\
Id
\end{cases} \\
Qu: \\
\begin{cases}
P\left(S^{t+k}|O^0\wedge\cdots\wedge O^t\right) \\
(k=0)\equiv Filtering \\
(k>0)\equiv Prediction \\
(k<0)\equiv Smoothing
\end{cases}
\end{cases}
\tag{13.50}
$$

13.3.4.2 Efferent copy

Usually, in robotics and control contexts, the state of the observed system can be not only observed but also acted upon. The previous models are enriched with control variables as additional inputs for state estimation. *Reading* the values of these control variables after they have been decided, in order to ameliorate state estimation or prediction, is one possible reason for there being efferent copies of motor variables in animal central nervous systems.

Let us recall this model (13.51) that has been already presented and commented upon (see Chapter 11 and Section 13.2.5).

Input/output HMMs [Bengio and Frasconi, 1995] have been introduced in the machine learning literature and applied as benchmarks in grammatical inference.

In the robotic localization context, the Markov localization model is the most common example of this model category [Thrun et al., 2005].

In life sciences, Laurens and Droulez [2007, 2008] applied Bayesian state estimation using efferent copies to the modeling of 3-D head orientation in space in humans and showed that such models could account for a variety of known perceptual illusions.

$$Pr\begin{cases} Ds\begin{cases} Sp(\pi)\begin{cases} Va: \\ S^0,\cdots,S^T,A^0,\cdots,A^T,O^0,\cdots,O^T \\ Dc: \\ \begin{cases} P\left(S^0\wedge\cdots\wedge O^T|\pi\right) \\ = \left[\begin{array}{c} P\left(S^0\wedge O^0|\pi\right)\times\prod_{t=0}^{T}\left[P\left(A^t|\pi\right)\right] \\ \prod_{t=1}^{T}\left[P\left(S^t|S^{t-1}\wedge A^{t-1}\wedge\pi\right)\times P\left(O^t|S^t\wedge\pi\right)\right]\end{array}\right]\end{cases} \\ Fo: \end{cases} \\ Id \end{cases} \\ Qu: \\ P\left(S^T|a^0\wedge\cdots\wedge a^{T-1}\wedge o^0\wedge\cdots\wedge o^{T-1}\wedge\pi\right)\end{cases} \tag{13.51}$$

13.3.4.3 Attention focusing and action selection

Instead of trying to find a structural decomposition of the state space automatically, alternative approaches can be pursued in order to reduce the computational complexity of the planning and control processes. It is assumed that the model already incorporates knowledge about the task or domain.

For instance, modeling cognitive systems requires including in the model knowledge about the environment structure or task progression structure, so as to separate elementary filters, and to reduce the complexity and dimensionality of the internal state space. Attention systems must also be defined so that computations are performed only in those subsets of the state space that are thought to be relevant. Finally, elementary motor tasks can be identified and modeled as behaviors, so that they are not planned completely through each time period.

Incorporating such knowledge into models makes them less generic and more sophisticated. Figure 13.2 shows the joint distribution of the full model from Koike [2008].

$$
\begin{aligned}
&P(A^{0:t}\ S^{0:t}\ O^{0:t}\ B^{0:t}\ C^{0:t}\ \lambda^{0:t}\ \beta^{0:t}\ \alpha^{0:t} \mid \pi)\\
&= \left[\begin{array}{l}
\prod_{j=1}^{t}\left[\begin{array}{l}
\prod_{i=1}^{N_i} P(S_i^j \mid S_i^{j-1}\ A_i^{j-1}\ \pi_i) \prod_{i=1}^{N_i} P(O^j \mid S_i^{j-1}\ C^j\ \pi_i)\\
P(B^j \mid \pi) \prod_{i=1}^{N_i} P(\beta^j \mid B^j\ S_i^j\ B^{j-1}\ \pi_i)\\
P(C^j \mid \pi) \prod_{i=1}^{N_i} P(\alpha^j \mid C^j\ S_i^j\ B^j\ \pi_i)\\
P(A^j \mid \pi) \prod_{i=1}^{N_i} P(\lambda^j \mid A^j\ B^j\ S_i^j\ A^{j-1}\ \pi_i)
\end{array}\right]\\
P(A^0\ S^0\ O^0\ B^0\ C^0\ \lambda^0\ \beta^0\ \alpha^0 \mid \pi)
\end{array}\right]
\end{aligned}
$$

FIGURE 13.2: Joint probability factorization for the full model as designed by Koike [2005].

In Figure 13.2, π_i refers to the i elementary filters, which are assumed to be independent (hence all the $\prod_{i=1}^{N_i}$ products). $P(S_i^j \mid S_i^{j-1}\ A_i^{j-1}\ \pi_i)$ are their dynamic models, $P(O^j \mid S_i^{j-1}\ C^j\ \pi_i)$ are their sensor models, $P(\beta^j \mid B^j\ S_i^j\ B^{j-1}\ \pi_i)$ are their behavior models, $P(\alpha^j \mid C^j\ S_i^j\ B^j\ \pi_i)$ are their attention models, and $P(\lambda^j \mid A^j\ B^j\ S_i^j\ A^{j-1}\ \pi_i)$ are their motor command models. Finally, $P(A^0\ S^0\ O^0\ B^0\ C^0\ \lambda^0\ \beta^0\ \alpha^0 \mid \pi)$ encodes the initial state of the system. This model is then used to compute the probability distribution over the next action to be performed, given the past history of observation and control variables: $P(A^t \mid O^{0:T}\ A^{0:T-1})$.

Koike [2005] and Koike et al. [2008] applied this approach in the domain of autonomous mobile sensorimotor systems. Koike showed how time and space limits on computations, and limited on-board processing power, could be accommodated, thanks to simplifying assumptions such as the stationarity of the temporal models, partial independence between sensory processes, domain-of-interest selection at the processing stages (attention focusing), and behavior selection.

13.3.4.4 Discussion

There is no clear definition of a loop. In this section, we have only presented the definition and examples of temporal loops. These loops can be compared to loops in the field of computer science, occurring when the execution flow gets several times through the same set of instructions. Such instructions are specified once for all the executions of the loop, and the global program is its replication through time.

This replication often occurs with fixed time spans. However, in biological systems, multiple loops may take place simultaneously with different and sometimes varying time constants. In robotics, many processes are run concurrently with different levels of priority. There is a need in Bayesian modeling for a proper way of integrating and synchronizing loops with different time scales. Finally, loops can also be considered without reference to time. Bayesian filters are a single model that is replicated at each time step, with an optional temporal dependency on preceding time steps. Models have also been proposed for spatial replication of models, with dependencies occurring over a neighborhood. One interesting difference is that temporal relations between instances are oriented according to the passage of time, whereas models of spatial loops, such as the Markov random field, rely on a symmetrical relation between neighbors.

Chapter 14

Bayesian Inference Algorithms Revisited

> "Five to one against and falling?" she said, "four to one against and falling...three to one...two...one...probability factor of one to one...we have normality, I repeat we have normality." She turned her microphone off, then turned it back on, with a slight smile and continued: "Anything you still can't cope with is therefore your own problem."
>
> *The Hitchhiker's Guide to the Galaxy*
> Douglas Adams [1995]

This chapter surveys the main available general purpose algorithms.

It is well known that general Bayesian inference is a very difficult problem, which may be practically intractable. Exact inference has been proved to be NP-hard [Cooper, 1990] as has the general problem of approximate inference [Dagum and Luby, 1993].[1] Numerous heuristics and restrictions to

[1] For more details, see Chapter 16, Section 16.7, "Computational complexity of Bayesian inference."

the generality of possible inferences have been proposed to achieve admissible computation time. The purpose of this chapter is to make a short review of these heuristics and techniques.

Before starting to crunch numbers, it is usually possible (and wise) to make some symbolic computations to reduce the amount of numerical computation required. The first section of this chapter presents the different possibilities. We will see that these symbolic computations can be either exact or approximate.

Once simplified, the expression obtained must be numerically evaluated. In a few cases exact (exhaustive) computation may be possible thanks to the previous symbolic simplification, but most of the time, even with the simplifications, only approximate calculations are possible. The second section of this chapter describes the principles of the main algorithms to do so.

Finally, in a third section we present the specific algorithms used in ProBT: the inference engine used to interpret the programs given as examples in this book.

14.1 Stating the problem

Given the joint distribution:

$$\begin{aligned} & P(X_1 \wedge X_2 \wedge \cdots \wedge X_N | \delta \wedge \pi) \\ = \; & P(L_1 | \delta \wedge \pi) \times P(L_2 | R_2 \wedge \delta \wedge \pi) \times \cdots \times P(L_K | R_K \wedge \delta \wedge \pi) \end{aligned} \tag{14.1}$$

it is always possible to compute any possible instantiated question:

$$P(Searched | known \wedge \delta \wedge \pi) \tag{14.2}$$

using the following general inference:

$$\begin{aligned} & P(Searched | known \wedge \delta \wedge \pi) \\ = \; & \sum_{Free} [P(Searched \wedge Free | known \wedge \delta \wedge \pi)] \\ = \; & \frac{\sum_{Free} [P(Searched \wedge Free \wedge known | \delta \wedge \pi)]}{P(known | \delta \wedge \pi)} \\ = \; & \frac{\sum_{Free} [P(Searched \wedge Free \wedge known | \delta \wedge \pi)]}{\sum_{Free \wedge Searched} [P(Searched \wedge Free \wedge known | \delta \wedge \pi)]} \\ = \; & \frac{1}{Z} \times \sum_{Free} [P(Searched \wedge Free | known \wedge \delta \wedge \pi)] \\ = \; & \frac{1}{Z} \times \sum_{Free} \left[P(L_1 | \delta \wedge \pi) \times \prod_{k=2}^{K} [P(L_k | R_k \wedge \delta \wedge \pi)] \right] \end{aligned} \tag{14.3}$$

where the first equality results from the marginalization rule (12.11), the second results from the conjunction rule (12.7), and the third corresponds to a second application of the marginalization rule.

The denominator appears to be a normalization term. Consequently, by convention, we will either replace it with Z or write a proportional equation ($\propto$) instead of an equality one ($=$).

Finally, it is possible to replace the joint distribution by its decomposition (14.1).

The problem of symbolic simplification can be stated very simply. How can we modify the expression:

$$\frac{1}{Z} \times \sum_{Free} \left[P(L_1|\delta \wedge \pi) \times \prod_{k=2}^{K} [P(L_k|R_k \wedge \delta \wedge \pi)] \right] \tag{14.4}$$

to produce a new expression requiring less computation that gives the same result or a good approximation of it?

Section 14.2 presents the different possibilities. We will see that these symbolic computations can be either exact (Section 14.2.1) or approximate (Section 14.2.2), in which case they lead to an expression that, while not mathematically equal to Equation 14.4, should be close enough.

Once simplified, the expression obtained is used to compute:

$$P(Searched|known \wedge \delta \wedge \pi) \tag{14.5}$$

must be evaluated numerically.

In a few cases, exact (exhaustive) computation may be possible, thanks to the previous symbolic simplification, but normally, even with the simplifications, only approximate calculation is possible. Section 14.3 describes the principles of the main algorithms used.

Two main problems must be solved: searching the modes in a high-dimensional space, and marginalizing in a high-dimensional space.

Because *Searched* may be a conjunction of numerous variables, each of them possibly having many values or even being continuous, it is seldom possible to compute exhaustively $P(Searched|known \wedge \delta \wedge \pi)$ and find the absolute most probable value for *Searched*. One may then decide to either build an approximate representation of this distribution or to directly sample from this distribution. In both cases, the challenge is to find the modes of:

$$P(Searched|known \wedge \delta \wedge \pi) \propto \sum_{Free} \left[P(L_1|\delta \wedge \pi) \times \prod_{k=2}^{K} [P(L_k|R_k \wedge \delta \wedge \pi)] \right]. \tag{14.6}$$

(on the search space defined by *Searched*), where most of the probability density is concentrated. This may be very difficult, as most of the probability may be concentrated in very small subspaces of the whole search space.

The situation is even worse, as computing the value of $P(Searched|known)$ for a given value of *Searched* (a single point of the search space of the preceding paragraph) is by itself a difficult problem. Indeed, it requires marginalizing the joint distribution on the space defined by *Free*. *Free* (like *Searched*) may be a conjunction of numerous variables, each of them possibly having many values or even being continuous. Consequently, the sum should also be either approximated or sampled. The challenge is then to find the modes of:

$$P(L_1|\delta \wedge \pi) \times \prod_{k=2}^{K} [P(L_k|R_k \wedge \delta \wedge \pi)] \tag{14.7}$$

(on the search space defined by *Free*), where most of the probability density is concentrated and which mostly contribute to the sum. Finally, marginalizing in a high-dimensional space appears to be a very similar problem to searching the modes in a high-dimensional space.

14.2 Symbolic computation

In this section, we give an overview of the principal techniques to simplify the calculation needed either to evaluate $P(Searched|known)$, or to find the most probable value for *Searched*, or to draw values for *Searched* according to this distribution.

The goal is to perform symbolic computation on the expression:

$$\frac{1}{Z} \times \sum_{Free} \left[P(L_1|\delta \wedge \pi) \times \prod_{k=2}^{K} [P(L_k|R_k \wedge \delta \wedge \pi)] \right] \tag{14.8}$$

to obtain another expression to compute the same result with far fewer elementary operations (sum and product). It is called symbolic computation because this can be done independently of the possible numerical values of the considered variables.

We will present these different algorithms as pure and simple algebraic manipulations of expression 14.8 above, even if most of them have been historically proposed from different points of view (especially in the form of manipulation of graphs and message passing along their arcs).

14.2.1 Exact symbolic computation

We first restrict our analysis to mathematically exact symbolic computations that lead to a simplified expression mathematically equivalent to the starting one.

14.2.1.1 Question-specific symbolic computation

It is seldom possible to solve analytically the question $P(Searched|known)$. Most of the time, the integral in expression 14.8 has no explicit solution.

However, this is possible for Kalman filters as defined by the Bayesian program in Equation 13.9. This explains their popularity and their importance in applications. Indeed, once analytically solved, the answer to the question may be computed very efficiently.

14.2.1.2 Question-dependent symbolic computation

We first take an example to introduce the different possible simplifications.

This example is defined by the following decomposition (Equation 14.9) of a joint distribution of nine variables:

$$\begin{aligned} & P(X_1 \wedge X_2 \wedge \cdots \wedge X_9) \\ = \; & P(X_1) \times P(X_2|X_1) \times P(X_3|X_1) \times P(X_4|X_2) \times P(X_5|X_2) \\ & \times P(X_6|X_3) \times P(X_7|X_3) \times P(X_8|X_6) \times P(X_9|X_6) \end{aligned} \quad (14.9)$$

corresponding to the Bayesian network defined in Figure 14.1:

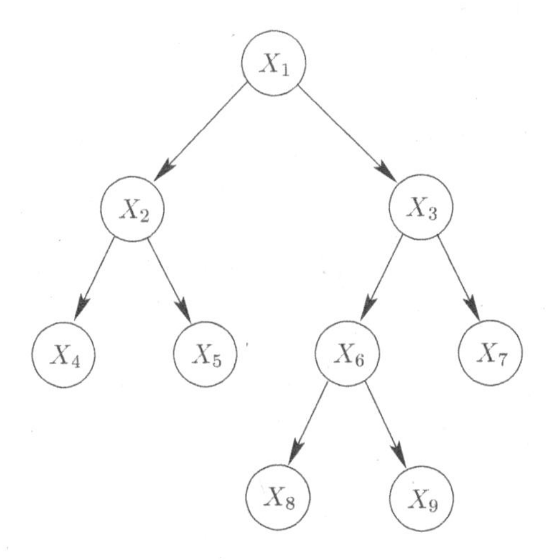

FIGURE 14.1: The Bayesian network corresponding to the decomposition in Equation 14.9.

Take, for instance, the instantiated question $P(X_1|x_5 \wedge x_7)$ where $Searched = X_1$, $known = x_5 \wedge x_7$, and $Free = X_2 \wedge X_3 \wedge X_4 \wedge X_6 \wedge X_8 \wedge X_9$. We know that:

$$
\begin{aligned}
&P(X_1|x_5 \wedge x_7) \\
&\propto \sum_{\substack{X_2 \wedge X_3 \\ X_4 \wedge X_6 \\ X_8 \wedge X_9}} \left[\begin{array}{l} P(X_1) \times P(X_2|X_1) \times P(X_3|X_1) \\ \times P(X_4|X_2) \times P(x_5|X_2) \\ \times P(X_6|X_3) \times P(x_7|X_3) \\ \times P(X_8|X_6) \times P(X_9|X_6) \end{array} \right]
\end{aligned} \tag{14.10}
$$

If each of the variables X_n may take 10 different possible values, then evaluating expression 14.10 for a given value of X_1 requires 9×10^6 elementary operations.

To reduce this number, we can first reorder the different summations the following way:

$$
\begin{aligned}
&P(X_1|x_5 \wedge x_7) \\
&\propto \sum_{X_2 \wedge X_3} \left[\begin{array}{l} P(X_1) \times P(X_2|X_1) \times P(X_3|X_1) \\ \times P(x_5|X_2) \times P(x_7|X_3) \\ \sum_{X_4} \left[\begin{array}{l} P(X_4|X_2) \\ \sum_{X_6} \left[\begin{array}{l} P(X_6|X_3) \\ \sum_{X_8} \left[\begin{array}{l} P(X_8|X_6) \\ \sum_{X_9} [P(X_9|X_6)] \end{array} \right] \end{array} \right] \end{array} \right] \end{array} \right]
\end{aligned} \tag{14.11}
$$

and we see that $\sum_{X_9} [P(X_9|X_6)]$ vanishes as it sums to one. We obtain:

$$
\begin{aligned}
&P(X_1|x_5 \wedge x_7) \\
&\propto \sum_{X_2 \wedge X_3} \left[\begin{array}{l} P(X_1) \times P(X_2|X_1) \times P(X_3|X_1) \\ \times P(x_5|X_2) \times P(x_7|X_3) \\ \sum_{X_4} \left[\begin{array}{l} P(X_4|X_2) \\ \sum_{X_6} \left[\begin{array}{l} P(X_6|X_3) \\ \sum_{X_8} [\ P(X_8|X_6)\ ] \end{array} \right] \end{array} \right] \end{array} \right]
\end{aligned} \tag{14.12}
$$

This same simplification can also be applied to the sums on X_8, X_6, and X_4 to yield:

$$
\begin{aligned}
&P(X_1|x_5 \wedge x_7) \\
&\propto \sum_{X_2 \wedge X_3} \left[\begin{array}{l} P(X_1) \times P(X_2|X_1) \times P(X_3|X_1) \\ \times P(x_5|X_2) \times P(x_7|X_3) \end{array} \right]
\end{aligned} \tag{14.13}
$$

Evaluating expression 14.13 requires 5×10^2 elementary operations. Eliminating the parts of the global sum that sum to one is indeed a very efficient simplification.

However, further rearranging the order of the sums in Equation 14.13 may lead to more gains.

First, $P(X_1)$ may be factorized out of the sum:

$$\begin{aligned} & P(X_1|x_5 \wedge x_7) \\ \propto \; & P(X_1) \times \sum_{X_2 \wedge X_3} \left[\begin{array}{l} P(X_2|X_1) \times P(X_3|X_1) \\ \times P(x_5|X_2) \times P(x_7|X_3) \end{array} \right] \end{aligned} \tag{14.14}$$

Then, the sum on X_2 and X_3 can be split, leading to:

$$\begin{aligned} & P(X_1|x_5 \wedge x_7) \\ \propto \; & P(X_1) \times \sum_{X_2} \left[\begin{array}{l} P(X_2|X_1) \times P(x_5|X_2) \\ \times \sum_{X_3} [P(X_3|X_1) \times P(x_7|X_3)] \end{array} \right] \end{aligned} \tag{14.15}$$

and as $\sum_{X_3} [P(X_3|X_1) \times P(x_7|X_3)]$ is only a function of X_1, it can be factored out of the sum on X_2, to finally have:

$$\begin{aligned} & P(X_1|x_5 \wedge x_7) \\ \propto \; & P(X_1) \times \sum_{X_2} [\, P(X_2|X_1) \times P(x_5|X_2) \,] \\ & \times \sum_{X_3} [P(X_3|X_1) \times P(x_7|X_3)] \end{aligned} \tag{14.16}$$

Evaluating expression 14.16 requires $19 + 19 + 2 = 40$ elementary operations, five orders of magnitude less than that required for the initial expression to perform the exact same calculation!

Other questions lead to similar symbolic simplifications. For instance, for the question $P(X_2|x_5 \wedge x_7)$, we have:

$$\begin{aligned} & P(X_2|x_5 \wedge x_7) \\ \propto \; & P(x_5|X_2) \times \sum_{X_1} \left[\begin{array}{l} P(X_1) \times P(X_2|X_1) \\ \times \sum_{X_3} [P(X_3|X_1) \times P(x_7|X_3)] \end{array} \right] \end{aligned} \tag{14.17}$$

However, the final simplification step is not similar to the simplification of $P(X_1|x_5 \wedge x_7)$ because $\sum_{X_3} [P(X_3|X_1) \times P(x_7|X_3)]$ depends on X_1 and consequently cannot be factored out of the sum on X_1.

Evaluating expression 14.17 requires $(21 \times 10) + 9 + 1 = 220$ elementary operations.

To perform these symbolic simplifications, it is only necessary to apply two fundamental rules: the distributive law and the normalization rule.

The simplification of:

$$= \frac{1}{Z} \times \sum_{Free} \left[P(L_1|\delta \wedge \pi) \times \prod_{k=2}^{K} [P(L_k|R_k \wedge \delta \wedge \pi)] \right]$$

may be done in three steps:

1. *Eliminate distributions that sum to one*: When a term

$$P(L_k|R_k \wedge \delta \wedge \pi)$$

appears in the sum, if all the variables appearing in L_k are summed and none of them appears in any of the other R_j, then $P(L_k|R_k \wedge \delta \wedge \pi)$ sums to one and vanishes out of the global sum. Of course, the list of summed variables, initialized to $Free$, must then be updated by removing the variables of L_k. This process can be recursively applied until no more terms of the product can be removed. It leads to an expression of the form:

$$\begin{aligned} & P(Searched|known \wedge \delta \wedge \pi) \\ = \ & \frac{1}{Z} \times \sum_{Summed} \left[\prod_{k} [P(L_k|R_k \wedge \delta \wedge \pi)] \right] \end{aligned} \tag{14.18}$$

where $Summed \subseteq Free$. An example of this was given in Equation 14.13.

2. *Factorize*: Each term of the remaining product $\prod_{k} [P(L_k|R_k \wedge \delta \wedge \pi)]$, where all the variables are either $Searched$ or $known$, is independent of the variables appearing in $Summed$, and consequently it can be factored out of the sum. We then obtain a new expression of the form:

$$\begin{aligned} & P(Searched|known \wedge \delta \wedge \pi) \\ = \ & \frac{1}{Z} \times \prod_{k} [P(L_k|R_k \wedge \delta \wedge \pi)] \times \sum_{Summed} \left[\prod_{l} [P(L_l|R_l \wedge \delta \wedge \pi)] \right] \end{aligned} \tag{14.19}$$

An example of this factorization was given in Equation 14.14.

3. *Order the sums cleverly*: Finally, the last type of simplification that can be made is to reorder the sums of $\sum_{Summed} \left[\prod_{l} [P(L_l|R_l \wedge \delta \wedge \pi)] \right]$ to minimize the number of operations required. This third step is much more complicated than the two previous ones: finding the optimal ordering is indeed NP-hard [Arnborg et al., 1987]. Only heuristics can be proposed but they are useful even if they do not find the optimal ordering. Any ordering helps to break the exponential complexity of the computation of the sum.

Numerous algorithms have been proposed to deal with these simplifications. Among the most interesting or most well known are the Symbolic Probabilistic Inference (SPI) algorithm [Shachter et al., 1990; Li and D'Ambrosio, 1994], the variable elimination family [Zhang and Poole, 1996], the bucket elimination family of algorithms [Dechter, 1999; Dechter and Rish, 1997], the Query-DAG framework [Darwiche and Provan, 1997], the general distributive law algorithm [Aji and McEliece, 2000], and the Successive Restriction Algorithm (SRA) [Mekhnacha et al., 2007].

14.2.1.3 Question-independent symbolic computation

Instead of trying to simplify only:

$$\begin{aligned} & P(Searched|known \wedge \delta \wedge \pi) \\ = & \frac{1}{Z} \times \sum_{Free} \left[P(L_1|\delta \wedge \pi) \times \prod_{k=2}^{K} \left[P(L_k|R_k \wedge \delta \wedge \pi) \right] \right] \end{aligned} \tag{14.20}$$

we can try to simplify for a family of such questions.

For instance, in Bayesian nets, where all the L_k of the decomposition are restricted to a single variable (13.1) it may be interesting to apply symbolic computation to minimize globally the number of numerical operations required for the family of questions:

$$\left\{ \begin{array}{l} P(X_1|known) \\ P(X_2|known) \\ \dots \\ P(X_N|known) \end{array} \right\} \tag{14.21}$$

Each of these questions is called a *belief*. The given value of $known$ is called the *evidence*.

We return to the example of the previous section. The family of interesting questions is:

$$\left\{ \begin{array}{l} P(X_1|x_5 \wedge x_7) \\ P(X_2|x_5 \wedge x_7) \\ P(X_3|x_5 \wedge x_7) \\ P(X_4|x_5 \wedge x_7) \\ P(X_6|x_5 \wedge x_7) \\ P(X_8|x_5 \wedge x_7) \\ P(X_9|x_5 \wedge x_7) \end{array} \right\} \tag{14.22}$$

Using the simplification scheme of the previous section for each of these seven questions, we obtain:

$$\begin{aligned} & P(X_1|x_5 \wedge x_7) \\ \propto & P(X_1) \times \sum_{X_2} \left[P(X_2|X_1) \times P(x_5|X_2) \right] \\ & \times \sum_{X_3} \left[P(X_3|X_1) \times P(x_7|X_3) \right] \end{aligned} \tag{14.23}$$

$$P(X_2|x_5 \wedge x_7) \propto P(x_5|X_2) \times \sum_{X_1} \left[\begin{array}{l} P(X_1) \times P(X_2|X_1) \\ \times \sum_{X_3} [P(X_3|X_1) \times P(x_7|X_3)] \end{array} \right] \quad (14.24)$$

$$P(X_3|x_5 \wedge x_7) \propto P(x_7|X_3) \times \sum_{X_1} \left[\begin{array}{l} P(X_1) \times P(X_3|X_1) \\ \times \sum_{X_2} [P(X_2|X_1) \times P(x_5|X_2)] \end{array} \right] \quad (14.25)$$

$$P(X_4|x_5 \wedge x_7) \propto \sum_{X_2} \left[\begin{array}{l} P(X_4|X_2) \times P(x_5|X_2) \\ \times \sum_{X_1} \left[\begin{array}{l} P(X_1) \times P(X_2|X_1) \\ \sum_{X_3} [P(X_3|X_1) \times P(x_7|X_3)] \end{array} \right] \end{array} \right] \quad (14.26)$$

$$P(X_6|x_5 \wedge x_7) \propto \sum_{X_3} \left[\begin{array}{l} P(X_6|X_3) \times P(x_7|X_3) \\ \times \sum_{X_1} \left[\begin{array}{l} P(X_1) \times P(X_3|X_1) \\ \sum_{X_2} [P(X_2|X_1) \times P(x_5|X_2)] \end{array} \right] \end{array} \right] \quad (14.27)$$

$$P(X_8|x_5 \wedge x_7) \propto \sum_{X_6} \left[\begin{array}{l} P(X_8|X_6) \times \sum_{X3} [P(X_6|X_3) \times P(x_7|X_3)] \\ \times \sum_{X_1} \left[\begin{array}{l} P(X_1) \times P(X_3|X_1) \\ \sum_{X_2} [P(X_2|X_1) \times P(x_5|X_2)] \end{array} \right] \end{array} \right] \quad (14.28)$$

$$P(X_9|x_5 \wedge x_7) \propto \sum_{X_6} \left[\begin{array}{l} P(X_9|X_6) \times \sum_{X3} [P(X_6|X_3) \times P(x_7|X_3)] \\ \times \sum_{X_1} \left[\begin{array}{l} P(X_1) \times P(X_3|X_1) \\ \sum_{X_2} [P(X_2|X_1) \times P(x_5|X_2)] \end{array} \right] \end{array} \right] \quad (14.29)$$

These seven expressions share many terms. To minimize the number of elementary numerical operations, they should not be computed several times but only once. This implies an obvious order in these computations:

1. *Step 0*: First, $P(x_5|X_2)$ and $P(x_7|X_3)$, which appear everywhere and can be computed immediately.

2. *Step 1*: Then $\sum_{X_2} [P(X_2|X_1) \times P(x_5|X_2)]$ and $\sum_{X3} [P(X_6|X_3) \times P(x_7|X_3)]$ can be computed directly.

3. *Step 2*: In the third step, the first belief can be evaluated:

$$\begin{aligned} & P(X_1|x_5 \wedge x_7) \\ \propto \; & P(X_1) \times \sum_{X_2} \left[P(X_2|X_1) \times P(x_5|X_2) \right] \\ & \times \sum_{X_3} \left[P(X_3|X_1) \times P(x_7|X_3) \right] \end{aligned} \tag{14.30}$$

4. *Step 3*: Then the two questions $P(X_2|x_5 \wedge x_7)$ and $P(X_3|x_5 \wedge x_7)$ can be solved:

$$\begin{aligned} & P(X_2|x_5 \wedge x_7) \\ \propto \; & P(x_5|X_2) \times \sum_{X_1} \left[\begin{array}{l} P(X_2|X_1) \\ \times \frac{P(X_1|x_5 \wedge x_7)}{\sum_{X_2} \left[P(X_2|X_1) \times P(x_5|X_2) \right]} \end{array} \right] \end{aligned} \tag{14.31}$$

$$\begin{aligned} & P(X_3|x_5 \wedge x_7) \\ \propto \; & P(x_7|X_3) \times \sum_{X_1} \left[\begin{array}{l} P(X_3|X_1) \\ \times \frac{P(X_1|x_5 \wedge x_7)}{\sum_{X3} \left[P(X_6|X_3) \times P(x_7|X_3) \right]} \end{array} \right] \end{aligned} \tag{14.32}$$

5. *Step 4*: The next two expressions, $P(X_4|x_5 \wedge x_7)$ and $P(X_6|x_5 \wedge x_7)$, can be deduced directly from the two previous ones as:

$$\begin{aligned} & P(X_4|x_5 \wedge x_7) \\ \propto \; & \sum_{X_2} \left[P(X_4|X_2) \times P(X_2|x_5 \wedge x_7) \right] \end{aligned} \tag{14.33}$$

$$\begin{aligned} & P(X_6|x_5 \wedge x_7) \\ \propto \; & \sum_{X_3} \left[P(X_6|X_3) \times P(X_3|x_5 \wedge x_7) \right] \end{aligned} \tag{14.34}$$

6. *Step 5*: Finally, the last two questions can be computed:

$$\begin{aligned} & P(X_8|x_5 \wedge x_7) \\ \propto \; & \sum_{X_6} \left[P(X_8|X_6) \times P(X_6|x_5 \wedge x_7) \right] \end{aligned} \tag{14.35}$$

$$\begin{aligned} & P(X_9|x_5 \wedge x_7) \\ \propto \; & \sum_{X_6} \left[P(X_9|X_6) \times P(X_6|x_5 \wedge x_7) \right] \end{aligned} \tag{14.36}$$

This order of computation may be interpreted as a message-passing algorithm in the Bayesian network (see Figure 14.2 below).

This algorithm was simultaneously and independently proposed by Judea

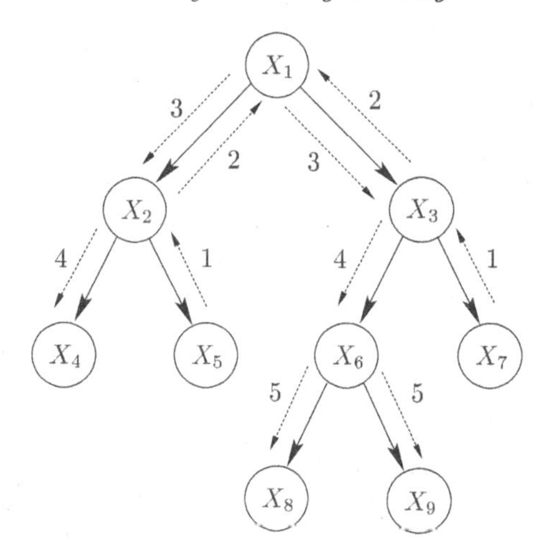

FIGURE 14.2: The order of computation described by steps 0 to 5 (above) may be interpreted as a message-passing algorithm in the Bayesian net.

Pearl [Pearl, 1988] under the name of *Belief Propagation*, and by Lauritzen and Spiegelhalter [Lauritzen and Spiegelhalter, 1988; Lauritzen, 1996] as the *Sum–Product* algorithm.

When the graph associated with the Bayesian network has no undirected cycles,[2] it is always possible to find this ordering, ensuring that each subexpression is evaluated once and only once.

On the other hand, when the graph of the Bayesian net has some undirected cycles the situation is trickier and such a clever ordering of the computation may not be found.

For instance, let us modify the above example by adding a dependency between X_2 and X_3. We then obtain the new decomposition:

$$
\begin{aligned}
& P(X_1 \wedge X_2 \wedge \cdots \wedge X_9) \\
= \; & P(X_1) \times P(X_2|X_1) \times P(X_3|X_2 \wedge X_1) \times P(X_4|X_2) \times P(X_5|X_2) \\
& \times P(X_6|X_3) \times P(X_7|X_3) \times P(X_8|X_6) \times P(X_9|X_6)
\end{aligned}
\tag{14.37}
$$

which corresponds to the graph of Figure 14.3 below.

Applying the simplification rules to the different questions, we obtain:

$$
\begin{aligned}
& P(X_1|x_5 \wedge x_7) \\
\propto \; & P(X_1) \times \sum_{X_2 \wedge X_3} \left[\begin{array}{l} P(X_2|X_1) \times P(X_3|X_2 \wedge X_1) \\ \times P(x_5|X_2) \times P(x_7|X_3) \end{array} \right]
\end{aligned}
\tag{14.38}
$$

[2]It is either a tree or a polytree.

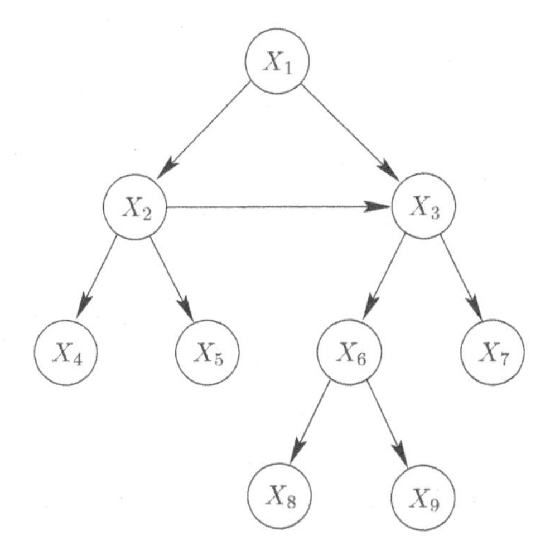

FIGURE 14.3: The Bayesian network corresponding to the joint distribution in Equation 14.37.

$$
\begin{aligned}
& P(X_2|x_5 \wedge x_7) \\
\propto \quad & P(x_5|X_2) \times \sum_{X_1 \wedge X_3} \left[\begin{array}{c} P(X_1) \times P(X_2|X_1) \\ \times P(X_3|X_2 \wedge X_1) \times P(x_7|X_3) \end{array} \right]
\end{aligned} \tag{14.39}
$$

$$
\begin{aligned}
& P(X_3|x_5 \wedge x_7) \\
\propto \quad & P(x_7|X_3) \times \sum_{X_1 \wedge X_2} \left[\begin{array}{c} P(X_1) \times P(X_2|X_1) \\ \times P(X_3|X_2 \wedge X_1) \times P(x_5|X_2) \end{array} \right]
\end{aligned} \tag{14.40}
$$

The four other cases are unchanged relative to these three (see Equations 14.33, 14.34, 14.35, and 14.36).

Obviously, the different elements appearing in these three expressions may not be neatly separated as in the previous case. The conjunction of variables $X_1 \wedge X_2 \wedge X_3$ must be considered as a whole: they form a new variable $A = X_1 \wedge X_2 \wedge X_3$. The decomposition (Equation 14.37) becomes:

$$
\begin{aligned}
& P(X_1 \wedge X_2 \wedge \cdots \wedge X_9) \\
\propto \quad & P(A) \times P(X_4|A) \times P(X_5|A) \times P(X_6|A) \times P(X_7|A) \\
& \times P(X_8|X_6) \times P(X_9|X_6)
\end{aligned} \tag{14.41}
$$

This corresponds to the graph in Figure 14.4 below, which again has a tree structure.

We have recreated the previous case, where the message-passing algorithms may be applied. However, this has not eliminated our troubles completely, because to compute $P(X_1|x_5 \wedge x_7)$, $P(X_2|x_5 \wedge x_7)$, and $P(X_3|x_5 \wedge x_7)$, we shall now require marginalization of the distribution $P(A|x_5 \wedge x_7)$:

$$
P(X_1|x_5 \wedge x_7) \propto \sum_{X_2 \wedge X_3} [P(A|x_5 \wedge x_7)] \tag{14.42}
$$

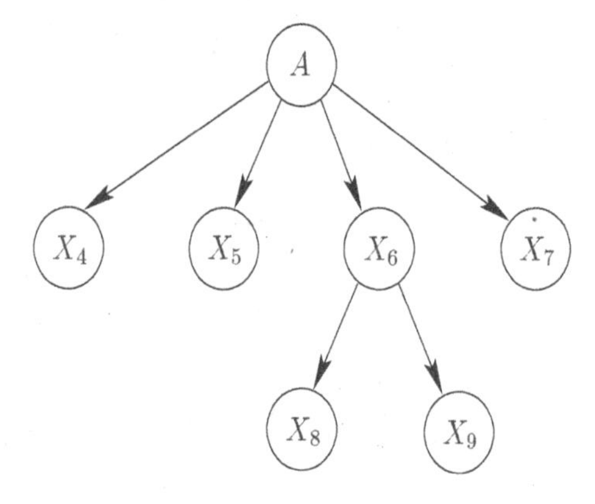

FIGURE 14.4: The Bayesian network resulting from the introduction of a new variable $A = X_1 \wedge X_2 \wedge X_3$ in the Bayesian network as shown in Figure 14.3.

$$P(X_2|x_5 \wedge x_7) \propto \sum_{X_1 \wedge X_3} [P(A|x_5 \wedge x_7)] \tag{14.43}$$

$$P(X_3|x_5 \wedge x_7) \propto \sum_{X_1 \wedge X_2} [P(A|x_5 \wedge x_7)] \tag{14.44}$$

Indeed, the computation of these sums may be very expensive.

The *junction tree* algorithm, also often called *JLO* after its inventors (Jensen, Lauritzen, and Olesen [Jensen et al., 1990]), searches for such a decomposition and implements the corresponding computation to solve $P(X_n|known)$.

The main idea of the junction tree algorithm is to convert the Bayesian network into a tree by clustering the nodes together. After building this tree of clusters, inference can be done efficiently by the single message-passing algorithm.

All the details of this popular algorithm may be found in the textbooks concerning graphical models cited previously (see Section 13.1.1) and also in Jensen's book [Jensen, 1996].

It is very important to note that this JLO algorithm may lead to very poor simplification, as, in some cases, the required marginalizations may be very expensive. The cost of these marginalizations grows exponentially with the number of variables in the conjunctions being considered.

Historically, another solution was proposed by Pearl [1988] to deal with graphs, with undirected cycles. It is called the *cut-set* algorithm. The principle of the cut-set algorithm is to break the cycles by fixing the values of some of the variables in these cycles and computing in turn the different questions for these different possible values.

14.2.1.4 Viterbi, max-product, and min-sum algorithms

When there are no $Free$ variables and if we are interested in the most probable value of *Searched*, the usual equation:

$$
\begin{aligned}
& P\left(Searched|known \wedge \delta \wedge \pi\right) \\
= \; & \frac{1}{Z} \times \sum_{Free} \left[P\left(L_1|\delta \wedge \pi\right) \times \prod_{k=2}^{K} \left[P\left(L_k|R_k \wedge \delta \wedge \pi\right) \right] \right]
\end{aligned} \tag{14.45}
$$

is transformed into:

$$
\begin{aligned}
& Max_{Searched} \left[P\left(Searched|known \wedge \delta \wedge \pi\right) \right] \\
= \; & Max_{Searched} \left[P\left(L_1|\delta \wedge \pi\right) \times \prod_{k=2}^{K} \left[P\left(L_k|R_k \wedge \delta \wedge \pi\right) \right] \right]
\end{aligned} \tag{14.46}
$$

The distributive law applies to the couple $\left(Max, \prod\right)$ in the same way as it applies to the couple $\left(\sum, \prod\right)$. Consequently, most of the previous simplifications are still valid with this new couple of operator.

The *sum-product* algorithm becomes the *max-product* algorithm, or more commonly, the *min-sum* algorithm, as it may be further transformed by operating on the inverse of the logarithm [MacKay, 2003].

It is also known as the *Viterbi* algorithm [Viterbi, 1967] and it is particularly used with hidden Markov models (HMMs) to find the most probable series of states that lead to the present state, knowing the past observations as stated in the Bayesian program in Equation 13.7.

14.2.1.5 Optimality considerations for exact inference

As previously mentioned, ordering the sums is NP-hard. As a consequence, all inference algorithms are doomed to select heuristics to obtain reasonable orderings. For optimality considerations, additional criteria concerning the computational cost of the inferred expression can be taken into account when selecting the heuristic.

For example, consider the Bayesian network in Figure 14.5, for which we are interested in constructing the target distribution $P(B \mid A)$. The d parameter represents the "depth" of the network.

The purpose of this example is to show different orderings of sums for the target distribution $P(B \mid A)$ depending on the chosen optimization criterion. The corresponding computational costs are quantified for each case.

First, we assume that the d parameter (the depth of the Bayesian network) is fixed to one. We will also assume that all variables of the network take values in a finite set with n elements.

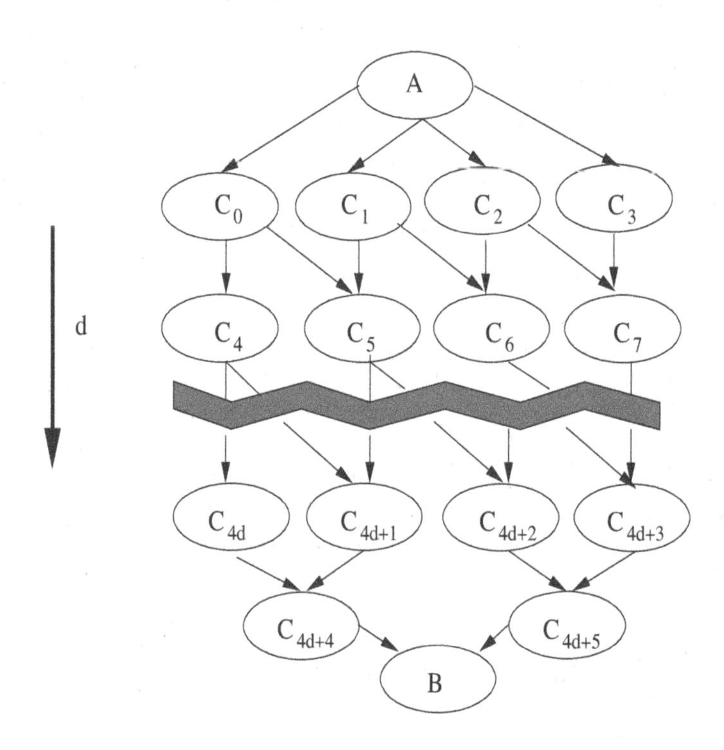

FIGURE 14.5: A Bayesian network from which $P(B \mid A)$ must be computed. In our example, two criteria concerning the computational cost are considered.

Using the "first compilation time minimization" criterion

Let us suppose that the time required to compile the target distribution $P(B \mid A)$ for a single evidence value a is the main issue. In this case, we will select the following ordering for sums:

$$P(B \mid a) = \sum_{C_3} \left\{ \begin{array}{l} P(C_3 \mid a) \times \\ \sum\limits_{C_2} \left\{ \begin{array}{l} P(C_2 \mid a) \times \\ \sum\limits_{C_7} \left\{ \begin{array}{l} P(C_7 \mid C_2 \wedge C_3) \times \\ \sum\limits_{C_1} \left\{ \begin{array}{l} P(C_1 \mid a) \times \\ \sum\limits_{C_6} \left\{ \begin{array}{l} P(C_6 \mid C_1 \wedge C_2) \times \\ \sum\limits_{C_9} \left\{ \begin{array}{l} P(C_9 \mid C_6 \wedge C_7) \times \\ \sum\limits_{C_0} \left\{ \begin{array}{l} P(C_0 \mid a) \times \\ \boxed{\sum\limits_{C_5} \left\{ \begin{array}{l} P(C_5 \mid C_0 \wedge C_1) \times \\ \sum\limits_{C_4} \left\{ \begin{array}{l} P(C_4 \mid C_0) \times \\ \sum\limits_{C_8} P(C_8 \mid C_4 \wedge C_5) \times P(B \mid C_8 \wedge C_9) \end{array} \right. \end{array} \right.} \end{array} \right. \end{array} \right. \end{array} \right. \end{array} \right. \end{array} \right. \end{array} \right. \end{array} \right. \tag{14.47}$$

Using this ordering, the number N_C of the arithmetic operations (additions and multiplications) required to compute $P(B \mid a)$ and the one number N_U required to update this table for a new evidence value of a' are respectively:

$$\begin{aligned} N_C &= 10n^5 + 6n^4 + 2n^3 + 2n^2 \\ N_U &= 4n^5 + 6n^4 + 2n^3 + 2n^2 \end{aligned} \tag{14.48}$$

Notice that the constant part of the sum (requiring no update when the evidence value of A changes) are quite deep in the nested loops.

Using the "update time minimization" criterion

Suppose now that we are interested in computing the target distribution

$P(B \mid a)$ in a program loop with a different observed evidence value of A at each iteration. In this case, minimizing the time required to update the corresponding probability table is the main issue, and we should use a different ordering of sums.

$$
\begin{aligned}
&P(B \mid a) = \\
&\sum_{C3} \left\{ \begin{array}{l}
P(C3 \mid a) \times \\
\sum_{C2} \left\{ \begin{array}{l}
P(C2 \mid a) \times \\
\sum_{C1} \left\{ \begin{array}{l}
P(C1 \mid a) \times \\
\sum_{C0} \left\{ \begin{array}{l}
P(C0 \mid a) \times \\
\boxed{\sum_{C7} \left\{ \begin{array}{l}
P(C7 \mid C2 \wedge C3) \times \\
\sum_{C6} \left\{ \begin{array}{l}
P(C6 \mid C1 \wedge C2) \times \\
\sum_{C9} \left\{ \begin{array}{l}
P(C9 \mid C6 \wedge C7) \times \\
\sum_{C5} \left\{ \begin{array}{l}
P(C5 \mid C0 \wedge C1) \times \\
\sum_{C4} \left\{ \begin{array}{l}
P(C4 \mid C0) \times \\
\sum_{C8} P(C8 \mid C4 \wedge C5) \times P(B \mid C8 \wedge C9)
\end{array} \right.
\end{array} \right.
\end{array} \right.
\end{array} \right.
\end{array} \right.}
\end{array} \right.
\end{array} \right.
\end{array} \right.
\end{array} \right.
\end{aligned}
\tag{14.49}
$$

Using this ordering, the number N_C of the arithmetic operations (additions and multiplications) required to compute $P(B \mid a)$ and the one number N_U required to update this table for a new evidence value of a' are respectively:

$$
\begin{aligned}
N_C &= 6n^6 + 8n^5 + 2n^4 + 2n^3 + 2n^2 \\
N_U &= 2n^5 + 2n^4 + 2n^3 + 2n^2
\end{aligned}
\tag{14.50}
$$

Notice that the constant part of the sum (requiring no update when the evidence value of A changes) is now much higher in the nested loops, which leads to fewer computations when changing the evidence A, but at a bigger initial cost.

Let us now assume that n is fixed to 2. Table 14.1 gives the number of arithmetic operations (addition and multiplication) required to compute

$P(B \mid a)$ (N_C) and to update it (N_U) as a function of the parameter d for the "first compilation time minimization" and "update time minimization" criteria.

d	Optimization criterion			
	Compilation		Update	
	N_C	N_U	N_C	N_U
1	**440**	248	696	**120**
2	**1112**	536	1592	**120**
3	**2072**	600	2488	**120**
6	**5016**	216	5176	**120**
10	**7064**	216	7224	**120**
100	**53144**	216	53304	**120**
300	**155544**	216	155704	**120**

TABLE 14.1: Computational Cost for Different Values of d When Using the "First Compilation Time Minimization" (Left) and "Update Time Minimization" (Right) Criteria (The results of the chosen optimization criterion are shown in bold.)

The file "chapter14/optimization.py" provides an example to change the optimization criterion.

```
question_S=model.ask(B,A,PL_OPTIMIZE_COMPILATION_TIME)
question_U=model.ask(B,A,PL_OPTIMIZE_UPDATE_TIME)
```

The optional key "PL_OPTIMIZE_COMPILATION_TIME" should be used for a single inference such as:

```
Va=plValues(A)
Va=0
pBka = question_S.instantiate(Va).compile()
```

while the key "PL_OPTIMIZE_UPDATE_TIME" should be in code such as:

```
for i in range(10):
        Va=i
        pBka = question_U.instantiate(Va).compile()
```

14.2.2 Approximate symbolic computation

Often, the exact symbolic simplification methods described previously are not sufficient. We must then use approximate symbolic simplifications that lead to mathematical expressions that approach the values of $P(Searched|known \wedge \delta \wedge \pi)$.

14.2.2.1 Variational methods

The key to variational methods is to convert a probabilistic inference problem into an optimization problem. In this way, the standard tools of constrained optimization can be used to solve the inference problem. The idea is to replace a joint distribution $P(X) = P(X_1 \wedge X_2 \wedge \cdots \wedge X_N)$ (represented by an acyclic graph in the case of a Bayesian net) by an approximation $Q(X)$, and to compute the Kullback–Leibler divergence between the two distributions.

The variational free energy (or Kullback–Leibler distance) is defined as:

$$F(Q, P) = \sum_{X} \left[Q(X) \times log\left(\frac{Q(X)}{P(X)}\right) - log(Z) \right] \tag{14.51}$$

This distance is minimized when $P(X) = Q(X)$.

Of course, minimizing F is as difficult as the original inference problem. However, by considering a different family of $Q(X)$, we obtain a different approximation of F and as a consequence, different variational methods.

For example, if one restricts oneself to the family of factorized independent distributions:

$$Q(X) = \prod_{i=1}^{I} \left[Q_i(X_i) \right] \tag{14.52}$$

the variational method boils down to the *mean field* approximation. Minimizing $F(Q, P)$ is greatly simplified using the acyclic graph structure of $P(X)$.

These approaches have been used successfully in a considerable number of specific models where exact inference becomes intractable, that is, when the graph is highly connected. A general introduction to variational methods may be found in introductory texts by Jordan [1999], Jordan and Weiss [2002], and Jaakkola and Jordan [1999].

14.3 Numerical computation

A major problem in probabilistic modeling with many variables is the computational complexity involved in typical calculations for inference. For sparsely connected probabilistic networks, this problem has been solved by the introduction of efficient algorithms for exact inference. However, in large or densely connected models, exact inference is often intractable. This means that the computation time increases exponentially with the problem size. In such cases, approximate inference techniques are alternative solutions that make the problem more tractable.

This section will briefly review the main approximate probabilistic inference techniques. These techniques are presented in two classes of approaches.

The first class groups the sampling-based techniques, while the second concerns the variational methods.

Sampling-based (or Monte Carlo) approaches for approximate Bayesian inference group together several stochastic simulation techniques that can be applied to solve optimization and numerical integration problems in large-dimensional spaces. Since their introduction in the physics literature in the 1950s, Monte Carlo methods have been at the center of the recent Bayesian revolution in applied statistics and related fields [Geweke, 1996]. They are applied in numerous other fields such as, for instance, image synthesis [Keller, 1996], CAD modeling [Mekhnacha et al., 2001], and mobile robotics [Dellaert et al., 1999; Fox et al., 1999].

The aim of this section is to present some of the most popular sampling-based techniques and their use in the problem of approximate Bayesian inference.

14.3.1 Sampling high-dimensional distributions

Sampling distributions is a central issue in approximate inference. Sampling is required when: (i) using Monte Carlo methods for numerical estimation of integrals (see Section 14.3.7), (ii) sampling a posteriori distributions.

The problem of drawing samples from a given distribution is still a challenging one, especially in high-dimensional spaces.

If we have an acceptable uniform random generator at our disposal, it is possible in some simple cases to use a "transformation function" to sample a given nonuniform parametric distribution [Rubinstein, 1981]. One of the more important and well-known transformation functions is the "Box–Muller transformation" [Box and Muller, 1958]. It permits generation of a set of random numbers drawn from a one-dimensional normal distribution using another set of random numbers drawn from a uniform random generator.

Therefore, direct sampling techniques are available for some standard simple distributions. However, for more complicated cases, indirect sampling methods such as "forward sampling", "importance sampling," "rejection sampling," and "Markov Chain Monte Carlo" (MCMC) are alternative solutions.

In this section we present some of the most popular variants of these algorithms. Two excellent starting points on Monte Carlo methods are the tutorials by Neal [1993] and MacKay [1996].

14.3.2 Forward sampling

Using the "forward sampling" algorithm to sample a joint distribution represented as a Bayesian network consists of drawing values of the variables one by one, starting with the root nodes and continuing in the implicit order defined by the Bayesian network graph. In other words, first each root variable X_i is drawn from $P(X_i)$. Then each nonroot variable X_j is drawn, in the ancestral ordering, from $P(X_j \mid pa(X_j))$, where $pa(X_j)$ are the parents of X_j, for which values have been already drawn.

Suppose for example that we are interested in drawing a point from the distribution $P(X_1 X_2) = P(X_1)\, P(X_2 \mid X_1)$ where $P(X_1)$ and $P(X_2 \mid X_1)$ are simple distributions for which direct sampling methods are available. Drawing a point $x^{(i)} = (x_1^{(i)}, x_2^{(i)})$ from $P(X_1 X_2)$ using forward sampling consists of: (i) drawing $x_1^{(i)}$ from $P(X_1)$, then (ii) drawing $x_2^{(i)}$ from $P(X_2 \mid X_1 = x_1^{(i)})$.

This sampling scheme may be used when no evidence values are available or when evidence concerns only the conditioning (right side) variables.

When evidence on the conditioned (left side) variables is available, forward sampling may also be used by introducing rejection of samples that are not consistent with the evidence. In this case, this algorithm may be very inefficient (have a high rejection rate) for evidence values with small probabilities of occurrence. Moreover, applying this algorithm is impossible when evidence concerns continuous variables.

14.3.3 Importance sampling

Suppose we are interested in sampling a distribution $P(X)$ for which no direct sampling method is available and that we are able to evaluate this distribution for each point x of the state space.

Suppose also that we have a simpler distribution $Q(X)$ (called the "proposal distribution") that we can evaluate for each point x and for which a direct sampling method is available.

Using "importance sampling" to sample $P(X)$ consists of generating n pairs $\{(w_i, x_q^{(i)})\}_{i=1}^n$ where $\{x_q^{(i)}\}_{i=1}^n$ are drawn from $Q(X)$ and $w_i = P(x_q^{(i)})/Q(x_q^{(i)})$.

The reader is referred to Yuan and Druzdzel [2006] for a more complete presentation of importance sampling techniques. It also presents a more sophisticated algorithm for importance sampling in Bayesian networks.

14.3.4 Rejection sampling

Suppose, as in the previous case, that we are interested in sampling a distribution $P(X)$ for which no direct sampling method is available and that we can evaluate this distribution for each point x of the state space.

Suppose also that we have a simpler distribution $Q(X)$ that we can evaluate for each point x_i and for which a direct sampling method is available, respecting the constraint: $\exists c,\ \forall x,\ c \times Q(x) > P(x)$.

Using rejection sampling to draw a point of $P(X)$ consists of drawing a point x_q from $Q(X)$ and accepting it with a probability of $\dfrac{c \times Q(x_q)}{P(x_q)}$:

1. draw a candidate point x_q from $Q(X)$,

2. evaluate $c \times Q(x_q)$,
3. generate a uniform random value u in $[0, c \times Q(x_q)]$,
4. if $P(x) > u$ then the point x_q is accepted. Otherwise, the point is rejected.

It is clear that this rejection sampling is efficient if the distribution $Q(X)$ is a good approximation of $P(X)$. Otherwise, the rejection rate will be very important.

14.3.5 Gibbs sampling

Gibbs sampling is an example of "Markov chain Monte Carlo" (MCMC) sampling techniques. MCMC methods use a Markovian process in which a sequence of states $\{x^{(t)}\}$ is generated. Each new state $x^{(t)}$ depends on the previous one $x^{(t-1)}$. These algorithms are based on the theory of Markov chains [Neal, 1993].

The Gibbs sampling method came into prominence with the work of Geman and Geman [1984] and Smith and Roberts [1993]. It is a method for sampling from distributions over at least two dimensions. It is assumed that while $P(X)$ is too complex to draw samples from directly, its conditional distributions $P(X_i|\{X_j\}_{i \neq j})$ are tractable to work with. In the general case of a system of N variables, a single iteration involves sampling one parameter at a time:

$$\begin{array}{rcl} x_1^{(t+1)} & \sim & P(X_1 \mid x_2^{(t)} x_3^{(t)} \cdots x_N^{(t)}) \\ x_2^{(t+1)} & \sim & P(X_2 \mid x_1^{(t)} x_3^{(t)} \cdots x_N^{(t)}) \\ & \vdots & \\ x_N^{(t+1)} & \sim & P(X_N \mid x_1^{(t)} x_2^{(t)} x_3^{(t)} \cdots x_{N-1}^{(t)}). \end{array}$$

14.3.6 Metropolis algorithm

The Metropolis algorithm [Metropolis et al., 1953] is another example of MCMC methods. It is one of the more widely used techniques for sampling high-dimensional probability distributions and densities. This algorithm only requires a way to evaluate the sampled expression for each point of the space.

The main idea of the Metropolis algorithm is to use a proposal distribution $Q(x_c, x^{(t)})$, which depends on the current state $x^{(t)}$. This proposal distribution can be a simple distribution (a normal distribution having $x^{(t)}$ as a mean value, for example).

Suppose the current state is $x^{(t)}$. A candidate x_c is generated from $Q(X_c, x^{(t)})$. To accept or reject this candidate we must compute

$$a = \frac{P(x_c)Q(x_c, x^{(t)})}{P(x^{(t)})Q(x^{(t)}, x_c)}.$$

If $a > 1$ then x_c is accepted, otherwise it is accepted with probability a. If x_c is accepted, we set $x^{(t+1)} = x_c$. If x_c is rejected, then we set $x^{(t+1)} = x^{(t)}$.

If the proposal distribution $Q(X_c, x^{(t)})$ is symmetrical (a normal distribution having $x^{(t)}$ as a mean value, for example), then $\frac{Q(x_c, x^{(t)})}{Q(x^{(t)}, x_c)} = 1$ and we obtain: $a = P(x_c)/P(x^{(t)})$.

One drawback of MCMC methods is that we must in general wait for the chain to reach equilibrium. This can take a long time and it is sometimes difficult to tell when it happens.

14.3.7 Numerical estimation of high-dimensional integrals

Integral (sums) calculus is a central issue in Bayesian inference. Unfortunately, analytic methods for integral evaluation seem very limited in real-world applications, where integrands may have complex shapes and integration spaces may have very high dimensionality. Furthermore, these techniques are not useful for general purpose inference, where the distributions may be simple probability tables.

In this section, we will generally assume that X is a k-dimensional vector with real or discrete components, or a mixture of real and discrete components. We will also use the symbol $\int$ as a generalized integration operator for both real integrals (over real components) and sums (over discrete components).

The aim of Monte Carlo methods for numerical integration is to approximate efficiently the k-dimensional (where k can be very large) integral

$$I = \int P(X)g(X)\, d^k X. \tag{14.53}$$

Assuming that we cannot visit every location x in the state (integration) space, the simplest solution we can imagine to estimate the integral 14.53 is to uniformly sample the integration space and then estimate I by $\hat{I}$:

$$\hat{I} = \frac{1}{N} \sum_i P(x^{(i)}) g(x^{(i)}).$$

where $\{x^{(i)}\}_{i=1}^N$ are randomly drawn in the integration space.

Because high-dimensional probability distributions are often concentrated on a small region T of the state (integration) space, known as its "typical set" [MacKay, 1996, 2003], the number N of points drawn uniformly for the state (integration) space must be sufficiently large to cover the region T containing most of the probability mass of $P(X)$.

Instead of exploring the integration space uniformly, Monte Carlo methods try to use the information provided by the distribution $P(X)$ to explore this space more efficiently. The main idea of these techniques is to approximate the integral 14.53 by estimating the expectation of the function $g(X)$ under the distribution $P(X)$:

$$I = \int P(X)g(X)\, d^k X = \langle g(X) \rangle.$$

Clearly, if we are able to generate a set of points (vectors) $\{x^{(i)}\}_{i=1}^N$ from $P(X)$, the expectation of $\hat{I}$ is I. As the number of samples N increases, the variance of the estimator $\hat{I}$ will decrease as $\frac{\sigma^2}{N}$ where σ^2 is the variance of g:

$$\sigma^2 = \int P(X)(g(X) - \langle g(X) \rangle)^2\, d^k X.$$

Suppose we are able to obtain a set of samples $\{x^{(i)}\}_{i=1}^N$ (k-vectors) from the distribution $P(X)$. We can use these samples to find the estimator

$$\hat{I} = \frac{1}{N} \sum_i g(x^{(i)}). \tag{14.54}$$

This Monte Carlo method assumes the capacity to sample the distribution $P(X)$ efficiently. It is called "perfect Monte Carlo integration". A a good survey of Monte Carlo sampling techniques can be found in Neal [1993].

14.4 Approximate inference in ProBT

Exact inference is often intractable for large and/or densely connected models. This is especially the case when the variables involved take values in huge sets of states. Moreover, exact inference is impossible for arbitrary models involving continuous variables.[3]

ProBT uses a set of sampling-based techniques to propose three levels of approximation. These techniques are used for variables taking values in finite sets as well as for continuous ones.

14.4.1 Approximation in computing marginalization

The first level of approximation proposed in ProBT concerns the problem of numerically estimating integrals. When evaluating an expression $E(X)$ for a given point of the target space, ProBT allows the computation of a more or less accurate numerical estimation of the integrals (sums) involved in this expression using perfect Monte Carlo methods (see Section 14.3.7). When the

[3]Exact inference using continuous variables is only possible for models allowing analytical calculation.

evaluated expression involves a marginalization over a conjunction of numerous variables that possibly take values in a huge (or infinite for continuous variables) set of states, its exact evaluation may have a very high computational cost. When using this approximation scheme, the expression is said to be a "Monte Carlo expression."

Let us take as an example the following decomposition:

$$\begin{aligned} & P(A \wedge B \wedge C \wedge D \wedge E \wedge F) \\ = \; & P(A)\, P(B)\, P(C|A \wedge B)\, P(D||B) \\ & P(E|C)\, P(F|D) \end{aligned} \tag{14.55}$$

The problem is to find an efficient way to approximate $P(a\ b \mid e\ f)$ using a Monte Carlo estimation of the integral:

$$I = \int P(C \mid a\ b)\ P(D \mid b)\ P(e \mid C)\ P(f \mid D)\ dC\ dD. \tag{14.56}$$

Is it more efficient to use:

$$\hat{I}_1 = \frac{1}{N} \sum_i P(e \mid c^{(i)})\ P(f \mid d^{(i)}), \tag{14.57}$$

where $\{c^{(i)}\}_{i=1}^{N}$ and $\{d^{(i)}\}_{i=1}^{N}$ are generated from $P(C \mid a\ b)$ and $P(D \mid b)$, or:

$$\hat{I}_2 = \Big(\frac{1}{N_C} \sum_j P(e \mid c^{(j)})\Big)\ \Big(\frac{1}{N_D} \sum_k P(f \mid d^{(k)})\Big), \tag{14.58}$$

where $\{c^{(j)}\}_{j=1}^{N_C}$ and $\{d^{(k)}\}_{k=1}^{N_D}$ are generated from $P(C \mid a\ b)$ and $P(D \mid b)$?

More generally, is it more efficient to use the sum/product evaluation tree built using the SRA algorithm (see Section 14.2.1.2) to estimate the integrals (sums) using Monte Carlo approximation?

To answer this question, we must consider error propagation in the estimation of intermediate terms and the convergence of this estimation. In ProBT, we use Equation 14.57 rather than 14.58 to estimate integrals (sums). In other words no elimination ordering is done. This choice is motivated as follows:

- It is more efficient to use the estimator in Equation 14.57 to avoid error propagation.
- Monte Carlo methods for integral estimation perform better in high-dimensional spaces [Neal, 1993].

ProBT allows two ways to control the cost/accuracy of the estimate.

The first way is to specify the number of sample points to be used for estimating the integral. This allows the user to express constraints on the computational cost and on the required accuracy of the estimate. This parameter (i.e., the number of sample points) is also used internally in the MCSEM algorithm (see Section 14.4.3) for a posteriori distributions optimization.

The second way to control the cost/accuracy of the estimate is to provide a convergence threshold. The convergence of the estimate is supposed to be reached when adding sampling points does not sensibly modify the value of the estimate. This is accomplished by checking the convergence criterion at each incremental estimation of the integral as follows. Starting with an initial number n_0 of sampling points, an estimate $\hat{E}_{n_0}(x)$ is computed. For each step s, the number of sampling points is increased according to a given scheduling scheme and the estimate $\hat{E}_{n_{s-1}}(x)$ is updated to give the estimate $\hat{E}_{n_s}(x)$. Then, given a threshold value ϵ_{estim}, the convergence criterion is checked. This convergence criterion is defined as:

$$\frac{|\hat{E}_{n_s}(x) - \hat{E}_{n_{s-1}}(x)|}{\hat{E}_{n_{s-1}}(x)} < \epsilon_{estim}.$$

14.4.2 Approximation in sampling distributions

ProBT implements a drawing method for each probability distribution (or density function), whether or not it is a standard built-in distribution or an inferred expression.

ProBT implements standard direct sampling methods for simple distributions such as normal (with one or many dimensions), Poisson, gamma, and so on.

For inferred expressions (exact or approximate), a direct sampling is possible if we construct an explicit numerical representation of the expression by compiling it. Compilation may be exhaustive, to obtain a corresponding probability table, or approximate, to obtain a corresponding MRBT[4] [Bessière, 2002]. In both cases, a direct drawing method is available. Unfortunately, this compilation process is often very expensive and seldom possible for high-dimensional distributions.

An alternative to compiling an expression before sampling it is to use an indirect sampling method that does not require a global numerical construction of the target distribution. The idea is to apply an indirect sampling method on the expression to be sampled. The following two cases must be considered.

14.4.2.1 Expressions with no evidence on the conditioned variables

For expressions containing no evidence or containing evidence exclusively on the conditioning variables, the simple (without rejection) forward sampling algorithm (see Section 14.3.2) is used. Suppose we have a given expression

$$E(X_1X_2) = P(X_1X_2 \mid e) = P(X_1 \mid e)\ P(X_2 \mid X_1),$$

where $P(X_1 \mid e)$ is a given probability table and $P(X_2 \mid X_1)$ is a normal distribution having the value of X_1 as mean and a fixed value as variance.

[4] Multi Resolution Binary Tree.

Drawing a point $x^{(i)} = (x_1^{(i)}, x_2^{(i)})$ using forward sampling consists of: (i) drawing $x_1^{(i)}$ from $P(X_1 \mid e)$ using the corresponding repartition function, (ii) drawing $x_2^{(i)}$ from the normal distribution $P(X_2 \mid X_1 = x_1^{(i)})$ using the Box–Muller algorithm.

The same method is used for expressions involving sums (integrals) and containing no evidence values in the left side of all components (i.e., simple marginalization expressions). This is done by drawing from the corresponding joint distribution and then simply projecting the drawn point on the target space. For example, suppose we have the expression

$$E(X) = P(X \mid e) \propto \int P(Y \mid e)\ P(X \mid Y)\ dY.$$

To draw from $E(X)$, we must draw $P(X\ Y \mid e)$ from the joint distribution using the forward sampling method, then take the X component as a sampling point of $E(X)$.

14.4.2.2 Expressions with evidence on the conditioned variables

Suppose now that we have the task of sampling the expression

$$E(X_1X_2) = P(X_1X_2 \mid e) \propto P(X_1)\ P(e \mid X_1)\ P(X_2 \mid X_1).$$

This expression contains evidence on the conditioned variable E. Using forward sampling with rejection in this case consists of drawing points from $P(X_1X_2\ E)$ using the standard forward sampling above, then rejecting points that are not coherent with the evidence (i.e., points where $e^{(i)} \neq e$). This algorithm may be very inefficient (high rejection rate), especially for evidence values having small probability of occurrence (small value of $P(e)$) and/or for continuous variables (i.e., when E is continuous).

Therefore, for expressions containing evidence on the conditioned variables of at least one component, ProBT uses the Metropolis algorithm (see Section 14.3.6).

14.4.3 Approximation in computing MAP

Searching a maximum a posteriori (MAP) solution for a given inference problem consists of maximizing the corresponding target distribution $P(X \mid E = e)$ (i.e., finding $x^* = \arg\max_X P(X \mid E = e)$), where e is a given evidence value.

X may be a conjunction of numerous variables that possibly take values in a huge (or infinite for continuous variables) set of states. Moreover, evaluating $P(X \mid E = e)$ for a given value x of the space may require evaluating an integral (or sum) over numerous variables. In these cases, evaluating the target distribution $P(X \mid E = e)$ for all possible states of the space is intractable and an appropriate numeric optimization technique must be used.

MAP is known to be a very hard problem. Its complexity has been investigated in Park [2002] and some approximation methods have been proposed for discrete Bayesian networks. However, we think that the continuous case is harder and needs more adapted algorithms.

For general purpose Bayesian inference problems, the optimization method to be used must satisfy a set of criteria in relation to the shape and nature of the objective function (target distribution) to optimize. The method must:

1. be global, because the distribution to optimize (objective function) is often multimodal,

2. allow multiprecision evaluation of expressions requiring integral (sums) computing; estimation with high accuracy may require long computation times,

3. allow parallel implementation to improve efficiency.

The resolution method used in ProBT to solve this double integration/optimization problem is based on an adaptive genetic algorithm. The accuracy of integral numerical estimation is controlled by the optimization process to reduce computation time.

Genetic algorithms (GAs) are stochastic optimization techniques inspired by the biological evolution of species. Since their introduction by Holland [Holland, 1975] in the 1970s, these techniques have been used for numerous global optimization problems, thanks to their ease of implementation and their relative independence of application fields. They are widely used in a large variety of domains including artificial intelligence and robotics [Mazer et al., 1998].

Biological and mathematical motivations of genetic algorithms and their principles are not discussed in this chapter. We only discuss the practical problems we face when using standard genetic algorithms in Bayesian inference. We give the required improvements and the corresponding algorithms.

In the following, we use $E(X)$ to denote the probability expression corresponding to the target distribution to be optimized. This expression is used as an evaluation function in our genetic algorithm. $E(X)$ may contain integrals (sums) to be evaluated exactly if the expression is an exact one, or approximately (using Monte Carlo) if the expression is an approximate one.

14.4.3.1 Narrowness of the objective function: Constraint relaxation

The objective function $E(X)$ may have a narrow support (the region where the value is not null) for very constrained problems. The initialization of the population with random individuals from the search space may give null values of the function $E(X)$ for most individuals. This will make the evolution of the algorithm very slow and its behavior will be similar to random exploration.

To deal with this problem, a concept inspired from classical simulated

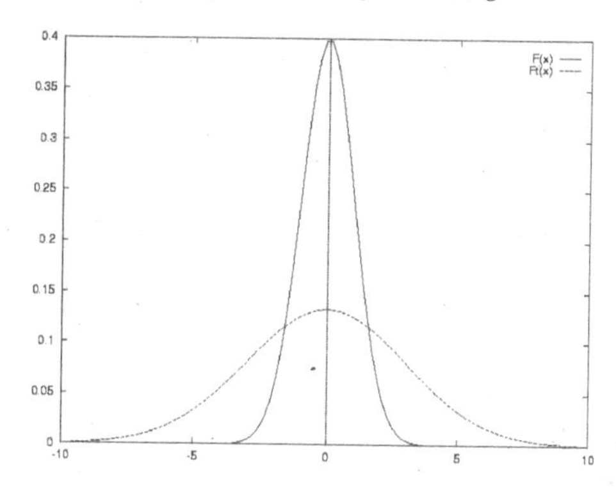

FIGURE 14.6: A normal (Gaussian) distribution at different temperature values.

annealing algorithms [Corana et al., 1987] consists of introducing a notion of "temperature." The principle is to first widen the support of the function by changing the original function to obtain nonnull values even for configurations that are not permitted (i.e., with probability zero). To do so, we introduce an additional parameter T (for temperature) in the objective function $E(X)$. Our goal is to obtain another function $E^T(X)$ that is smoother and has wider support, with

$$\lim_{T \to 0} E^T(X) = E(X).$$

To widen the support of $E(X)$, all elementary terms (distributions) of $E(X)$ are widened.

Because the main idea is to perform successive optimization cycles for successive values of the temperature T, we must respect the following additional condition:

$$\forall x_1, x_2 \in \mathcal{S}, \quad \forall t_1, t_2 \in I\!R^+, \qquad E^{t_1}(x_1) \leq E^{t_1}(x_2) \Rightarrow E^{t_2}(x_1) \leq E^{t_2}(x_2).$$

To do so, all elementary distributions must accept an additional temperature parameter and must themselves satisfy the following condition:

$$\forall x_1, x_2 \in \mathcal{S}, \quad \forall t_1, t_2 \in I\!R^+, \qquad f^{t_1}(x_1) \leq f^{t_1}(x_2) \Rightarrow f^{t_2}(x_1) \leq f^{t_2}(x_2).$$

For example, for a normal (Gaussian) distribution (see Figure 14.6) we have:

$$f(x) = \frac{1}{\sqrt{2\pi}\sigma} e^{-\frac{1}{2}\frac{(x-\mu)^2}{\sigma^2}}, \quad \text{and} \quad f^T(x) = \frac{1}{\sqrt{2\pi}\sigma(1+T)} e^{-\frac{1}{2}\frac{(x-\mu)^2}{[\sigma(1+T)]^2}}.$$

For nonparametric distributions (probability tables for example), this additional parameter may be simply ignored.

14.4.3.2 Accuracy of the estimates: Multiprecision computing

When solving problems involving approximate expressions (i.e., Monte Carlo numerical estimation of integrals), the second problem we face is that only an approximation $\hat{E}(X)$ of $E(X)$ is available, of unknown accuracy.

This accuracy depends on the number of points N used for the estimation: more points mean more accurate estimations. However, using a large number of points to obtain sufficient accuracy in the whole optimization process may be very expensive in computation time and is seldom possible for complicated problems. The main idea we propose to solve this problem is to introduce N as an additional parameter to define a new function $\hat{E}_N(X)$.

The main assumption in using this additional parameter is that the final population of a GA initialized and run for some cycles with $\hat{E}_{N_1}(X)$ as its evaluation function is a good initialization for another GA having $\hat{E}_{N_2}(X)$ as evaluation function with $N_2 > N_1$.

14.4.3.3 General optimization algorithm

In the following, we label the evaluation function (the objective function) by the temperature T and the number N of points used for estimation. It is denoted by $E_N^T(X)$.

Our optimization algorithm may be described by the following three phases:

1. Initialization and initial temperature determination.
2. Reduction of temperature to recreate the original objective function.
3. Augmentation of the number of points to increase the accuracy of the estimates.

14.4.3.4 Initialization

The population of the GA is initialized at random from the search space. To minimize computing time in this initialization phase, we use a small number N_0 of points to estimate integrals. We propose the following algorithm as an automatic initialization procedure for the initial temperature T_0, able to adapt to the complexity of the problem.

```
INITIALIZATION()
  FOR each population[i] DO
      REPEAT
              population[i] = random(Space)
              value[i] = E^T_{N0}(population[i])
              if (value[i] == 0.0)
                T = T + ΔT
      UNTIL (value[i]> 0.0)
  END FOR
  Reevaluate_population()
```

where ΔT is a small increment value.

This phase of the algorithm is schematized in Figure 14.7.

14.4.3.5 Temperature reduction

To obtain the original objective function ($T = 0.0$), a possible scheduling procedure consists of multiplying the temperature, after running the GA for a given number of cycles nc_1, by a factor α ($0 < \alpha < 1$). A small value for α may cause the divergence of the algorithm, while a value too close to 1.0 may considerably increase the computation time. In ProBT, the value of α has been experimentally fixed to 0.8 as the default; however, it can be fixed by the user. We summarize the algorithm as follows:

```
TEMPERATURE_REDUCTION()
   WHILE (T > T_ϵ) DO
            FOR i=1 TO nc_1 DO
                  Run_GA()
            END FOR
            T = T * α
            Reevaluate_population()
   END WHILE
   T = 0.0
   Reevaluate_population()
```

where T_ϵ is a small threshold value.

This phase of the algorithm is schematized in Figure 14.8.

14.4.3.6 Increasing the number of points N

At the end of the temperature reduction phase, the population may contain several possible solutions for the problem. To decide between these solutions, we must increase the accuracy of the estimates. One approach is to increase N, after running the GA for a given number of cycles nc_2, by a factor β ($\beta > 1$) so that the variance of the estimate is divided by β:

$$Var(E^0_{\beta * N}(X)) = \frac{1}{\beta} Var(E^0_N(X)).$$

We can describe this phase by the following algorithm.

```
NUMBER_OF_POINTS_INCREASING()
   WHILE (N < N_max) DO
            FOR i=1 TO nc_2 DO
                  Run_GA()
            END FOR
            N = N * β
            Reevaluate_population()
   END WHILE
```

where N_{max} is the number of points that allow convergence of the estimates $\hat{E}_N^0(X)$ for all individuals of the population.

This phase of the algorithm is schematized in Figure 14.9.

ProBT also provides an anytime version of the MCSEM algorithm. In this version, the user is allowed to fix the maximum number of evaluations of the objective function or the maximum time to be used to maximize it.

A preliminary implementation of the MCSEM algorithm and its use in high-dimensional inference problems has been presented in Mekhnacha et al. [2001, 2000] in which this algorithm is used as a resolution module in a probabilistic CAD system.

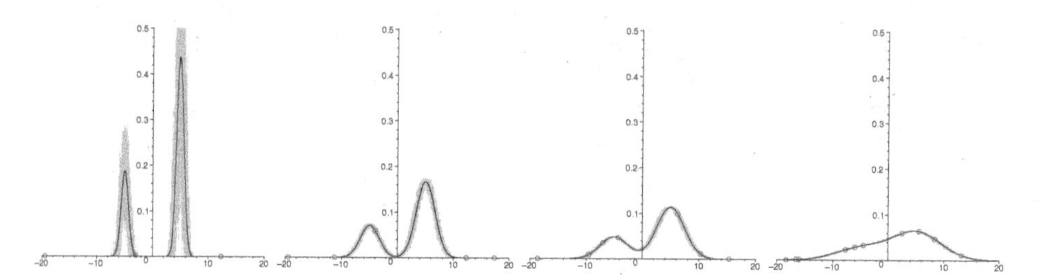

FIGURE 14.7: The initialization phase of the MCSEM algorithm. In black, the theoretical distribution to maximize and in gray, the estimated one using Monte Carlo numerical integration. From left to right, the T (temperature) parameter is increased starting from zero (i.e., the initial distribution).

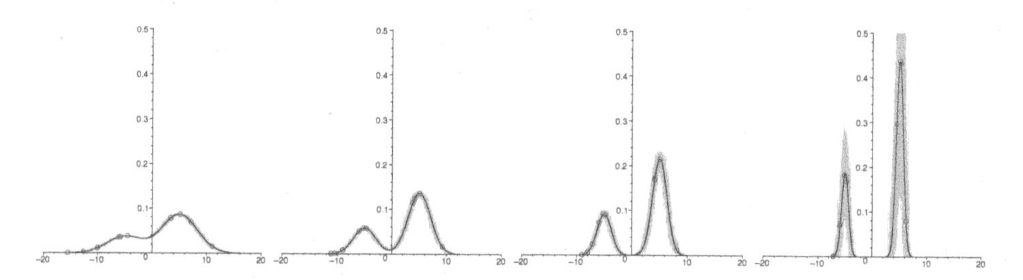

FIGURE 14.8: The "temperature reduction" phase of the MCSEM algorithm. In black, the theoretical distribution to maximize and in gray, the estimated one using Monte Carlo numerical integration. From left to right, the T (temperature) parameter is decreased to obtain the original distribution (i.e., $T = 0.0$).

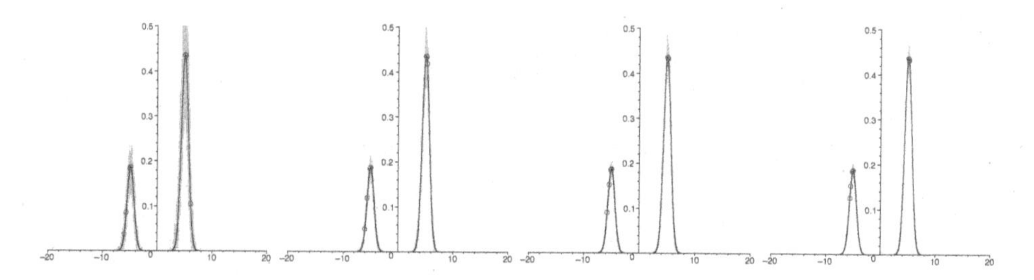

FIGURE 14.9: The "increasing number of points" phase of the MCSEM algorithm. In black, the theoretical distribution to maximize and in gray, the estimated one using Monte Carlo numerical integration. From left to right, the parameter N (number of sampling points used to estimate the integral) is increased to obtain increasingly accurate estimations of the distribution.

Chapter 15

Bayesian Learning Revisited

> The scientist shall organize; science is done with facts as a house with stones, but an accumulation of facts is no more a science than a heap of stones is a house.[1]
>
> *La Science et l'Hypothèse*
> Henri Poincaré [1902]

In Chapter 4 we have seen how data are used to transform a "specification" into a "description": the free parameters of the distributions are instantiated with the data, making the joint distribution computable for any value of the variables. This identification process may be considered as a learning mechanism allowing the data to shape the description before any inferences could be made. In this chapter, we consider learning problems in more detail and show how some of them may be expressed as special instances of Bayesian programs.

[1] Le savant doit ordonner ; on fait la science avec des faits comme une maison avec des pierres ; mais une accumulation de faits n'est pas plus une science qu'un tas de pierres n'est une maison.

15.1 Parameter identification

15.1.1 Problem statement

Let's recall the general form of a Bayesian program as it is presented in Chapter 2:

$$Program \begin{cases} Description. \begin{cases} Specification(\pi) \begin{cases} Variables \\ Decomposition \\ Forms \end{cases} \\ Identification\ (based\ on\ \delta) \end{cases} \\ Question \end{cases}$$

The identification part is based on a data set δ. For example, in Chapter 4 we used data to instantiate $P(Vrot \mid Dir \wedge Prox \wedge \pi \wedge \delta)$. The data are obtained either by driving the robot to perform a "pushing" (δ_{push}) or a "following" (δ_{follow}) behavior. As a result, we obtained with the same specifications π two distinct descriptions depending on the data set $\delta_{behavior}$.

$$Pr \begin{cases} Ds \begin{cases} Sp(\pi) \begin{cases} Va: \\ Vrot, Dir, Prox \\ Dc: \\ \begin{cases} P(Vrot \wedge Dir \wedge Prox \wedge \pi) \\ = P(Dir \wedge Prox \mid \pi) \\ \times P(Vrot \mid Dir \wedge Prox \wedge \pi) \end{cases} \\ Fo: \\ \begin{cases} P(Dir \wedge Prox \wedge \pi) = Uniform \\ P(Vrot \mid Dir \wedge Prox \wedge \pi) = \\ B(\mu = f(\delta)), \sigma = g(\delta)) \end{cases} \end{cases} \\ Id(\delta_{behavior}) : \mu_{dir,prox} = f(\delta_{behavior}), \sigma_{dir,prox} = g(\delta_{behavior}) \end{cases} \\ Qu: \\ P(Vrot \mid Dir \wedge Prox \wedge \pi \wedge \delta_{behavior}) \end{cases} \tag{15.1}$$

For this particular identification, $k = card(Dir) \times card(Prox)$ probability distributions $P(Vrot)_{1,\ldots,k}$ are computed from the set of observations $\delta_{behavior} = \{vrot_i, dir_i, prox_i\} : i \in \{1, l\}$. These distributions are indexed by the values dir and $prox$. The parameters of each distribution $\mu_{dir,prox}$ and $\sigma_{dir,prox}$ are obtained by pruning the set $\{vrot_i, dir_i, prox_i\} : i \in \{1, \ldots, l\}$ to obtain a subset $\{vrot_j, dir_j, prox_j\}$ with $dir_j = dir$ and $prox_j = prox$ and by computing the experimental mean and standard deviation:

$$\mu_{dir,prox} = E(vrot_j)$$
$$\sigma_{dir,prox} = E((vrot_j - mu_{dir,prox})^2)$$

This shows that multiple descriptions may be obtained with multiple data sets, however, many hypotheses remain hidden in that implementation.

15.1.2 Bayesian parametric estimation

Our goal is to obtain a probability distribution on a given variable O which summarizes some previous observations $\delta = o_1, \ldots, o_N$ on this variable. We denote this distribution $P(O \mid \delta \wedge \pi)$. The parametric approach consists of choosing a family of probability distributions (for example Normal laws) parameterized by a set of variables $\Lambda = \lambda_1, \wedge \ldots, \wedge \lambda_M$ (for example the mean and/or the variance). The choice of the distribution and of the parameters are part of the modeling process and belong to the prior knowledge of the programmer: π. We discuss the Bayesian approach to the parametric estimation by considering the program in Equation 15.2.

$$Pr\begin{cases} Ds\begin{cases} Sp(\pi_0)\begin{cases} Va: \\ O, O_1, \ldots, O_n, \Lambda \\ Dc: \\ \begin{cases} P(O \wedge O_1 \wedge \ldots \wedge O_N \wedge \Lambda \wedge \pi_0) \\ = P(\Lambda \mid \pi_0) \\ \times P(O_1 \mid \Lambda \wedge \pi_0) \times \ldots P(O_N \mid \Lambda \wedge \pi_0) \end{cases} \\ Fo: \\ \begin{cases} P(\Lambda \mid \pi_0) = \text{prior knowledge on the distribution } \Lambda \\ P(O \mid \Lambda \wedge \pi_0) = \text{prior on the distribution on } O \\ P(O_1 \mid \Lambda \wedge \pi_0) = \text{prior on the distribution on } O \\ \ldots \\ P(O_N \mid \Lambda \wedge \pi_0) = \text{prior on the distribution on } O \end{cases} \end{cases} \\ Id: \end{cases} \\ Qu: P(O \mid O_1 \ldots\ldots\ldots \wedge O_N \wedge \pi_0) \end{cases} \tag{15.2}$$

$P(O \mid O_1 = o_1 \ldots \wedge O_N = o_N \wedge \pi_0)$ is the probability distribution obtained by instantiating the question of the Bayesian program 15.2 with the data set δ. It is the result of several modeling choices:

1. The parametric forms and the parameters are modeling choices.

2. The model assumes δ is the result of independent trials.

This distribution may also be obtained with another approach: let's consider the following subprogram.

$$Pr \begin{cases} Ds \begin{cases} Sp(\pi_\lambda) \\ \begin{cases} Va: \\ O_1, \ldots, O_n, \Lambda \\ Dc: \\ \begin{cases} P(O_1 \wedge \ldots \wedge O_N \wedge \Lambda \wedge \pi_\lambda) \\ = P(\Lambda \mid \pi_\lambda) \\ \times P(O_1 \mid \Lambda \wedge \pi_\lambda) \times \ldots P(O_N \mid \Lambda \wedge \pi_\lambda) \end{cases} \\ Fo: \\ \begin{cases} P(\Lambda \mid \pi_\lambda) = \text{prior knowledge on the distribution } \Lambda \\ P(O_1 \mid \Lambda \wedge \pi_\lambda) = \text{prior on the distribution on } O \\ \ldots \\ P(O_N \mid \Lambda \wedge \pi_\lambda) = \text{prior on the distribution on } O \end{cases} \end{cases} \\ Id: \end{cases} \\ Qu: P(\Lambda \mid O_1 \ldots \wedge O_N \wedge \pi_\lambda) \end{cases} \tag{15.3}$$

We denote by $P(\Lambda \mid \delta \wedge \pi_\lambda) = P(\Lambda \mid O_1 = o_1 \ldots \wedge O_N = o_n \wedge \pi_\lambda)$ the probability distribution obtained by instantiating the question of program 15.3 with the previous readings δ. We can now define a simple program:

$$Pr \begin{cases} Ds \begin{cases} Sp(\delta \wedge \pi) \\ \begin{cases} Va: \\ O, \Lambda \\ Dc: \\ \begin{cases} P(\Lambda \wedge O \wedge \pi) = P(\Lambda \mid \delta \wedge \pi) \times P(O \mid \Lambda \wedge \pi) \end{cases} \\ Fo: \\ \begin{cases} P(\Lambda \mid \delta \wedge \pi) : \text{instantiated question of program 15.3} \\ P(O \mid \Lambda \wedge \pi) : \text{prior distribution on } O \end{cases} \end{cases} \\ Id: \end{cases} \\ Qu: P(O \mid \delta \wedge \pi) \end{cases} \tag{15.4}$$

We can show that:

$$P(O \mid \delta \wedge \pi) = P(O \mid O_1 = o_1 \ldots \wedge O_N = o_n \wedge \pi_0) \tag{15.5}$$

On the one hand, the question of the program in Equation 15.3 can be computed as:

$$P(\Lambda \mid O_1 \ldots \wedge O_N \wedge \pi_\lambda) = \frac{P(\Lambda \mid \pi_\lambda) \times P(O_1 \mid \Lambda \wedge \pi_\lambda) \times \ldots P(O_n \mid \Lambda \wedge \pi_\lambda)}{\sum_{\Lambda} P(\Lambda \mid \pi_\lambda) \times P(O_1 \mid \Lambda \wedge \pi_\lambda) \times \ldots P(O_n \mid \Lambda \wedge \pi_\lambda)} \tag{15.6}$$

while the question of the program in Equation 15.4 can be computed as:

$$P(O \mid \delta \wedge \pi) = \sum_{\Lambda} P(\Lambda \mid \delta \wedge \pi) \times P(O \mid \Lambda \wedge \pi) \tag{15.7}$$

which can also be written as:

$$\frac{\sum_{\Lambda} P(\Lambda \mid \pi_\lambda) \times P(O_1 = o_1 \mid \Lambda \wedge \pi_\lambda) \times \ldots P(O_n = o_n \mid \lambda \wedge \pi_\lambda) \times P(O \mid \Lambda \wedge \pi)}{\sum_{\Lambda} P(\Lambda \mid \pi_\lambda) \times P(O_1 = o_1 \mid \Lambda \wedge \pi_\lambda) \times \ldots P(O_n = o_n \mid \Lambda \wedge \pi_\lambda)} \tag{15.8}$$

On the other hand, the question in Equation 15.2 can be computed as:

$$\begin{aligned} & P(O \mid O_1 \ldots \wedge O_N \wedge \pi_0) \\ &= \frac{\sum_{\Lambda} P(\Lambda \mid \pi_0) \times P(O \mid \Lambda \wedge \pi_0) \times P(O_1 \mid \Lambda \wedge \pi_0) \times \ldots P(O_n \mid \Lambda \wedge \pi_0)}{\sum_{\Lambda} \left(P(\Lambda \mid \pi_0) \times P(O_1 \mid \Lambda \wedge \pi_0) \times \ldots P(O_n \mid \Lambda \wedge \pi_0) \times \sum_{O} P(O \mid \Lambda \wedge \pi_0) \right)} \\ &= \frac{\sum_{\Lambda} P(\Lambda \mid \pi_0) \times P(O \mid \Lambda \wedge \pi_0) \times P(O_1 \mid \Lambda \wedge \pi_0) \times \ldots P(O_n \mid \Lambda \wedge \pi_0)}{\sum_{\Lambda} P(\Lambda \mid \pi_0) \times P(O_1 \mid \Lambda \wedge \pi) \times \ldots P(O_n \mid \Lambda \wedge \pi)} \end{aligned} \tag{15.9}$$

leading to the desired result:

$$\begin{aligned} & P(O \mid O_1 = o_1 \ldots \wedge O_N = o_n \wedge \pi_0) = \\ & \frac{\sum_{\Lambda} P(\Lambda \mid \pi_\lambda) \times P(O_1 = o_1 \mid \Lambda \wedge \pi_\lambda) \times \ldots P(O_n = o_n \mid \Lambda \wedge \pi_\lambda) \times P(O \mid \Lambda \wedge \pi_\lambda)}{\sum_{\Lambda} P(\Lambda \mid \pi_\lambda) \times P(O_1 = o_1 \mid \Lambda \wedge \pi_\lambda) \times \ldots P(O_n = o_n \mid \Lambda \wedge \pi_\lambda)} \end{aligned} \tag{15.10}$$

The distribution $P(\Lambda \mid \delta \wedge \pi)$ is a means to summarize and to reuse the previous experiments and the initial priors. Several methods and hypothesis are used to set the initial priors $P(\Lambda \mid \pi)$ and to compute $P(\Lambda \mid O_1 = o_1 \ldots \wedge O_N = o_n \wedge \pi)$ leading to several types of parametric estimations. We describe two of them in the next two sections.

15.1.3 Maximum likelihood (ML)

The most common approach to parametric estimation is to assume a uniform distribution for the prior: $P(\Lambda \mid \pi) = \text{Uniform}$ and to replace the posterior distribution $P(\Lambda \mid \delta \wedge \pi)$ by a Dirac distribution:

$$\delta_{\Lambda^*} : \lambda^* = \max_{\lambda \in \Lambda} : P(\lambda \mid \delta \wedge \pi)$$

For example, if we consider a Gaussian distribution for $P(O \mid \Lambda \wedge \pi)$ with a given variance σ, the maximum likelihood estimator for the mean μ^* is obtained by solving a least square $\mu^* = \min_{\mu} : \sum_i ((\mu - o_i)^2)$ leading to a simplified Bayesian program:

$$
Pr \begin{cases} Ds \begin{cases} Sp(\delta \wedge \pi) \begin{cases} Va: \\ O, \Lambda \\ Dc: \\ \left\{ P(O \wedge \pi) = P(\Lambda \mid \delta \wedge \pi) \times P(O \mid \Lambda \wedge \pi) \right. \\ Fo: \\ \begin{cases} P(\Lambda \mid \delta \wedge \pi) : \delta_{\mu^*} \\ P(O \mid \lambda \wedge \pi) : N(\lambda, \sigma) \end{cases} \end{cases} \\ Id: \end{cases} \\ Qu: \\ P(O \mid \delta \wedge \pi) \end{cases} \tag{15.11}
$$

and a simple solution for the question: $P(O \mid \delta \wedge \pi) = N(\mu^*, \sigma)$.

The program in Equation 15.11 is implemented in the file: "chapter15/ML.py". DataDescriptors are used to connect with data. They offer a convenient way to parse the data and to interface with database, csv files, streams, and so forth:

```
O = plSymbol('O',plRealType(-100,100))

file = 'C:/Users/mazer/Documents/Publications/\
BPbook/Chapters/chapter15/code/previous_O.csv'
#define the data source ignoring unknown fields
previous_O=plCSVDataDescriptor(file,O^O_I)
previous_O.ignore_unknown_variables()
#define the type of ML learner
learner_O=plLearn1dNormal(O)
#use data
i= learner_O.learn_using_data_descriptor(previous_O)
#retrieve the distribution
distrib = learner_O.get_distribution()
#print it
print distrib
```

Several distributions may be learned with ML from a single source: here a histogram on the variable "O_I".

```
#learning another distribution from the same source
previous_O.rewind()
learner_O_I=plLearnHistogram(O_I)
i= learner_O_I.learn_using_data_descriptor(previous_O)
distrib_I = learner_O_I.get_distribution()
print distrib_I
```

To speed up the learning process, these distributions may be learned at the same time from a single data source.

```
#same as above but brows the data only once
#define a vector of distributions to learn
global_learner = plLearnDistribVector([learner_O_I,learner_O],O_I^O)
i=global_learner.learn_using_data_descriptor(previous_O)
list_distrib=global_learner.get_computable_object_list()
print 'global learning \n',list_distrib[0],'\n',list_distrib[1]
```

ML learners are incremental: new data may be added to update the current estimators.

```
previous_O1=plCSVDataDescriptor(file1,O)
previous_O1.ignore_unknown_variables()
i= learner_O.learn_using_data_descriptor(previous_O)
i= learner_O.learn_using_data_descriptor(previous_O1)
distrib = learner_O.get_distribution()
```

15.1.4 Bayesian estimator and conjugate laws

The Bayesian estimator is obtained when choosing a nonuniform distribution for the initial prior $P(\Lambda \mid \pi)$ and by using a Dirac δ_{λ^*} for the posterior with $\lambda^* = E(P(\Lambda \mid \delta \wedge \pi))$.

In some cases, it is possible to find $P(\Lambda \mid \pi)$ and $P(O_i \mid \Lambda \wedge \pi)$ such that the posterior $P(\Lambda \mid O_i \wedge \pi)$ has the same parametric form as the prior $P(\Lambda \mid \pi)$: then $P(\Lambda \mid \pi)$ and $P(O_i \mid \Lambda \wedge \pi)$ are said to be conjugate.

For example if $P(O_i \mid \Lambda = \lambda \wedge \pi)$ is a binomial distribution:

$$
\begin{aligned}
P(O_i = 1) &= \lambda \\
P(O_i = 0) &= 1 - \lambda
\end{aligned}
$$

and if the prior on $P(\Lambda \mid \pi)$ is a beta distribution having α and β as parameters:

$$P(\lambda) = \frac{\lambda^{\alpha-1}(1-\lambda)^{\beta-1}}{beta(\alpha,\beta)}$$
$$beta(\alpha,\beta) = \int_0^1 t^{\alpha-1}(1-t)^{\beta-1}dt$$

then $P(\Lambda \mid O_i = o_i \wedge \pi)$ follows a beta distribution $beta(\alpha + o_i, \beta + 1)$.

The conjugate property allows us to obtain a new prior after each observation and to obtain a close formula for N observations. Following our previous example we have:

$$\begin{aligned} P(\Lambda \mid O_1 = o_1 \ldots \wedge O_N = o_n \wedge \pi) &= beta(\alpha + \sum_{i=1}^{N} o_i, \beta + N) \\ \lambda^* = E(P(\Lambda \mid \delta \wedge \pi) &= \frac{\alpha + \sum_{i=1}^{N} o_i}{\beta + N} \end{aligned} \tag{15.12}$$

leading to:

$$P(O = 1 \mid \delta \wedge \pi) = \lambda^*$$

The file "chapter15/BE.py" contains an example showing how to obtain the Bayesian estimator for a binomial distribution. Here, the priors α and β are set to 1 (corresponding to a uniform prior).

```
from pypl import *
O = plSymbol('O',plIntegerType(0,1))
file = 'ExDir+'chapter15/data/B_O.csv'
#define the data source ignoring unknown fields
previous_O=plCSVDataDescriptor(file,O)
#define the type of Bayesian learner
#the prior beta distribution (here uniform : alpha=1 beta=1)
learner_O=plBayesLearnBinomial(O,1,1)
#print the distribution before learning
distrib = learner_O.get_distribution()
print distrib
#use data
i= learner_O.learn_using_data_descriptor(previous_O)
#retrieve the distribution
distrib = learner_O.get_distribution()
#print it
print distrib
```

Using a Bayesian estimator is a way to reduce the computation load to infer $P(O \mid \delta \wedge \pi)$. In fact, replacing the program in Equation 15.13 with the program in Equation 15.14 is a way to avoid computing:

$$\int_{\lambda} P(\Lambda \mid \delta \wedge \pi) \times P(O \mid \Lambda \wedge \pi)$$

by considering only a single value $\lambda^* = E(P(\Lambda \mid \delta \wedge \pi))$. On some occasions, it is worthwhile considering to keep all the information at hand on Λ and to perform an approximate inference.

$$
Pr \begin{cases} Ds \begin{cases} Sp(\delta \wedge \pi) \begin{cases} Va: \\ O, \Lambda \\ Dc: \\ \left\{ P(O \wedge \Lambda \wedge \pi) = P(\Lambda \mid \delta \wedge \pi) \times P(O \mid \Lambda \wedge \pi) \right. \\ Fo: \\ \begin{cases} P(\Lambda \mid \delta \wedge \pi) := beta(\alpha + \sum_{i=1}^{N} o_i, \beta + N) \\ P(O \mid \Lambda \wedge \pi) : \text{Binomial distribution} \end{cases} \end{cases} \\ Id: \end{cases} \\ Qu: P(O \mid \delta \wedge \pi) \end{cases} \tag{15.13}
$$

$$
Pr \begin{cases} Ds \begin{cases} Sp(\delta \wedge \pi) \begin{cases} Va: \\ O, \Lambda \\ Dc: \\ \left\{ P(O \wedge \pi) = P(\Lambda \mid \delta \wedge \pi) \times P(O \mid \Lambda \wedge \pi) \right. \\ Fo: \\ \begin{cases} P(\Lambda \mid \delta \wedge \pi) := \delta_{E(beta(\alpha + \sum_{i=1}^{N} o_i, \beta + N))} \\ P(O \mid \Lambda \wedge \pi) : \text{Binomial distribution} \end{cases} \end{cases} \\ Id: \end{cases} \\ Qu: P(O \mid \delta \wedge \pi) \end{cases} \tag{15.14}
$$

The file "chapter15/BE.py" shows how to implement the program in Equation 15.13 to use the information contained in $P(\Lambda \mid \delta \wedge \pi)$. To build the decomposition the posterior distribution is obtained from the learner. Approximate inference is used to compute $P(O \mid \delta \wedge \pi)$. Note that it is possible to control either the precision of the integration or the number of steps used to approximate the sum.

```
#get the posterior distribution
posterior=learner_O.get_aposteriori_distribution(L)
#write the decomposition
decomposition= posterior*plCndBinomial(O,L)
#define de joint distribution
joint = plJointDistribution(decomposition)
#tell the interpreter to use 1000 sampling points
#to appoximate the integral
qu=joint.ask_mc_sample(O,1000)
print qu.compile()

#another way is to tell the interpreter to stop integrating
#when the precision is below 0.0001
qu=joint.ask_mc_threshold(O,0.0001)
print qu.compile()
```

15.2 Expectation–Maximization (EM)

The expectation–maximization (EM) is a class of iterative algorithms that generalize the parameter estimation to models with latent or partially observable variables (see Dempster et al. [1977]). It is used in a wide variety of situations best described as incomplete-data problems. The idea behind the EM algorithms is intuitive and natural and it is used as a common basis in many classical learning algorithms. The EM algorithms are named from their two phases: the expectation and the maximization steps. Both steps may be seen as Bayesian programs. The two programs in Equation 15.15 could be interleaved in a loop to produce a version of the EM algorithm.

$$\begin{cases} \text{PrE : Program for E Step} \\ Ds \\ \begin{cases} Sp(\pi \wedge \pi_i) \\ \begin{cases} Va: \\ O, Z, \Lambda \\ Dc: \\ \begin{cases} P(O \wedge Z \wedge \Lambda \mid \pi \wedge \pi_i) \\ = P(\Lambda \mid \pi_i) \\ \times P(O \wedge Z \mid \Lambda \wedge \pi) \end{cases} \\ Fo: \\ \begin{cases} P(\Lambda \mid \pi_i) : \text{(i)} \\ P(O \wedge Z \mid \Lambda \wedge \pi) : \text{Model} \end{cases} \end{cases} \\ Id: \end{cases} \\ Qu : P(Z \mid O \wedge \pi_i) \end{cases} \quad \begin{cases} \text{PrM : Program for the M step} \\ Ds \\ \begin{cases} Sp(\pi) \\ \begin{cases} Va: \\ O, Z, \Lambda \\ Dc: \\ \begin{cases} P(O \wedge Z \wedge \Lambda \mid \pi \wedge \pi_i) \\ = P(\Lambda \mid \pi) \\ \times P(O \wedge Z \mid \Lambda \wedge \pi) \end{cases} \\ Fo: \\ \begin{cases} P(\Lambda \mid \pi) : \text{(ii)} \\ P(O \wedge Z \mid \Lambda \wedge \pi) : \text{Model} \end{cases} \end{cases} \\ Id: \end{cases} \\ Qu : P(\Lambda \mid O \wedge Z \wedge \pi) \end{cases} \tag{15.15}$$

Both programs share the same variables:

- O: The conjunction of all observable variables.
- Z: The conjunction of all latent variables.
- Λ: The conjunction of all the parameters used in the parametric forms to define the probability distributions.

They also share the same model:

$$P(O \wedge Z \mid \Lambda \wedge \pi)$$

On this basis, it is possible to design a variety of EM algorithms. We describe one particular version among many:

```
Kullback-Leibler-distance = +∞
define P(Λ | π_0){E step Prior}
i=0
while Kullback-Leibler-distance > ϵ do
  {E Step}
  define PrE with P(Λ | π_i) as prior
  infer P(Z | O ∧ π_i)
  Instantiate with readings P(Z | O = δ ∧ π_i)
  {M Step}
  define PrM with P(Λ | π) {Initial Prior}
  infer P(Λ | O ∧ Z ∧ π)
  Compute the soft evidence
  P(Λ | π_{i+1}) = Σ_{z∈Z} P(z | O = δ ∧ π_i) × P(Λ | O = δ ∧ Z = z ∧ π)
  i=i+1
  compute Kullback-Leibler-distance(P(Λ | π_{i+1}), P(Λ | π_i))
end while
return P(Λ | π_{i+1})
```

The algorithm starts with the E step: a prior distribution on the parameters $P(\Lambda \mid \pi_0)$ is given to initialize the process. The result of the inference $P(Z \mid O \wedge \pi_i)$ is instantiated with the observed data $P(Z \mid \delta \wedge \pi_i) = P(Z \mid O = o_1^i ... o_n^i \wedge \pi_i)$. The algorithm then proceeds with the program designed for the M step. The distribution $P(\Lambda \mid \pi)$ is set with the prior on Λ before considering any data. The program is used to compute $P(\Lambda \mid O \wedge Z \wedge \pi)$. In this version, we use soft evidence 8.5 to compute the new prior $P(\Lambda \mid \pi_{i+1})$ for the next E step:

$$P(\Lambda \mid \pi_{i+1}) = \sum_{z \in Z} P(z \mid O = \delta \wedge \pi_i) \times P(\Lambda \mid O = \delta \wedge Z = z \wedge \pi)$$

The Kullback–Leibler distance between $P(\Lambda \mid \pi_{i+1})$ and $P(\Lambda \mid \pi_i)$ could be used as a stopping condition: k denotes the value of $i+1$ when the condition is reached.

The result $P(\Lambda \mid \pi_k)$ could be used to define a parametric form in a classification program in Equation 15.16, where R and X are limited to the observable and latent variables attached to a single occurrence of the phenomenon.

$$Pr\begin{cases} Ds\begin{cases} Sp(\pi\wedge\delta)\begin{cases} Va: \\ \Lambda,R,X \\ Dc: \\ \begin{cases} P(R\wedge X\wedge\Lambda\wedge\pi\wedge\delta) \\ = P(\Lambda\mid\pi\wedge\delta) \\ \times P(R\wedge X\mid\Lambda\wedge\pi) \end{cases} \\ Fo: \\ \begin{cases} P(\Lambda\mid\pi\wedge\delta)=P(\Lambda\mid\pi_k) \\ P(R\wedge X\mid\Lambda\wedge\pi):\text{model} \end{cases} \end{cases} \\ Id: \end{cases} \\ Qu: \\ P(X\mid R\wedge\pi\wedge\delta) \end{cases} \tag{15.16}$$

Given an initial template for the EM algorithm we can design further variations leading to different results, computing times, and convergence properties. For example, a Dirac distribution δ_{λ^*} may be used in the E step to describe $P(\Lambda \mid \pi_{i+1})$.

$$\begin{aligned} &P(\Lambda\mid\pi_{i+1})=\sum_{z\in Z}P(z\mid O=\delta\wedge\pi_i)\times P(\Lambda\mid O=\delta\wedge Z=z\wedge\pi) \\ &\lambda^*=\max_{\lambda}P(\Lambda=\lambda\mid\pi_{i+1}) \\ &P(\Lambda\mid\pi_{i+1})=\delta_{\lambda^*} \end{aligned} \tag{15.17}$$

Also, the EM algorithms may be seen as filters. The result $P(\Lambda \mid \pi \wedge \delta) = P(\Lambda \mid \pi_k)$ is a summary of the information given by a data set δ. If a new data set comes along, it can be used as an initial prior in the program for the M step. Note that regrouping the two data sets may lead to a different result.

Nothing forbids continuously redefining the latent and the observable variables. Special models may also be used to ease the computation, for example, using conjugate laws in the M step.

At this point, no assumption have been made concerning $P(O \wedge Z \mid \Lambda \wedge \pi)$. For this reason, the EM algorithm applies to a variety of applications: classification, HMM, model selection, and so forth.

15.2.1 EM and classification

One instance of the classification problem is a model where the distributions on the observations i on the attributes A depend on a given class C. In general, the problem is to learn and then to infer the class from the observations. In this particular case the learning problem is defined as in the program in Equation 15.18.

$$
Pr \begin{cases} Ds \begin{cases} Sp(\pi) \begin{cases} Va: \\ A_1, \ldots, A_n; A_i : \text{observations} ; n \text{ cardinality of learning set} \\ C_1, \ldots, C_n; C_i \in [1, \ldots\ k] : \text{k number of classes} \\ \Lambda \\ Dc: \\ \begin{cases} P(A_1 \wedge C_1 \ldots A_N \wedge C_N \wedge \Lambda \wedge \pi) \\ = P(\Lambda \mid \pi) \\ \times P(A_1 \wedge C_1 \mid \Lambda \wedge \pi) \times \ldots \times P(A_N \wedge C_N \mid \Lambda \wedge \pi) \end{cases} \\ Fo: \\ \begin{cases} P(\Lambda \mid \pi) = \text{prior knowledge on the distribution } \Lambda \\ P(A_i \wedge C_i \mid \Lambda \wedge \pi) = \text{ Observational model} \end{cases} \end{cases} \\ Id: \end{cases} \\ Qu: P(\Lambda \mid A_1 \wedge C_1 \ldots A_N \wedge C_N \wedge \pi) \end{cases} \tag{15.18}
$$

The distribution on Λ is then given by

$$
P(\Lambda \mid \delta \wedge \pi) = P(\Lambda \mid A_1 = a_1 \wedge C_1 = c_1 \ldots A_N = a_n \wedge C_N = c_n \wedge \pi) \tag{15.19}
$$

When the class is not observed in the data, it is an instance of the unsupervised learning problem. Unsupervised learning may be used to classify multidimensional data, to discretize variables, to approximate a complex distribution, or to learn with an incomplete data set. For example, we may want to study the weight of a species. One possible assumption is to consider that the weight is dependent on the gender of each individual. The EM algorithm will be used to obtain a classifier (female or male) while only being able to observe the weight. Let's describe a version of this classifier.

We consider a population of N individuals. The descriptions will use the following variables:

- W_j: stands for weight of the individual j: $O = W_1 \wedge \ldots \wedge W_N$.
- $C_j = \{0, 1\}$: stands for the class of the individual j: $Z = C_1 \wedge \ldots \wedge C_N$.
- $\Lambda_f^\sigma, \Lambda_f^\mu$ are the parameters of the Normal distributions used to represent the probability on the weights for females.
- $\Lambda_m^\sigma, \Lambda_m^\mu$ are the parameters of the Normal distributions used to represent the probability on the weights for males.
- Λ_g is the parameter of the binomial distribution used to model the probability to have a female or a male.

- $\Lambda = \Lambda_g \wedge \Lambda_f^\sigma \wedge \Lambda_f^\mu \wedge \Lambda_m^\sigma \wedge \Lambda_m^\mu$

The model $P(O \wedge Z \mid \Lambda \wedge \pi)$ for the E and M steps is defined as a mixture of Normal distributions:

$$\begin{cases} P(O \wedge Z \mid \Lambda \wedge \pi) = \\ P\left(W_j \mid C_j \wedge \Lambda_f^\sigma \wedge \Lambda_f^\mu \wedge \Lambda_m^\sigma \wedge \Lambda_m^\mu \wedge \pi\right) = \\ \begin{cases} C_j = 0 : \\ P\left(W_j \mid \Lambda_f^\sigma \wedge \Lambda_f^\mu\right) = N(\Lambda_f^\sigma, \Lambda_f^\mu) \\ C_j = 1 : \\ P(W_j \mid \Lambda_m^\sigma \wedge \Lambda_m^\mu) = N(\Lambda_m^\sigma, \Lambda_m^\mu) \end{cases} \end{cases} \tag{15.20}$$

The prior for the E step is defined as a Dirac (see Equation 15.17):

$$\text{(i) } P(\Lambda \mid \pi_{i+1}) = \delta_{\lambda^*}$$

The initial prior for the E step is defined according to some background knowledge: $\lambda_g^* = 0.7, \lambda_f^{\mu *} < \lambda_m^{\mu *}$

The prior for the M step is defined as a uniform distribution:

$$\text{(ii) } P(\Lambda \mid \pi) = \textbf{Uniform}$$

The result of this EM algorithm will give the following result:

$$P(\Lambda \mid \pi_k) = \lambda_g^k, \Lambda_f^{\sigma_k}, \Lambda_f^{\mu_k}, \Lambda_m^{\sigma_k}, \Lambda_m^{\mu_k}$$

which are the parameters of the distributions on the gender and on weights knowing the gender. These parameters may be used to classify any individual knowing its weight using the program in Equation 15.21.

$$
Pr \begin{cases}
Ds \begin{cases}
Sp(\pi \wedge \delta) \\
\begin{cases}
Va : C, W \\
Dc : \\
\begin{cases}
P(C \wedge W \wedge \pi \wedge \delta) \\
= P(C \mid \pi \wedge \delta) \\
\times P(W \mid C \wedge \pi \wedge delta)
\end{cases} \\
Fo : \\
\begin{cases}
P(C \mid \pi \wedge \delta) = \text{Binomial}(\lambda_g^k) \\
P(W \mid C \wedge \pi \wedge \delta) = \\
\begin{cases}
C = 0 : \\
P(W \mid C \wedge \pi \wedge \delta) = N(\Lambda_f^{\sigma_k}, \Lambda_f^{\mu_k}) \\
C_j = 1 : \\
P(W \mid C \wedge \pi \wedge \delta) = N(\Lambda_m^{\sigma_k}, \Lambda_m^{\mu_k})
\end{cases}
\end{cases}
\end{cases} \\
Id :
\end{cases} \\
Qu : \\
P(C \mid W \wedge \pi \wedge \delta)
\end{cases}
\tag{15.21}
$$

Figure 15.1 represents the joint distribution of a Gaussian mixture with two components and unknown parameters. Table 15.1 shows how an EM algorithm may retrieve the parameters with 1000 samples drawn out from this distribution.

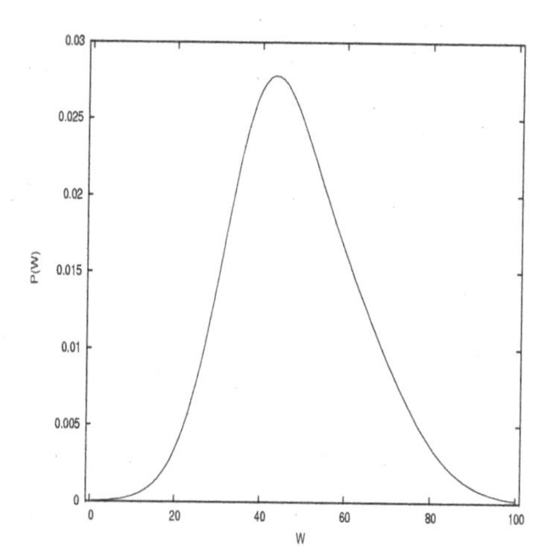

FIGURE 15.1: A Gaussian mixture with two components.

TABLE 15.1: The EM Algorithm Could Retrieve a Given Mixture with a Good Precision While Being Intialized with Relatively Poor Initial Condition

Λ	**Used for generation**	**Prior E step**	**Found by EM**
λ_g	0.55	0.7	0.66
λ_f^{μ}	45.0	20.0	45.1
λ_f^{σ}	10	40	9.9
λ_m^{μ}	55	70	52.8
λ_m^{σ}	15	40	13.7

The program in file: "chapter15/genweights.py" generates a set of data based on known parameters (see Table 15.1). It samples the distribution represented in Figure 15.1. The file "chapter15/getlambda.py" contains a program to retrieve these parameters with an EM algorithm (see Table 15.1).

```
from pypl import *
C = plSymbol('C',plIntegerType(0,1))
W = plSymbol('W',plRealType(0,100))
#define a ML learner for a binomial law
pC_learner = plLearnHistogram(C)
pW_learner = plCndLearn1dNormal(W,C)
#define intial guess : P(Lambda|pi_0)
pC_init = plBinomial(C,0.70)
pWkC_init = plDistributionTable(W,C)
pWkC_init.push(plNormal(W,20.0,40.0),0)
pWkC_init.push(plNormal(W,70.0,40.0),1)
#define the learner
learner=plEMLearner(pC_init*pWkC_init,
                    [pC_learner,pW_learner])
#define the data source
data = plCSVDataDescriptor('weights.csv',W)
#perform learning stop with a threshold
learner.run(data,10e-9)
#get the prameters
print learner.get_distribution(0)
print learner.get_distribution(1)
```

15.2.2 EM and HMM

Hidden Markov models are special cases of generic Bayesian filters (see Equation 11.2) designed to learn and to label temporal sequences of sensor readings. They are used in many applications, more notably in applications related to speech and behavior recognition. We describe the model of HMM

(Section 15.2.2.1), how to learn the model parameters of an HMM (Section 15.2.2.2), how to use this model to infer the distribution on states from a series of observations (Section 15.2.2.3), and finaly how to detect a behavior from time series among N, by using N HMMs running in parallel, each indexed by u (Section 15.2.2.4).

15.2.2.1 HMM

A HMM indexed with u is defined as Bayesian program 15.22. The variables O_t^u, S_t^u respectively denote the observations and the states at time t and $\Lambda_0^u, \Lambda_M^u, \Lambda_S^u$ are the model parameters. Λ_0^u is the set of parameters for the initial condition on states, Λ_M^u is the set of parameters for the sensor model, and

$$\Lambda_s^u = \Lambda_1^u \wedge \Lambda_2^u \ldots \wedge \Lambda_j^u \ldots \wedge \Lambda_{a(u)-1}^u$$

are the parameters for the state transitions, with $P(S_t \mid S_{t-1} = j)$ parameterized by Λ_j^u. The cardinality $a(u)$ of the states $S_{0,\ldots,T}^u$ may vary from one HMM to another.

$$Pr\begin{cases} Ds\begin{cases} Sp(\pi^u) \\ \begin{cases} Va: \begin{cases} \Lambda_0^u, \Lambda_M^u, \Lambda_S^u, \\ S_t^u, \forall t \in [0, \ldots, T] : S_t^u \in 1, \ldots., a(u) \\ O_t^u, \forall t \in [1, \ldots, T] : O_t^u \in D \end{cases} \\ Dc: \begin{cases} P(S_o^u \wedge O_1, \ldots S_t^u \wedge O_t^u \ldots S_T^u \wedge O_T^u \wedge \Lambda_0^u \wedge \Lambda_M^u \wedge \Lambda_S^u \mid \pi^u) = \\ P(\Lambda_0^u \wedge \Lambda_M^u \wedge \Lambda_S^u \mid \pi^u) \\ P(S_0^u \mid \Lambda_0^u \wedge \pi^u) \\ \prod_{t\in[1\ldots T]} P(S_t^u \mid S_{t-1}^u \wedge \Lambda_S^u \wedge \pi^u) P(O_t^u \mid S_t \wedge \Lambda_M^u \wedge \pi^u)) \end{cases} \\ Fo: \begin{cases} P(\Lambda_0^u \wedge \Lambda_0^u \wedge \Lambda_S^u \wedge \mid \pi^u) \\ P(S_0 \mid \Lambda_0^u \wedge \pi^u) = \text{multinomial} \\ P\left(S_t^u \mid S_{t-1}^u \wedge \Lambda_S^u \wedge \pi^u\right) = \text{multinomial} \\ P(O_t^u \mid S_t^u \wedge \Lambda_M^u \wedge \pi^u) = \text{Sensor Model} \end{cases} \end{cases} \\ Id: P(\Lambda_0^u \wedge \Lambda_0^u \wedge \Lambda_S^u \mid \pi^u) = \text{identified with EM} \end{cases} \\ Qu: P(S_T^u \mid O_1^u \wedge \ldots \wedge O_T^u) \end{cases} \tag{15.22}$$

15.2.2.2 Learning HMM with EM

In this framework, we have to find $\Lambda = \Lambda_0^u \wedge \Lambda_O^u \wedge \Lambda_S^u$ given several occurrences of the sequence u.

We use the EM algorithm by stating:

- $Z = S_0^u \wedge S_1^u \wedge \ldots \wedge S_{T-1}^u$

- $O = O_0^u \wedge O_1^u \wedge \ldots \wedge O_T^u \wedge S_T^u$
- $\Lambda = \Lambda_0^u \wedge \Lambda_O^u \wedge \Lambda_S^u$

In practice, we perform supervised learning for each sequence by considering several sequences of observations $j \in 1, \ldots, L : O^j$, each leading to the desired type u. The following algorithm is used:

$P(\Lambda \mid \pi)$=Uniform
for $j = 1 \to L$ **do**
  set $\delta = \{o_1^j, \ldots o_{n(j)}^j, a(u)\}$
  RUN EM
  Update $P(\Lambda \mid \pi)$ {Use EM as a filter}
end for
return $P(\Lambda \mid \pi)$ as $P(\Lambda \mid \pi^u)$

Here, we choose to constrain the parametric space by asserting that each learning sequence of type u should reach the state $a(u)$ at the end of the sequence: $S_T^u = a(u)$. The algorithm accommodates sequences of any length and uses the EM as a filter which updates the prior on $P(\Lambda \mid \pi)$ after each sequence is recognized as belonging to the same class.

15.2.2.3 State estimation with HMM

Given $P(\Lambda_0^u \wedge \Lambda_M^u \wedge \Lambda_S^u \mid \pi^u)$ and a sequence of observations $\{o_1, \ldots o_T\}$, the question $P(S_T \mid O_1 \wedge \ldots O_T \pi^u)$ is used to assert:

$$P(S_T \mid O_1 = o_0 \wedge \ldots O_T = o_T \wedge \pi^u)$$

In practice, the recursive version of Bayesian filters is used at each time step using the program in Equation 15.23.

$$Pr \begin{cases} Ds \begin{cases} Sp(\pi_t^u) \\ \begin{cases} Va: \begin{cases} S_t^u, S_{t-1}^u \in 1, ..., a(u) \\ \Lambda \\ O_t^u, \in D \end{cases} \\ Dc: \begin{cases} P(\Lambda \wedge S_{t-1}^u \wedge S_t^u \wedge O_t^u \wedge \pi_t^u) = \\ P(\Lambda \mid \pi_t^u) P(S_{t-1}^u \mid \pi_t^u) \\ P(S_t^u \mid S_{t-1}^u \wedge \Lambda_S \wedge \pi_t^u) P(O_t^u \mid S_t \wedge \Lambda_M \wedge \pi_t^u) \end{cases} \\ Fo: \begin{cases} P(\Lambda \mid \pi_t^u) = \text{obtained during learning} \\ P(S_{t-1} \mid \pi_t^u) = P(S_{t-1}^u \mid O_{t-1}^u = o_{t-1} \wedge \pi_{t-1}^u) \\ P(S_t^u \mid S_{t-1}^u \wedge \Lambda_S^u \wedge \pi_t^u) = \text{transition Model} \\ P(O_t^u \mid S_t^u \wedge \Lambda_M \wedge \pi_t^u) = \text{Sensor Model} \end{cases} \\ Id: \end{cases} \end{cases} \\ Qu: P(S_t^u \mid O_t^u \wedge \pi_t^u) \end{cases} \tag{15.23}$$

The question is used to obtain a probability distribution over the S_t^u given an observation o_t^u. In turn, it can be used in a prediction step at time t to be compared with other sequences u'.

$$
Pr\begin{cases} Ds\begin{cases} Sp(\pi_t'^u) \begin{cases} Va: \begin{cases} S_t^u, \in 1, ..., a(u) \\ O_t^u, \in D \end{cases} \\ Dc: \begin{cases} P\left(S_t^u \wedge O_t^u \wedge \pi_t'^u\right) = \\ P\left(S_t^u \mid \pi_t'^u\right) P\left(O_t^u \mid S_t \wedge \pi_t'^u\right) \end{cases} \\ Fo: \begin{cases} P\left(S_t \mid \pi_t'^u\right) = P\left(S_t^u \mid O_t^u = o_t \wedge \pi_t^u\right) \text{ see 15.23} \\ P\left(O_t^u \mid S_t^u \wedge \pi_t'^u\right) = \text{Sensor Model} \end{cases} \end{cases} \\ Id: \end{cases} \\ Qu: P\left(O_t^u \mid \pi_t'^u\right) \end{cases} \tag{15.24}
$$

15.2.2.4 Using a bank of HMMs

To detect a type of sequence u among m we use the program in Equation 15.25 as a filter to deliver a probability distribution over each type of sequence after each reading. The transition probability distribution may be given or also identified with the EM algorithm.

$$
Pr\begin{cases} Ds\begin{cases} Sp(\pi_T^u) \begin{cases} Va: \begin{cases} U_t, U_{t-1} \in 1, ..., m; \text{current and previous sequence} \\ O_t = \bigcup\limits_{u \in [1...m]} \{O_t^u\} : O_t^u \in D; \text{current observation vector} \end{cases} \\ Dc: \begin{cases} P\left(U_{t-1} \wedge U_t \wedge O_t \wedge \pi_t^u\right) = \\ P\left(U_{t-1} \mid \pi_t\right) P\left(U_t \mid U_{t-1} \wedge \pi_t\right) P\left(O_t \mid U_t \wedge \pi_t\right) \end{cases} \\ Fo: \begin{cases} P\left(U_{t-1} \mid \pi_t\right) = \text{previous estimation} \\ P\left(U_t \mid U_{t-1}\right) = \text{transition model} \\ P\left(O_t \mid U_t = u \wedge \pi_t\right) \\ = \begin{cases} P\left(O_t^u \mid \pi_t'^u\right); \text{prediction with model } u \text{ (15.24)} \\ \times \\ Uniform(O_t - \bigcup\limits_{u \in [1...m]} \{O_t^u\}) \end{cases} \end{cases} \end{cases} \\ Id: \end{cases} \\ Qu: P\left(U_t \mid O_t \wedge \pi_T^u\right) \end{cases} \tag{15.25}
$$

15.2.3 Model selection

The model selection is of paramount importance in machine learning. Bayesian programs help to deal with a limited set of known models.

One simple version of the model selection problem has already been presented in Chapter 10. A variable H is used to select among several existing models $P(S \wedge I\pi_i)$ based on some input I. We recall the general program in Equation 15.26 for mixing models and show how to learn $P(H \mid I \wedge \pi_M)$ with the EM algorithm.

$$Pr: \begin{cases} Ds: \begin{cases} Sp(\pi_M): \\ \begin{cases} Va: H \in [1,\ldots,n], I, S_1,\ldots,S_n \\ Dc: \begin{cases} P(H \wedge I \wedge S \mid \pi_M) \\ = P(I \mid \pi_M) \\ \times P(H \mid I \wedge \pi_M) \\ \times P(S \mid H \wedge I \wedge \pi_M) \end{cases} \\ Fo: \begin{cases} P(I \mid \pi_M) = \text{Uniform} \\ P(H \mid I \wedge \pi_M) = \text{Given or learned with EM} \\ P(S \mid H \wedge I \wedge \pi_M) \\ = \begin{cases} H = 1: \\ P(S \mid I \wedge \pi_1) \\ \ldots \\ P(S \mid I \wedge \pi_i) = \text{Question to model } \pi_i \\ \ldots \\ H = n: \\ P(S \mid I \wedge \pi_n) \end{cases} \end{cases} \end{cases} \\ Id: \end{cases} \\ Qu: P(S \mid I \wedge \pi_M) \end{cases} \tag{15.26}$$

To learn $P(H \mid I \wedge \pi_M)$, we use the EM algorithm by stating:

- $Z = H$
- $O = S$
- $\Lambda =$ parameters of a multinomial law for H

This approach is used in Chapter 10 to learn select and mixed models attached to the behaviors "Phototaxy" and "Avoidance."

However, in many applications the analyst has to consider families of models. For example, the models may be parameterized by the number of classes or by the number of states as in the classification or as in the hidden Markov models. Choosing the right models among many is a problem which requires the analyst to consider the bias-variance trade-off. In the next section we consider selecting a model among a huge number of possible models: given a set of variables and a data set on these variables, we select the most appropriate decomposition.

15.3 Learning structure of Bayesian networks

Learning the decomposition from a data set can be seen as a model selection problem where the number of possible decompositions $d(n)$ is a function of the number of variables n. A lower bound of $d(n)$ may be found by considering the cardinality of the set of Bayesian networks or acyclic graphs with n nodes: $\mathcal{B}_n$. The cardinality $Card(\mathcal{B}_n) = b(n)$ is given by Equation 15.27:

$$\begin{aligned} & b(1) = 1 \\ & \sum_{i=1}^{n} (-1)^{i+1} \begin{pmatrix} n \\ i \end{pmatrix} 2^{i(n-i)} b(n-i) = n^{2^{O(n)}} \end{aligned} \tag{15.27}$$

For example, the number of acyclic graphs for 10 variables is of the order of 10^{18}. It may not be necessary to consider all the possibilities since several descriptions may lead to the same results no matter which data are given as a learning example: they may belong to the same Markov equivalence class.

For example, if we could equally learn $P(A)$, $P(B)$, $P(C)$, $P(A \mid B)$, $P(B \mid A)$, $P(C \mid B)$, and $P(B \mid C)$ from a set of triplets a_i, b_i, C_i, then the three decompositions $P(C)\,P(B \mid C)\,P(A \mid B)$, $P(A)\,P(B \mid A)\,P(C \mid B)$, and $P(B)\,P(C \mid B)\,P(A \mid B)$ will lead to the same joint distribution $P(A \wedge B \wedge C)$ while the decomposition $P(A)\,P(C)\,P(B \mid C \wedge A)$ will lead to another decomposition since $P(B \mid C \wedge A) \neq P(B \mid C)$.

The double exponential number of possible models makes the problem intractable, and methods such as the model selection algorithm presented in Section 15.2.3 cannot be applied. The existing algorithms rely on heuristics, and could be classified into two classes:

- The algorithms based on mutual information.

- The algorithms based on scores.

We use the work of Leray [2006] to briefly present the main approaches.

In the program in the file: "chapter15/structure_learning.py" the function "generate_data" generates a data set from a known joint distribution, here the distribution corresponding to the famous *Asia* model which is built with the function make_model_asia(). The goal of the following algorithms is to reconstruct a good approximation of the decomposition used to build this model from the produced data set.

```
def make_model_asia():
    # VARIABLES SPECIFICATION
    A = plSymbol ("A", PL_BINARY_TYPE); # visit to Asia?
    S = plSymbol ("S", PL_BINARY_TYPE); # Smoker?
    T = plSymbol ("T", PL_BINARY_TYPE); # has Tuberculosis
    L = plSymbol ("L", PL_BINARY_TYPE); # has Lung cancer
    B = plSymbol ("B", PL_BINARY_TYPE); # has Bronchitis
    O = plSymbol ("O", PL_BINARY_TYPE); # has tuberculosis \
                                          or cancer
    X = plSymbol ("X", PL_BINARY_TYPE); # positive X-Ray
    D = plSymbol ("D", PL_BINARY_TYPE); # Dyspnoea?
    # PARAMETRIC FORM SPECIFICATION
    tableA = [ 0.99, 0.01 ];
    tableS = [ 0.5,  0.5  ];
    tableT = [ 0.99, 0.01,     # P(T | [A=f])
               0.95, 0.05 ];   # P(T | [A=t])
    tableL = [ 0.99,  0.01,
               0.9,   0.1  ];
    tableB = [ 0.7,  0.3,
               0.4,  0.6  ];
    tableO = [ 1,    0,       # P(O | [T=f]^[L=f])
               0,    1,       # P(O | [T=f]^[L=t])
               0,    1,       # P(O | [T=t]^[L=f])
               0,    1    ];  # P(O | [T=t]^[L=t])
    tableX = [ 0.95, 0.05,
               0.02, 0.98 ];
    tableD = [ 0.9,  0.1,
               0.3,  0.7,
               0.2,  0.8,
               0.1,  0.9  ];
    # DECOMPOSITION
    P_A = plProbTable(A, tableA);
    P_S = plProbTable(S, tableS);
    P_T_k_A = plDistributionTable (T, A, tableT);
    P_L_k_S = plDistributionTable (L, S, tableL);
    P_B_k_S = plDistributionTable (B, S, tableB);
    P_O_k_T_L = plDistributionTable (O, T^L, tableO);
    P_X_k_O = plDistributionTable (X, O, tableX);
    P_D_k_O_B = plDistributionTable (D, O^B, tableD);
    variables = A^S^T^L^B^O^X^D;
    jd = plJointDistribution(variables, P_A * P_S * \
                              P_T_k_A * P_L_k_S *
                              P_B_k_S * P_O_k_T_L * \
                              P_X_k_O * P_D_k_O_B);
    return jd
```

15.3.1 Directed minimum spanning tree algorithm: DMST

The search space is limited to valid decompositions having the format used in the program in Equation 15.28, in other words, to any valid oriented tree having X_1 as its root.

$$Pr:\begin{cases}Ds:\begin{cases}Sp(\pi):\begin{cases}Va: X_1,\ldots,X_n\\ Dc:\begin{cases}P(X_1\wedge\ldots X_n\mid\pi)\\ =P(X_1\mid\pi)\\ \prod P(X_k\mid X_l\wedge\pi)\end{cases}\\ Fo:\begin{cases}P(X_1\mid\pi)=\text{Laplace}\\ P(X_k\mid X_l\wedge\pi)=\text{Laplace}\end{cases}\end{cases}\\ Id:\end{cases}\\ Qu:\end{cases} \tag{15.28}$$

The search space is reduced further by considering nonoriented trees since all the oriented trees built from this initial structure belong to the same Markov equivalent class. Intuitively, we should select $P(X_k \mid X_l)$ or $P(X_l \mid X_k)$ as a possible member of the decomposition if X_l and X_k are strongly correlated. Starting from this idea and a correlation measure to weight each pair X_k, X_l the DMST algorithm builds the spanning tree covering all the variables of interest and maximizing the sum of weights (measure of correlation) on the selected edges (using the Kruskal algorithm). This tree is then oriented starting from the variable having the highest rank, and standard learning could take place on each distribution to obtain the description.

Various correlation measures may be used, for example:

- Mutual information:

$$MI(X_l, X_k) = \sum_{x_l \in X_l} \sum_{x_k \in X_k} P(x_l \wedge x_k) \log\left(\frac{P(x_l \wedge x_k)}{P(x_l)\,P(x_k)}\right)$$

  where $P(x_l \wedge x_k)$,$P(x_l)$,$P(x_k)$ are computed as histograms from the data set.

- Laplace mutual information: as the MI measure but where $P(x_l \wedge x_k)$, $P(x_l)$, $P(x_k)$ are computed as Laplace laws from the data set.

- Normalize mutual information:

$$NMI(X_l, X_k) = -\frac{MI(X_l, X_k)}{H(X_l, X_k)}$$

  where the entropy $H(X_l, X_k)$ is computed as

$$\sum_{x_l \in X_l} \sum_{x_k \in X_k} -P(x_l \wedge x_k) \log\left(P(x_l \wedge x_k)\right)$$

In the program in the file: "chapter15/structure_learning.py" we use the MDL score to call the DMST algorithm.

```
score_dmst = plEdgeScoreMDL_( dataset)
result = learner.DMST( score_dmst, order, root_index )
result_dmst = learner.get_joint_distribution( dataset )
```

The vector order will be used by the K2 algorithm.

15.3.2 Score-based algorithms

Unlike the previous method that attempts to use conditional independence between variables, these approaches will find the structure that maximizes a certain score on the decomposition.

For these approaches to be feasible, it requires the score to be locally decomposable, that is to say equal to the sum of the local scores at each node. As an exhaustive search is not possible, the algorithms proposed work in a smaller space (scheduling nodes), or perform a greedy search (GS) in the search space.

15.3.2.1 Possible scores

Most of the existing scores are based on the Occam Razor principle: they tend to favor simple descriptions (Ds) while trying to best explain the available data. As such, most of the scores are made of two terms. A first term maximizes the likelihood $P(D \mid \Theta \wedge Dc)$ where Θ are the parameters necessary to instantiate the decomposition Dc. A second term penalizes too complex models as the ones having many parameters: $\dim \Theta$.

For example, if we have N independent observations $D = \bigcup_i^N D_i$, the $BIC(Ds, D)$ (Bayesian information criterion) score for a description Ds given a data set D is defined as follows:

$$BIC(Ds, D) = \log\left(\sum_i^n P(D_i \mid \Theta \wedge Dc)\right) - \frac{1}{2}\dim(\Theta)\log(N)$$

In the same spirit the score based on the minimum description length, MDL, penalizes descriptions having a large number (N_{Dc}) of distributions in their decomposition:

$$MDL(Ds, D) = \log\left(\sum_i^n P(D_i \mid \Theta \wedge Dc)\right) - N_{Dc}\log(N) - c\dim(\Theta)\log(N)$$

Many other scores are used to compare one description to another for example, Akaike's information criterion (AIC) and Bayesian Dirichlet equivalent uniform criterion (BDEU). Once a score has been selected it is used to locally navigate among the large space of descriptions.

15.3.2.2 The K2 algorithm

As MWST, the K2 algorithm considers a subspace of the decomposition set. An ordered set of variables $OS : \{X_1 < X_2 \ldots < X_n\}$ implicitly defines a subset of valid decompositions. Intuitively we will use the K2 algorithm if we have an idea of this ordering (for example if we believe they are causal relationships between the variables). We define the subspace as the valid decomposition which could be written as in Equation 15.29 (Bayes network) and having the following property:

$$\forall P\left(X_k \mid X_1^k \wedge \ldots \wedge X_p^k \wedge \pi\right) \in Dc : \forall j, k : X_k > X_j^k$$

$$Pr : \begin{cases} Ds : \begin{cases} Sp(\pi) : \\ \begin{cases} Va : X_1, \ldots, X_n \\ Dc : \begin{cases} P\left(X_1 \wedge \ldots X_n \mid \pi\right) \\ = P\left(X_1 \mid \pi\right) \\ \prod P\left(X_k \mid X_1^k \wedge \ldots \wedge X_p^k \wedge \pi\right) \end{cases} \\ Fo : \begin{cases} P\left(X_1 \mid \pi\right) = \text{Histograms} \\ P\left(X_k \mid X_1^k \wedge \ldots \wedge X_p^k \wedge \pi\right) = \text{Histograms} \end{cases} \end{cases} \\ Id : \end{cases} \\ Qu : \end{cases} \tag{15.29}$$

Given a score function $score()$, a training and an evaluation sets D' and D, as well as an upper limit $UMax$ for the number of conditioning variables, one version of the K2 algorithm could be defined as:

$Dc = P(X_1)P(X_2)...P(X_n)$
$Ds = identify(Dc, D')$ {use data to identify each term of the decomposition}
$Smax = score(Ds, D)$ {score it with another set}
for $i = 1 \rightarrow N$ **do**
  $A = \Phi$ {Reset the ancestor of variable i}
  U=0 {Reset the number of ancestor}
  for $j = i - 1 \rightarrow 1$ **do**
    {Find suitable ancestors}
    while $U < Umax$ **do**
      $Dc' = Dc$
      $A' = A \wedge X_j$
      Replace $P(X_i \mid A)$ by $P(X_i \mid A') inDc'$
      $Ds' = identify(Dc', D')$
      if $score(Ds', D) > Smax$ **then**
        $A = A'$
        $Smax = score(Ds', D)$
        $U = U + 1$
        $Dc = Dc'$
      end if
    end while
  end for
end for
return Dc

The main idea of the algorithm is to condition the probability distribution on a given variable by a limited number of variables with a higher rank in the given order OS.

The K2 algorithm uses the output of the DMST algorithm to define the distribution order.

```
# Apply the K2 algorithm with BDeu score on the same dataset.
Nprime = 100.0; #BDeu parameter
score_k2 =plNodeScoreBDeu_ (dataset, Nprime);
max_parents = 2; #K2 parameter
learner.K2(score_k2, order, max_parents);
result_k2 = learner.get_joint_distribution(dataset);
```

15.3.2.3 The greedy search

The greedy search algorithms are based on elementary operations to change the decomposition. In the context of Bayesian networks, these operations are ways to walk in the search space moving from one graph to another by modifying edges. The chosen score is used to evaluate the quality of a move and a decision is made selecting the best alternative. The algorithm never goes back and stops when no move leads to a better solution than the current one.

For example, we may consider the following operations on Bayesian networks provided they lead to a valid decomposition:

- **remove**:
  $P\left(X_k \mid X_1^k \wedge \ldots X_j^k \ldots \wedge X_p^k\right) \rightarrow P\left(X_k \mid X_1^k \wedge \ldots \wedge X_p^k\right)$
- **add**:
  $P\left(X_k \mid X_1^k \wedge \ldots \wedge X_p^k\right) \rightarrow P\left(X_k \mid X_1^k \wedge \ldots X_j^k \ldots \wedge X_p^k\right)$
- **reverse**:
  $P\left(X_k \mid X_1^k \wedge \ldots X_j^k \ldots \wedge X_p^k\right) \rightarrow P\left(X_k \mid X_1^k \wedge \ldots \wedge X_p^k\right) P\left(X_k^j \mid X_k\right)$

For a decomposition Dc, we define by $V(Dc)$ the neighborhood of Dc as the set of the valid decompositions which may be obtained my applying one of the above transformations on the decomposition Dc. Given a score, one version of the greedy search algorithm may be sketched as follows:

$Dc = P(X_1)\,P(X_2)\ldots P(X_n)$ {or any initial decomposition}
$Ds - I(Dc, D')$ {use data to identify each term of the decomposition}
$Smax = score(Ds, D)$ {score it with another data set}
while true **do**
  $Dc' : \forall dc \in V(Dc) score(I(Dc', D'), D) > score(I(dc, D'), D)$
  if $score(I(Dc', D'), D) < Smax$ **then**
    $Return(Ds = I(Dc, D'))$ {return local maximum}
  end if
end while

The variations among this type of algorithm are numerous. They could also be used to explore very small subspaces of the initial search space by only considering a small subset of the conditional distribution which may be modified.

The greedy search (GS) algorithm uses the output of the DMST algorithm to define the initial structure.

```
learner = plStructureLearner(result_dmst);
score_gs = plNodeScoreBIC_ (dataset);
learner.GS(score_gs);
result_gs = learner.get_joint_distribution(dataset);
```

Part IV

Frequently Asked Questions — Frequently Argued Matters

Chapter 16

Frequently Asked Questions and Frequently Argued Matters

> The modern scientist is a more apt recipient than any one else of Kipling's austere advice: "If you can see your life's work suddenly collapse and then start work again, if you can suffer, struggle and die without complaint, you will be a man my son." Only the work of science can make you love what you destroy, only here can you continue the past by repudiating it, only here can you honor your professor by contradicting him.[1]
>
> *La Formation de l'Esprit Scientifique: Contribution a une Psychanalyse de la Connaissance Objective*
> Gaston Bachelard [1938]

[1] C'est au savant moderne que convient, plus qu'à tout autre, l'austère conseil de Kipling: "Si tu peux voir s'écrouler soudain l'ouvrage de ta vie, et te remettre au travail, si tu peux souffrir, lutter, mourir sans murmurer, tu seras un homme, mon fils." Dans l'oeuvre de la science seulement on peut aimer ce qu'on détruit, on peut continuer le passé en le niant, on peut vénérer son maître en le contredisant.

We believe that Bayesian modeling is an elegant matter that can be presented simply, intuitively, and with mathematical rigor. We hope that we succeed in doing so in this book. However, the subjectivist approach to probability has been and still is a subject of countless controversies.

To make the main exposition as clear and simple as possible, none of these controversies, historical notes, epistemological debates, and tricky technical questions are discussed in the body of the book. We have made the didactic choice to develop all these questions in this chapter titled "FAQ and FAM" ("Frequently Asked Questions and Frequently Argued Matters").

This chapter is organized as a collection of "record cards," at most three pages long, presented in alphabetical order. Cross references to these subjects are included in the main text for readers who are interested in going further than a simple presentation of the principles of Bayesian modeling.

16.1 Alternative Bayesian inference engines

ProBT is of course not the only probabilistic inference engine available but it is the only one designed to implement Bayesian programming. Around 70 of them are present in the inventory by Kevin Murphy, made available on the Web at the following URL:

`http://www.cs.ubc.ca/~murphyk/Software/bnsoft.html`

This list is in constant evolution. It gathers very different software, either industrial or academic, either commercial or free, either open-source or not. Some are very specialized for a given type of inference or graphical model, some are more general. They are also at very different states of development as some are industrial products or advanced collaborative projects resulting from many years of work while some others are more "experimental." Among the most well known, we will cite in alphabetical order and without the probabilistic programming languages addressed in the specific Section 16.6:

- BayesiaLab: an industrial product implementing essentially Bayesian networks.
- BUGS: a very active open-source project presented in detail in the book titled *The BUGS Book: A Practical Introduction to Bayesian Analysis* [Lunn et al., 2012].
- Hugin expert: another industrial product dedicated to Bayesian nets which was the very first one available on the market 20 years ago.
- Infer.NET: an inference engine under development by Microsoft Research and combining graphical models and probabilistic programming.
- Netica: an historical reference for the inference in Bayesian nets.

- Stan: an academic inference engine running in C++ and R.

16.2 Bayesian programming applications

In this section we list the different applications of Bayesian programming. On one hand we have the academic applications, essentially 23 PhD theses defended in different universities around Europe during the past 15 years and, on the other hand, industrial applications mainly developed by the ProbaYes company.

The academic applications are summarized in the following list in chronological order. Twelve of them are described in the book titled *Probabilistic Reasoning and Decision Making in Sensory-Motor Systems* [Bessière et al., 2008].

- Eric Dedieu, PhD thesis titled "Contingent Representation," which identifies the intrinsic difficulty of incompleteness in robot programming and first proposed to use the Jaynes' "Probability as Logic" approach to deal with it (see [Dedieu, 1995](in French) and [Dedieu and Mazer, 1992; Bessiere et al., 1994, 1997]).

- Olivier Lebeltel, PhD thesis titled "Bayesian Robot Programming," where the original proposition of the Bayesian programming methodology can be found (see [Lebeltel, 1999](in French) and [Lebeltel et al., 2000, 2004]).

- Kamel Mekhnacha, PhD thesis titled "A Robotic CAD System Using a Bayesian Framework," which presents a methodology based on Bayesian formalism to represent and to handle geometric uncertainties in robotics and CAD systems (see [Mekhnacha, 1999](in French) and [Mekhnacha et al., 2000, 2001; Mekhnacha and Bessière, 2008]).

- Christophe Coué, PhD thesis titled "Bayesian Occupancy Filtering (BOF) for Multi-Target Tracking: An Automotive Application," where a new approach for robust perception and risk assessment in highly dynamic environments is proposed. This approach is called Bayesian Occupancy Filtering (BOF). Tt basically combines a four-dimensional occupancy grid representation of the obstacle state space with Bayesian filtering techniques (see [Coué, 2003](in French) and Coué et al. [2006]).

- Ruben Seren Garcia Ramirez, PhD thesis titled "Bayesian Programming of Robotic Arm," in which a pick and place task has been entirely probabilistically programmed (see [Garcia-Ramirez, 2003](in French)).

- Julien Diard, PhD thesis titled "The Bayesian Map: A Hierarchical Probabilistic Model for Mobile Robot Navigation," where a representation of space using hierarchies of Bayesian behavioral representations is proposed (see [Diard, 2003](in French) and [Diard et al., 2004; Diard and Bessière, 2008; Diard et al., 2010]).

- Céderic Pradalier, PhD thesis titled "Perceptual Navigation Around a Sensory-Motor Trajectory," where behavioral replay of a sensory-motor trajectory has been completely modeled using Bayesian programming (see [Pradalier, 2004](in French) and [Pradalier et al., 2003, 2004, 2005; Pradalier and Bessière, 2008]).

- Jihene Serkhane, PhD thesis titled "Building a Talking Baby Robot: A Contribution to Speech Acquisition and Evolution," in which a Bayesian model of babies speech early acquisition is proposed (see [Serkhane, 2005](in French) and [Serkhane et al., 2002, 2003, 2005; Schwartz et al., 2004; Serkhane et al., 2007, 2008; Boë et al., 2008]).

- Guy Ramel, PhD thesis which proposes a new approach for objects recognition that incorporates visual and range information with spatial arrangement between objects (see [Ramel, 2006](in French) and [Ramel and Siegwart, 2008]).

- Adriana Tapus, PhD thesis titled "Topological SLAM—Simultaneous Localization and Mapping with Fingerprints of Places" (see [Tapus, 2005] and [Tapus, 2008]).

- Carla Maria Chagas e Cavalcante Koike, PhD thesis titled "Bayesian Approach to Action Selection and Attention Focusing. An Application in Autonomous Robot Programming" (see [Koike, 2005] and [Koike et al., 2008]).

- Miriam Amavizca, PhD thesis titled "3D Human Hip Volume Reconstruction with Incomplete Multimodal Medical Images" (see [Amavizca, 2005](in French) and [Amavizca, 2008]).

- Jean Laurens, PhD thesis titled "Bayesian Model of Visuo-Vestibular Interaction," which proposes a Bayesian model of motion estimation using both visual and vestibular information and reproduces different classical human illusions (see [Laurens, 2006](in French) and [Laurens and Droulez, 2007, 2008]).

- Francis Colas, PhD thesis titled "Perception of Shape from Motion," where a unique Bayesian model of human perception of planes from the optical flow is proposed justifying the results of six different psychophysic experiments from the literature (see [Colas, 2006](in French) and [Colas et al., 2008a,b]).

- Ronan Le Hy, PhD thesis titled "Playing to Train Your Video Game Avatar," where it is demonstrated how a player of an FPS video game can teach an avatar how to play (see [Le Hy, 2007](in French) and [Le Hy et al., 2004; Le Hy and Bessière, 2008]).
- Pierre-Charles Dangauthier, PhD thesis titled "Bayesian Learning: Foundations, Method and Applications," which deals with different aspects of learning and especially addresses the automatic selection and creation of relevant variables to build a model (see [Dangauthier, 2007](in French) and [Dangauthier et al., 2004, 2005, 2007]).
- Shrihari Vasudevan, PhD thesis titled "Spatial Cognition for Mobile Robots: A Hierarchical Probabilistic Concept-Oriented Representation of Space" (see [Vasudevan, 2008] and [Vasudevan and Siegwart, 2008]).
- Francis Colas investigated the role of position uncertainty in the peripheral visual field to guide eye movement saccades (see [Colas et al., 2009]).
- Jorg Rett, PhD thesis titled "Robot-Human Interface Using Laban Movement Analysis Inside a Bayesian Framework" (see [Rett, 2008] and [Rett et al., 2010]).
- Estelle Gilet, PhD thesis titled "Bayesian Modeling of Sensory-Motor Loop: An Application to Handwriting," where a Bayesian Action Perception (BAP) model of the reading-writing sensory motor loop is proposed (see [Gilet, 2009](in French) and [Gilet et al., 2011]).
- Xavier Perrin, PhD thesis titled "Semi-Autonomous Navigation of an Assistive Robot Using Low Throughput Interfaces," where a Bayesian strategy to help a disabled person to drive a wheelchair using an EEG signal is proposed (see [Perrin, 2009] and [Perrin et al., 2010]).
- Joao Filipe Ferreira, PhD thesis titled "Bayesian Cognitive Models for 3D Structure and Motion Multimodal Perception" (see [Ferreira, 2011] and [Ferreira et al., 2012]).
- Clement Moulin-Frier, PhD thesis titled "Emergence of Articulatory-Acoustic Systems from Deictic Interaction Games in a 'Vocalize to Localize' Framework"(see [Moulin-Frier, 2011](in French) and [Moulin-Frier et al., 2011, 2012]).
- Gabriel Synnaeve, PhD thesis titled "Bayesian Programming and Learning for Multi-Player Video Games: Application to RTS AI," where a probabilistic model of a "bot" to automatically play Starcraft is proposed (see [Synnaeve, 2012] and also [Synnaeve and Bessière, 2010, 2011a,b,c]).

For an up-to-date description of the industrial applications please consult the Web site of ProbaYes (http://probayes.com).

16.3 Bayesian programming versus Bayesian networks

At first, the Bayesian programming syntax presented in this book may seem less convenient than the graphical presentation of standard Bayesian network software. The absence of an evident human-machine interface is not an oversight but a choice. This choice was made for four main reasons:

- We think that graphical representations impose supplementary constraints which do not result from the rules of probability nor from the logic of the problem. For instance, the rules of probability allow us to specify a decomposition including a distribution with two or more variables on the left part of the conditioning mark as, for example, $P(B \wedge C|A \wedge \pi)$. This is not possible in a Bayesian network graphical representation without introducing an intermediate variable. This limitation becomes even more bothersome as seen in Equation 16.1. The two variables B and C are defined with the same joint distribution $P(B \wedge C|\pi)$, while D is conditionally dependent on B and E on C.

$$P(B \wedge C|\pi) P(D|B \wedge \pi) P(E|C \wedge \pi) \tag{16.1}$$

  This decomposition becomes really difficult to represent in a graphical way. Any Bayesian network can be represented in the Bayesian programming formalism, but the opposite is not true. Indeed, the graphical representation with the same power of expression as Bayesian programming is probabilistic factor graphs [Loeliger, 2004].

- The algebraic notation used in Bayesian programming is very convenient for expressing iteration or recurrences. This greatly simplifies the specification of models that include the same submodel duplicated several times, such as Bayesian filters or hidden Markov models where $\prod_{t=1}^{T} \left[P(S^t|S^{t-1}) P(O^t|S^t)\right]$ specifies everything in a very compact and rigorous manner.

- Bayesian programming, using the subroutine call mechanism described in Chapter 9, offers a very simple way to build hierarchical complex probabilistic models built from simpler elementary bricks. Much simpler and rigorous than the attempt to do the same thing with graphical notation such as for instance the network fragment approach [Laskey and Mahoney, 1997] or the Object Oriented Bayesian Network (OOBN) [Koller and Pfeffer, 1997] to cite only the oldest ones.

- As shown in Chapter 14, the algebraic notation used by Bayesian programing may be used to revisit the different inference algorithms with a simpler and more systematic point of view, demonstrating that all these algorithms finally reduce to clever applications of the distributive and normalization laws.

16.4 Bayesian programming versus Bayesian modeling

Why did we call this book Bayesian "Programming"?

As stated in the very first paragraph, the use of computers makes the difference between "programming" and "modeling":

"Computers have brought a new dimension to modeling. A model, once translated into a program and run on a computer, may be used to understand, measure, simulate, mimic, optimize, predict, and control. During the last 50 years, science, industry, finance, medicine, entertainment, transport, and communication have been completely transformed by this revolution."

What is proposed in this book is a programming language and methodology.

The goal is to have probabilistic models running on computers. Bayesian programming goes one step further than modeling by providing the adequate tools to do so. Even if other approaches may be explored (see, for instance, the discussion "Bayesian programming versus probabilistic programming" in Section 16.6 just below) Bayesian programming offers all the features required from a programming language (as presented in Part II of this book).

A parallel can be made, for instance, with geometry. Geometry modeling was of course very interesting and has been very useful to humans since they owned land and had to compute the surface of their fields to buy or sell them. However, since geometry programming has been implemented on computers and especially since specific hardware has been developed to allow for incredibly efficient geometrical computation, revolutionary applications have appeared such as CAD systems to conceive 3D objects and buildings, mapping and localization systems, special effects for moving pictures and video-game animation.

Will our computers offer us embedded capabilities for perception, action, learning, and decision with incomplete and uncertain knowledge? Will they offer us completely new applications? Only the near future will provide the answer.

16.5 Bayesian programming versus possibility theories

The comparison between probabilistic approaches (not only Bayesian programming) and possibility theories has been debated for a long time and is, unfortunately, a very controversial matter.

Possibility theories (like, for instance, fuzzy sets [Zadeh, 1965], fuzzy logic [Zadeh, 1974, 1975] and possibility theory [Dubois and Prade, 2001]) propose different alternatives to probability to model uncertainty. They argue that probability is insufficient or inconvenient to model certain aspects of incomplete and uncertain knowledge.

The defense of probability is mainly based on Cox's theorem (see Section 16.8 below) which, starting from four postulates concerning rational reasoning in the presence of uncertainty, demonstrates that the only mathematical framework that satisfies these postulates is probability theory. The argument then goes like this: if you use a different approach than probability, then you necessarily infringe on one of these postulates. Let us see which one and discuss its utility. The debate is still open and it is beyond the scope of this book to discuss it in detail.

16.6 Bayesian programming versus probabilistic programming

The purpose of probabilistic programming is to unify the scope of classical programming languages with probabilistic modeling (especially Bayesian networks) in order to be able to deal with uncertainty but still profit from the power of expression of programming languages to describe complex models.

The extended classical programming languages can be logical languages as proposed in Probabilistic Horn Abduction [Poole, 1993], Independent Choice Logic [Poole, 1997], PRISM [Sato and Kameya, 2001], and ProbLog [Raedt et al., 2007] which propose an extension of Prolog. It can also be extensions of functional programming languages (essentially Lisp and Scheme) such as IBAL [Pfeffer, 2001] or CHURCH [Goodman et al., 2008]. The inspiring programming languages can even be object oriented like in BLOG [Milch et al., 2004] and FACTORIE [Mccallum et al., 2009] or more standard like in CES [Thrun, 2000a] and FIGARO [Pfeffer, 2012].

An interesting paper presenting a synthesis of the probabilistic programming approach and especially the semantics of these languages is proposed by Poole [2010].

The purpose of Bayesian programming is different. Following Jaynes' precept of "probability as logic" we defend that probability is an extension of and alternative to logic above which a complete theory of rationality, computation, and programming can be rebuilt. We do not search to extend classical languages but rather to replace them by a new programming approach based on probability and taking fully into account incompleteness and uncertainty. As stated in the Introduction, "the next step in this direction would be to develop specific hardware, a probabilistic computer, with Bayesian gates instead of Boolean gates and a Bayesian algebra to formalize probabilistic computation instead of the Boolean algebra used for logic-based calculus." We are working in this direction, but this is not in the scope of this book.

The precise comparison between the semantic and power of expression of Bayesian and probabilistic programming is still an open question. It would be very interesting, for instance, to exhibit examples of problems that can be addressed with one of the approaches and not with the other, or, at least that are much more simple to describe in one of the approaches than the other.

16.7 Computational complexity of Bayesian inference

As already stated several times, Bayesian inference is a very time consuming problem. The major problems of probabilistic inference have been proved NP-hard or worse. This is the case, for instance, for exact inference in singly connected Bayesian nets [Wu and Butz, 2005] and in multiply connected Bayesian nets [Cooper, 1990], for approximate inference [Dagum and Luby, 1993]), for optimal triangulation [Jordan and Weiss, 2002], and for optimal ordering [Arnborg et al., 1987]. Darwiche [2009] proposes a synthetic review of these questions.

The only way out is to simplify the problem. There are two approaches.

The first one is problem independent and consists of using generic approximation algorithms based on heuristics and sampling. Some of these have been described in Chapter 14. Markov chain Monte Carlo (MCMC) algorithms are obviously an immense success in this category. However, until proven otherwise, NP-hard problems stay NP-hard and this implies that there exist necessary cases where these algorithms will fail.

The second one is problem dependent and consists of dividing the problem into a combination of more simple ones using an additional conditional independence assumption. This is the central purpose of the decomposition step in the Bayesian programming approach. The efficacy of this process has been demonstrated numerous times in this book. However, any new conditional independence hypothesis is paid by a certain loss of information and transforms the initial problem into a different one. The very difficult challenge the programmer has to take is to find the conditional independence hypotheses that make sense and lead to acceptable results.

16.8 Cox theorem

The Cox theorem [Cox, 1946, 1961] demonstrates how the intuitive notion of plausibility can be and should be formalized by the mathematical concept of probability.

Let us note $(a|c)$ the plausibility of a given proposition a knowing a set of knowledge c.

The following postulates explicitly the plausibility notion:

1. Plausibilities are represented by real numbers[2]

$$(a|c) \in \mathbb{R} \tag{16.2}$$

2. Plausibilities are consistent: if there exist several correct calculi for the same plausibility, they should all lead to the same result.

3. If some new information c' replaces c and increases the plausibility of a, then the plausibility of the negation $\bar{a}$ should decrease.

$$[(a|c') > (a|c)] \Rightarrow [(\bar{a}|c') < (\bar{a}|c)] \tag{16.3}$$

4. If some new information c' increases the plausibility of a but does not concern in any way the plausibility of b, then the plausibility of the conjunction $a \wedge b$ should increase:

$$[[(a|c') > (a|c)] \wedge [(b|a \wedge c') = (b|a \wedge c)]] \Rightarrow [(a \wedge b|c') > (a \wedge b|c)] \tag{16.4}$$

Starting from these postulates, Richard T. Cox has demonstrated that plausible reasoning should follow the two rules from which all the theories can be rebuilt:

1. The normalization rule:

$$P(a|c) + P(\bar{a}|c) = 1 \tag{16.5}$$

2. The conjunction rule:

$$P(a \wedge b|c) = P(a|c)\,P(b|a \wedge c) = P(b|c)\,P(a|b \wedge c) \tag{16.6}$$

Furthermore, this theorem shows that any technic for plausibility calculus that would not respect these two rules would contradict at least one of the

[2]See an interesting discussion in Appendix A.3, p. 656 of Jaynes' book [2003] arguing that rational numbers are sufficient.

preceding postulates. Consequently, if we accept these postulates, probability calculus is the *only* means to do plausible reasoning.[3]

Chapter 2 of Jaynes' book [2003] is completely devoted to the demonstration and discussion of Cox's theorem and is a reference on this matter.

Cox's theorem has been partially disputed by Halpern [1999a; 1999b], himself contradicted by Arnborg and Sjödin [2000].

16.9 Discrete versus continuous variables

In this book we mainly restricted ourselves to the use of discrete variables.

The main reason for this is that we believe that continuous variables are a pure fiction on the computer, at worst a quite dangerous one. For instance, real numbers coded by types "float," "double," or "long double" are all coded using discrete values either on 32, 64, or 80 bits. Unfortunately, discretization of continuous values may lead to numerous problems would this discretization be using 2 bits or 80 bits. For instance, two different roots of a polynomial function may be indistinguishable as they may take the same value once discretized. Well-posed mathematical methods can lead to ill-behaved computer algorithms due to the effect of discretization.

We think that "discretization" is a very important and difficult modeling choice that deserves careful thinking and cannot be blindly left to computer float encoding.

For instance, if you have a signal with a high dynamic either in energy as light for optical sensors or sound for auditory ones, or in range as distance for laser devices, a logarithmic discretization is often wise to preserve accuracy for small values yet cover the whole range of the signal.

Another example is given by dynamical environment (with rapid change in time) where rough (with few values) discretization is often a clever choice. Indeed, there is a trade-off between the precision of discretization and the required computing time for inference. In some cases, lowering the accuracy in order to speed up computation may lead to better results by improving the reactivity to rapidly changing environments.

In Bayesian programming, the first and most difficult modeling choice is the selection of relevant variables. To be more complete and exact, we should say that this first modeling choice is the selection of the relevant variables and *their encoding as discrete variables.* This encoding choice is a very important component of the model which is too often neglected as it is a delicate question.

Furthermore, considering continuous variables obliges us to use measure

[3]See the corresponding discussion about "Bayesian programming versus possibility theories" in section 16.5.

theory as defined by Emile Borel and Henri-Léon Lebesgue and leads to Andrey Kolmogorov's axiomatization of probability. Of course this approach to probability is of primary importance and has countless applications. However, it is a very different concept and viewpoint on probability than the epistemological position adopted in this book where probability is considered as an alternative and extension of logic (see Jaynes [2003] for an extended discussion and comparison between his approach and Kolmogov's).

16.10 Incompleteness irreducibility

Incompleteness (the fact that a model neglects some factors influencing the phenomenon it is modeling) has been proved all along in this book to be the major difficulty encountered by the classical approaches to modeling (logic and usual programming). This difficulty is the central justification for the adoption of Bayesian programming proposed as a way to take into account this incompleteness.

However, a legitimate question is: "Is this incompleteness reducible?" or stated in other terms: "Is it possible, by adding more and more variables to our model, to obtain a model with no hidden variables?"

If the studied phenomenon is not formal the answer to these questions is *NO.*

There are three fundamental arguments in favor of this negative answer:

- The very idea of exhaustivity is in contradiction to the concept of the model. A model is interesting if, and only if, it is (much) simpler than the studied phenomenon.
- The fundamental laws of physics are boundless in distance: any single particle is in interaction with all the others in the universe. Consequently, the entire universe should be necessary to describe *completely* any physical phenomenon even the simplest.
- "Chaotic" systems prove that it is impossible to provide a "simplified" model of some phenomenon, even some elementary ones, as it is necessary to know "exactly" their initial conditions to be able to predict their evolution.

Henri Poincaré summarized this in beautiful words:

> To find a better definition of hazard, we have to look for facts that we agree to qualify as fortuitous and for which probability calculus applies. We will then look for their common characteristics. The first instance that we can choose in unstable equilibrium.

If a cone rests on its point we know that it will fall down but we don't know on which side. It seems that only hazard will decide. Would the cone be perfectly symmetric, would its axis be perfectly vertical, would there be absolutely no other force than gravity, it will not fall. But the slightest symmetry break will tilt it on one side or another and, as soon as it will be tilted, so little that it is, it will completely fall on that side. Even with a perfect symmetry, an infinitesimal juddering, a breath of air will tilt it of a few arc seconds and it will be sufficient to cause its fall and to determine the direction of this fall toward the initial inclination.

An infinitesimal cause that we overlook may determine a major effect that we cannot miss. We then say that this effect is due to hazard. Would we know exactly the laws of nature and the state of the universe at the initial instant, we could exactly predict the state of this same universe at the next moment. But, even with this perfect knowledge of the laws of nature, we have only an approximate knowledge of the initial state. If we can predict the next state with the same approximation, it's all what we need, the phenomenon has been forecast, it is ruled by laws. However, it is not always the case, it may happen that slight differences in initial conditions generate huge ones in final phenomenon. The prediction becomes impossible and we are facing a fortuitous phenomenon.[4]

Calcul des Probabilités
Henri Poincaré [1912]

[4]Pour trouver une meilleure définition du hasard, il nous faut examiner quelques-uns des faits qu'on s'accorde à regarder comme fortuits, et auxquels le calcul des probabilités paraît s'appliquer; nous rechercherons ensuite quels sont leurs caractères communs. Le premier exemple que nous allons choisir est celui de l'équilibre instable; si un cône repose sur sa pointe, nous savons bien qu'il va tomber, mais nous ne savons pas de quel côté; il nous semble que le hasard seul va en décider. Si le cône était parfaitement symétrique, si son axe était parfaitement vertical, s'il n'était soumis à aucune autre force que la pesanteur, il ne tomberait pas du tout. Mais le moindre défaut de symétrie va le faire pencher légèrement d'un côté ou de l'autre, et dès qu'il penchera, si peu que ce soit, il tombera tout à fait de ce côté. Si même la symétrie est parfaite, une trépidation très légère, un souffle d'air pourra le faire incliner de quelques secondes d'arc; ce sera assez pour déterminer sa chute et même le sens de sa chute qui sera celui de l'inclinaison initiale. Une cause très petite, qui nous échappe, détermine un effet considérable que nous ne pouvons pas ne pas voir, et alors nous disons que cet effet est dû au hasard. Si nous connaissions exactement les lois de la nature et la situation de l'univers à l'instant initial, nous pourrions prédire exactement la situation de ce même univers à un instant ultérieur. Mais, lors même que les lois naturelles n'auraient plus de secret pour nous, nous ne pourrions connaître la situation qu'approximativement. Si cela nous permet de prévoir la situation ultérieure avec la même approximation, c'est tout ce qu'il nous faut, nous disons que le phénomène a été prévu, qu'il est régi par des lois; mais il n'en est pas toujours ainsi, il peut arriver que de petites différences dans les conditions initiales en engendrent de très grandes dans les phénomènes finaux; une petite erreur sur les premières produirait une erreur énorme sur les derniers. La prédiction devient impossible et nous avons le phénomène fortuit.

In practice, however, it is not necessary to invoke these fundamental reasons to justify the irreducibility of incompleteness. A sensory motor system, either living or artificial, should evidently be able to take decisions with only a very partial knowledge of its interaction with its environment. Can we imagine that a bee has a complete model of its aerodynamic interaction with the environment to fly around without running into obstacles?

16.11 Maximum entropy principle justifications

The most intuitive justification of maximum entropy approaches consists in a combinatory argument coming directly from its statistical mechanic origins. This was first proposed by Boltzmann.

Let us suppose that we have N identical particles and that each of these particles can be in Q different microscopic states. We can define a macroscopic state ν_k as a set $\{n_1, \cdots, n_Q\}$ of Q numbers such that each n_q is the number of particles in microscopic state q.

We must have, of course:

$$\sum_{q=1}^{Q} [n_q] = N \tag{16.7}$$

The system must also verify some energetic constraints:

$$\sum_{q=1}^{Q} [n_q \times e_q] = E \tag{16.8}$$

where E is the global energy of the system and where e_q is the energy of state q.

If $W(\nu_k)$ is the number of permutations of microscopic states that realize the macroscopic state ν_k ,

$$W(\nu_k) = \frac{N!}{n_1! \times \cdots \times n_Q!} \tag{16.9}$$

For Boltzmann, the most probable macroscopic state is the one that can be realized by the highest number of possible permutations of microscopic states. In other words, the macroscopic state that maximizes $W(\nu_k)$.

Using the Stirling formula to approximate the factorial for large n:

$$log\,(n!) = n \times log(n) - n + \sqrt{2\pi n} + \frac{1}{12n} + o\left(\frac{1}{n^2}\right) \tag{16.10}$$

we get:

$$log\left(W\left(\nu_k\right)\right) \simeq -N \times \sum_{q=1}^{Q}\left[\frac{n_q}{N} \times log\left(\frac{n_q}{N}\right)\right] \tag{16.11}$$

and the most probable macroscopic state is the one that maximizes the entropy yet respects the constraints of Equations 16.7 and 16.8.

An exact parallel reasoning can be made for probability distributions.

Let us suppose that we have N observations (analogous to particles) and that each of these observations consist in observing that one of the Q mutually exclusive propositions is true (analogous to the microscopic state). We can define a description (analogous to the macroscopic state) δ_k as a set $\{n_1, \cdots, n_Q\}$ of Q numbers such that each n_q is the number of observations of the proposition q.

If we want this description to be a normalized probability distribution, we must have:

$$\sum_{q=1}^{Q}\left[\frac{n_q}{N}\right] = \sum_{q=1}^{Q}\left[p_q\right] = 1 \tag{16.12}$$

On the one hand, suppose we have M functions $f_m(q)$ (called "observable functions") and assume further we can impose a constraint on each of these functions by setting the value of its expectation to F_M (called "constraint levels"):

$$\forall m, m \in \{1, \cdots, M\} \sum_{q=1}^{Q}\left[p_q \times f_m\left(q\right)\right] = F_m \tag{16.13}$$

On the other hand, if $W(\delta_k)$ is the number of permutations of observations that realize the decomposition δ_k, again using the Stirling formula, we get:

$$log\left(W\left(\delta_k\right)\right) \simeq -N \times \sum_{q=1}^{Q}\left[p_q \times log\left(p_q\right)\right] \tag{16.14}$$

then the most probable probability distribution is the one that maximizes the entropy yet respects the M constraints in Equation 16.13. The most probable probability distribution is the one that corresponds to the highest number of possible permutations of the observations compatible with the imposed constraints.

Different rigorous justifications of the entropy principles exist in the literature. The more interesting, may be the entropy concentration theorems as initially demonstrated by Jaynes [1982] for the discrete case and by Robert [1990] for the continuous case.

16.12 Noise or ignorance?

What is noise?

Let us consider the throw of a dice.

It is a very "complex" physical phenomenon. Complex in the sense that many factors ought to be taken into account as, for instance, initial forces, gravity, aerodynamics, bouncing forces, and so forth. Initial forces themselves depend on the musculature, the shape, the health of the body, on the roughness of the skin, and precise shape of the hands and fingers, on the motor control and even "will" of the thrower. Gravity depends on the balance of the dice and the main gravity attraction of earth, but also on local perturbations due to geological constraints and eventually the presence of heavy objects in the vicinity. It also depends on far away attractions as, to cite only the most preeminent one, the position of the moon at the time of the throw. Aerodynamics depend on anything that may generate a breath of air, from the neighbor's sneeze to the butterfly effect on the local climate.

Is the trajectory of the dice a deterministic physical process? A very argued question but absolutely not relevant to our problem. Would it be completely deterministic, it will not help us to predict the outcome as, obviously, we lack boundless information to be able to reconstruct the course of events.

If the dice is "loaded," for instance, toward an outcome of 6 and if we consider only a standard vertical gravity, we may predict the probability of this outcome. Any deviation from this prediction can be interpreted as "noise." However, it is not related to any physical property of the system, this deviation is only the result of our ignorance (or simplification choices) that makes us discard all the relevant information enumerated above.

To ensure that a throw is "honest", you withdraw any observable information using, for instance, a balanced dice that prevents you from using any macroscopic gravity prediction and a cup that deprives you from any reachable information about the initial forces. "Noise" is then so preeminent, your ignorance is so complete, that the best prediction you can reach is a uniform distribution on the six possible outcomes.

16.13 Objectivism versus subjectivism controversy and the "mind projection fallacy"

There are two extreme positions in the epistemology of probability: the objectivist and the subjectivist ones. On the one hand, the objectivists consider that probability is a mathematical tool to model the real world. On the other hand, the subjectivists consider probabilities as a way to model the reasoning of a given *subject* about the world. It is considered by many as only philosophical quibbling, but they are wrong as it has very important practical consequences in the way probabilities are used and in how the obtained results may be interpreted.

For the objectivists, one should use probabilities to build models of the world as *objective* as possible, meaning that these models should depend on the observable data and only on them and should be independent of their own knowledge of any possible observers.[5] It is a praiseworthy goal, a direct heritage of the idea that science can provide an objective, an *exact*, or even a *true* description of the world.

The subjectivists consider that probability is a model of reasoning. As such, the *subject* who is reasoning and who's own knowledge about the world is central for the model; is at least as important as the data he collects by making observations. Subjectivists even deny the possibility of building a model of the world independent of any preliminary knowledge to interpret these data. They propose probability as an alternative and an extension of logic to formalize rational reasoning when information is incomplete and uncertain. Preliminary knowledge plays the same role for probability that axiomatic plays for logic. Starting from "wrong" axioms (not true in the world) will lead to "wrong" conclusions (not describing, explaining, or predicting the behavior of the world) even with exact logical inferences. Starting from "wrong" preliminary probabilistic knowledge will also lead to "wrong" conclusions whatever the data and even with perfectly valid probabilistic calculus. The "objectivity" of the subjectivist approach then lies in the fact that two different subjects with same preliminary knowledge and same observations will inevitably reach the same conclusions. A quite different meaning of objectivity than the one adopted by the objectivists.

The Laplace succession law controversy which has made for an exciting debate for the last 150 years is an interesting example of the two different points of view. Laplace proposed to model a series of experiments using the following law:

$$P(x) = \frac{1 + n_x}{\Omega + n} \tag{16.15}$$

where n_x is the number of times the x value appears in the series, Ω is the cardinal of variable X, and n is the total number of observations in the series.

If the observed series is the life of an individual and the variable X stands for "the individual survives this year," then $\Omega = 2$ and we get for a 14-year-old boy a probability of surviving one more year equal to $15/16$ when for his 75-year-old grandfather we get a probability of $76/77$.

Using this kind of argument, the objectivists have been making fun of Laplace and his succession law, saying that they were both stupid.

The subjectivists' position is that the Laplace succession law is just one

[5]They even often deny the existence or, at least, the necessity of these observers.

element of the reasoning subject to preliminary knowledge, and if the obtained result is in contradiction with "common sense" it just means that "common sense" has more information to make its judgment than solely this rule. Adding the knowledge that human life has an upper limit, indeed easily solves the question of the probability of survival of a given individual.

Edwin T. Jaynes in his book *Probability Theory: The Logic of Science* [Jaynes, 2003] presents a fervent plea for the subjectivist point of view. He warns us again of what he calls the "mind projection fallacy":

> Common language — or at least, the English language — has an almost universal tendency to disguise epistemological statements by putting them into a grammatical form which suggests to the unwary an ontological statement. A major source of error in current probability theory arises from an unthinking failure to perceive this. To interpret the first kind of statement in the ontological sense is to assert that one's own private thoughts and sensations are realities existing externally in Nature. We call this the "Mind Projection Fallacy," and note the trouble it causes many times in what follows. But this trouble is hardly confined to probability theory; as soon as it is pointed out, it becomes evident that much of the discourse of philosophers and Gestalt psychologists, and the attempts of physicists to explain quantum theory, are reduced to nonsense by the author falling repeatedly into the Mind Projection Fallacy.
>
> *Probability Theory: The Logic of Science*
> Edwin T. Jaynes [2003]

You can find in Jaynes' book many examples of misuses of probability due to an objectivist interpretation and especially a review of apparent paradoxes that can be easily solved with a subjectivist point of view.

Of course we presented here only the two extreme positions, when a lot of intermediary approaches exist. For instance, a usual definition for "Bayesianism" refers to probabilists that accept the use of priors as reasoning subjects' knowledge. Even this position has been largely attacked by objectivists with endless discussions on the relevance of the used priors. From a subjectivist position, the subject is free and takes his own risks when using a given prior. If he makes a wrong choice then he will get an inappropriate model.

In this book we went much further in the subjectivist direction. We do not only use priors but "preliminary knowledge." Priors are limited to the specification of a few parametric forms to summarize subject preliminary knowledge, when, in contrast, preliminary knowledge is made of the specification part of the Bayesian program made of (i) the choice of the relevant variables, (ii) the choice of the decomposition assuming conditional independences, and (iii) the choice of the parametric forms for each of the distributions appearing in the decomposition.

A major contribution of this book is precisely this formalization of the preliminary knowledge which, we hoped, has been proved in these pages to be general and generic enough to model a lot of different problems.

16.14 Unknown distribution

In a lot of models we stated that some distributions are $Uniform$.

For instance, this is the case in the model of a water treatment unit as stated in the Bayesian program in Equation 4.29 reproduced below:

$$
Pr\begin{cases}
Ds\begin{cases}
Sp(\pi)\begin{cases}
Va: \\
I_0, I_1, F, S, C, O \\
Dc: \\
\begin{cases}
& P(I_0 \wedge I_1 \wedge F \wedge S \wedge C \wedge O) \\
= & P(I_0) \times P(I_1) \times P(F) \times P(S|I_0 \wedge F) \\
& \times P(C) \times P(O|I_0 \wedge I_1 \wedge S \wedge C)
\end{cases} \\
Fo: \\
P(I_0) = Uniform \\
P(I_1) = Uniform \\
P(F) = Uniform \\
P(S|I_0 \wedge F) = \delta_{S=Int\left(\frac{I_0+F}{2}\right)} \\
P(C) = Uniform \\
P(O|I_0 \wedge I_1 \wedge S \wedge C) = Histograms
\end{cases} \\
Id
\end{cases} \\
Qu:
\end{cases}
\tag{16.16}
$$

However, these hypotheses are not always necessary. If for a given model, you are sure that some of the variables appearing in your model will always be known (appearing only in the right part of a question), then you do not need to specify prior distributions for these variables as these distributions will be canceled out in the inference by appearing both at the numerator and denominator of the expression required to be computed to solve any of the possible questions.

This is the case in the water treatment example for variables I_0 and I_1, which are always known. The answer to any question of the form $P(Search|known \wedge i_0 \wedge i_1)$ is obtained by:

$$
\begin{aligned}
& P\left(Search \mid known \wedge i_0 \wedge i_1\right) \\
= & \frac{\sum_{Free}\left[P\left(i_0\right) P\left(i_1\right) P(F) P\left(S \mid i_0 \wedge F\right) P(C) P\left(O \mid i_0 \wedge i_1 \wedge S \wedge C\right)\right]}{\sum_{Free \wedge Search}\left[P\left(i_0\right) P\left(i_1\right) P(F) P\left(S \mid i_0 \wedge F\right) P(C) P\left(O \mid i_0 \wedge i_1 \wedge S \wedge C\right)\right]} \\
= & \frac{\sum_{Free}\left[P(F) P\left(S \mid i_0 \wedge F\right) P(C) P\left(O \mid i_0 \wedge i_1 \wedge S \wedge C\right)\right]}{\sum_{Free \wedge Search}\left[P(F) P\left(S \mid i_0 \wedge F\right) P(C) P\left(O \mid i_0 \wedge i_1 \wedge S \wedge C\right)\right]}
\end{aligned} \tag{16.17}
$$

The answer depends on neither $P\left(I_0\right)$ nor on $P\left(I_1\right)$.

The situation is not the same for variables F and C that are, for some interesting questions, either searched or let free. For them, $P(F)$ and $P(C)$ must be specified.

Consequently, in Bayesian programming you can specify a distribution as *Unknown* and ProBT will provide an error message if you try to ask a question that supposes to use this distribution.

Chapter 17

Glossary

Knowledge advances integrating uncertainty, not exorcising it.[1]

La Méthode
Edgar Morin [1981]

This chapter is a very short summary of the book where the central concepts are recalled as an extended glossary.

17.1 Bayesian filter

Bayesian filters are a particular case of Bayesian programs (see Section 17.4) defined as follows:

[1] La connaissance progresse en intègrant en elle l'incertitude, non en l'exorcisant.

$$
Pr\begin{cases}
Ds\begin{cases}
Sp(\pi)\begin{cases}
Va: \\
S^0,\cdots,S^T,O^0,\cdots,O^T \\
Dc: \\
\begin{cases}
 & P\left(S^0\wedge\cdots\wedge S^T\wedge O^0\wedge\cdots\wedge O^T|\pi\right) \\
= & P\left(S^0\wedge O^0\right)\times\prod_{t=1}^{T}\left[P\left(S^t|S^{t-1}\right)\times P\left(O^t|S^t\right)\right]
\end{cases} \\
Fo: \\
\begin{cases}
P\left(S^0\wedge O^0\right) \\
P\left(S^t|S^{t-1}\right) \\
P\left(O^t|S^t\right)
\end{cases}
\end{cases} \\
Id
\end{cases} \\
Qu: \\
\begin{cases}
P\left(S^{t+k}|O^0\wedge\cdots\wedge O^t\right) \\
(k=0)\equiv Filtering \\
(k>0)\equiv Prediction \\
(k<0)\equiv Smoothing
\end{cases}
\end{cases}
\tag{17.1}
$$

See Section 13.1.2 for details and special cases like hidden Markov models (HMMs), Kalman filters, and particle filters.

17.2 Bayesian inference

Bayesian inference consists of computing the probability distribution corresponding to a *question* (17.14) knowing a *description* (17.8). The computation to be made is the following:

$$
\begin{aligned}
& P(Searched|known \wedge \delta \wedge \pi) \\
= & \sum_{Free} [P(Searched \wedge Free|known \wedge \delta \wedge \pi)] \\
= & \frac{\sum_{Free} [P(Searched \wedge Free \wedge known|\delta \wedge \pi)]}{P(known|\delta \wedge \pi)} \\
= & \frac{\sum_{Free} [P(Searched \wedge Free \wedge known|\delta \wedge \pi)]}{\sum_{Free \wedge Searched} [P(Searched \wedge Free \wedge known|\delta \wedge \pi)]} \\
= & \frac{1}{Z} \times \sum_{Free} [P(Searched \wedge Free|known \wedge \delta \wedge \pi)] \\
= & \frac{1}{Z} \times \sum_{Free} \left[P(L_1|\delta \wedge \pi) \times \prod_{k=2}^{K} [P(L_k|R_k \wedge \delta \wedge \pi)] \right]
\end{aligned}
\tag{17.2}
$$

The difficulty in this elementary computation is the dimension of the search space (variable *Searched*) and the dimension of the integration space (variable *Free*).

Most of the time only approximate inference is acceptable and all the algorithms tried to solve these optimization problems in clever ways. See Chapter 14 for details.

However, the main means of keeping Bayesian inference tractable is to make the right simplification choices in the *decomposition* (see Section 17.7) by assuming conditional independencies between variables numerous enough to reduce dimensionality, yet sparse enough, to keep the model relevant.

17.3 Bayesian network

Bayesian networks are a particular case of *Bayesian programs* (see Section 17.4) defined as follows:

$$Pr\begin{cases} Ds\begin{cases} Sp(\pi)\begin{cases} Va: \\ X_1,\cdots,X_N \\ Dc: \\ \begin{cases} & P(X_1\wedge\cdots\wedge X_N|\pi) \\ = & \prod_{n=1}^{N}\left[P(X_n|R_n\wedge\pi)\right] \end{cases} \\ Fo: \\ any \end{cases} \\ Id \end{cases} \\ Qu: \\ P(X_n|known) \end{cases} \tag{17.3}$$

Note the particularity of the *decomposition* (see Section 17.7) where one and only one variable X_n is appearing left of the conditioning sign and depending of its "antecedents" R_n.

See Section 13.1.1 for details and Section 16.3 for a discussion on "Bayesian programming versus Bayesian networks."

17.4 Bayesian program

Bayesian program is a central notion of this book. It is the framework implementing the Bayesian programming modeling methodology.

$$Program\begin{cases} Description\begin{cases} Specification(\pi)\begin{cases} Variables \\ Decomposition \\ Forms \end{cases} \\ Identification\ (based\ on\ \delta) \end{cases} \\ Question \end{cases}$$

A Bayesian program is made of two parts: the *description* (see Section 17.8) which formalizes the probabilistic models and the *question* (see Section 17.14), which specifies what problem has to be solved.

The description itself is made of two parts: the *specification* (see Section 17.15), which formalizes the *preliminary knowledge* (see Section 17.13) of the modeler, and the *identification* where learning of the nonspecified parameters is made.

Finally, the specification is made of three parts: the choice of the *variables* (see Section 17.17), the *decomposition* (see Section 17.7) where the modeler adds knowledge to simplify the model, and the *forms* (see Section 17.9) where the modeler specifies the mathematical means to compute and learn the elements of the decomposition.

17.5 Coherence variable

A *coherence variable* is a Boolean variable which, when true, imposes that two other variables are "coherent" which means that they should share the same probability distribution knowing the same premises.

A coherence variable Λ appears in a *decomposition* under the basic form:

$$P([\Lambda = 1] | A \wedge A') = \delta_{A=A'} \tag{17.4}$$

Chapter 8 discusses the semantic of coherence variables and details their numerous usages as, for instance, to reason with soft evidence, to activate or deactivate part of a model (switch), and to deal with cycles.

17.6 Conditional statement

Conditional statements in Bayesian programs are implemented using a choice variable which conditions the use of different probabilistic models.

It is very similar to the use of a conditional statement in classical programming, where the value of the choice variable decides which branch of the program should be executed.

However, in Bayesian programming we usually do not know the value of the choice variable but rather a probability distribution on this choice variable. Consequently, the choice variable has to be marginalized out, leading to taking in accounting for all the different models that it conditioned but with a focus on the importance proportional to the probability of the choice variable for this branch.

Details and examples can be found in Chapter 10.

17.7 Decomposition

Decomposition is an essential modeling step. It consists in expressing the joint distribution of the model as a product of simpler distributions assuming conditional independencies between variables.

$$\begin{aligned} & P(X_1 \wedge X_2 \wedge \cdots \wedge X_N | \delta \wedge \pi) \\ = \; & P(L_1 | \delta \wedge \pi) \times P(L_2 | R_2 \wedge \delta \wedge \pi) \times \cdots \times P(L_K | R_K \wedge \delta \wedge \pi) \end{aligned} \tag{17.5}$$

It is the algebraic analogous of defining a Bayesian net's graph even if more general (see Chapter 12 for a complete formal definition).

The art of decomposing is delicate. Indeed, the assumed conditional independencies between variables shall be carefully chosen in order to satisfy three contradictory constraints: (i) reducing the dimensionality to have tractable computation, (ii) not oversimplifying to keep most of the relevant information, and (iii) selecting elementary distributions that can be easily learned.

17.8 Description

Description in a Bayesian program is the probabilistic model considered.

This model results from, on one hand, knowledge provided by the programmer called *specification* (see Section 17.15) and, on the other hand, knowledge coming from the "environment" and learned during *identification*.

Any valid probabilistic model can be a description. In particular, descriptions are more general than Bayesian nets (see Section 17.3) as they are an algebraic way to express probabilistic factor graphs.

17.9 Forms

Forms are the third necessary ingredient in the *specification* (see Section 17.15) part of a *Bayesian program* to be able to compute the joint distribution. A form can either be a parametric form or a *question* (see Section 17.14) to another *Bayesian program*.

17.10 Incompleteness

Incompleteness is the fundamental problem that Bayesian programming addresses.

We believe that it is impossible to build a complete model of any "real" (nonformal) phenomenon. There are always some hidden variables that influence the phenomenon and are not taken into account by the model. The effect of these hidden variables is that the model cannot completely account for the behavior of the phenomenon. Uncertainty appears as a direct consequence of this incompleteness.

Using probability is then a necessity to describe these nonformal phenomenon.

Details may be found in Chapter 3 and a discussion about "incompleteness irreducibility" in Section 16.10.

17.11 Mixture

Mixtures are usually presented in the literature as an approximation function problem where a given probability distribution is modeled by a weighted sum of simpler distributions.

We propose to present systematically these approaches using a hidden variable H that could explain the obtained mixture but is ignored most of the time. We then obtain the following Bayesian program:

$$Pr \begin{cases} Ds \begin{cases} Sp(\pi) \begin{cases} Va: \\ X_1, \cdots, X_N, H \\ Dc: \\ \begin{cases} & P(X_1 \wedge \cdots \wedge X_N \wedge H|\pi) \\ = & P(H|\pi) \times P(X_1 \wedge \cdots \wedge X_N|H \wedge \pi) \end{cases} \\ Fo: \\ \begin{cases} P(H|\pi) \equiv Table \\ P(X_1 \wedge \cdots \wedge X_N|[H=m] \wedge \pi) \\ \equiv P(X_1 \wedge \cdots \wedge X_N|\pi_m) \end{cases} \end{cases} \\ Id \end{cases} \\ Qu: \\ \begin{cases} & P(X_1 \wedge \cdots \wedge X_N|\pi) \\ = & \sum_{m=1}^{M} [P([H=m]|\pi) \times P(X_1 \wedge \cdots \wedge X_N|\pi_m)] \end{cases} \end{cases} \tag{17.6}$$

The usual form of the mixture (the weighted sum) appears as the result of the inference done by ignoring and marginalizing out the variable H.

It is then easy to go one step further in generalization by considering that the variable H could be conditioned by some of the variables. Doing this we obtain the standard probabilistic conditional statement form (see Section 17.6).

17.12 Noise

Noise is anything that is not music, anything not written in the score, anything not specified in your model, everything that you are ignoring, which may be much.

See the discussion "Noise or ignorance?" in Section 16.12.

17.13 Preliminary knowledge

Preliminary knowledge is the set of knowledge brought to the system by the programmer. It is very strictly defined by *specifications* (see Section 17.15) made of three components, the choice of the *variables* (see Section 17.17), the *decomposition* (see Section 17.7), and the *forms* (see Section 17.9).

Preliminary knowledge made of these three components is sufficient to define any probabilistic model.

Preliminary knowledge is really what is defining our *subjectivist* approach to probability (see the discussion "Objectivism versus subjectivism" in Section 16.13). It goes much further than some definitions of "Bayesanism" restricted to models using priors. To specify a probabilistic model using preliminary knowledge the programmer must answer three fundamental questions:

- What are you talking about?: the *variables*;
- How is your knowledge structured?: the *decomposition*;
- How is your knowledge mathematically represented?: the *forms*.

Priors are limited to the values of some parameters of some of the forms, a very tiny part of the preliminary knowledge indeed.

17.14 Question

Questions are formally defined as a partition of the set of variables in three sub-sets:

- The *known* variables for which a value is imposed.
- The *searched* variables for which you are computing a probability distribution knowing the known variables.
- The *free* variables that you are ignoring and that have to be marginalized out.

A question is a family of probability distributions defined $P(Searched|Known)$ made up of as many distributions as the possible values of $Known$. Each instantiated question is defined by $P(Searched|known)$ and the answer is given by the computation:

$$P\left(Searched|known\right) = \frac{1}{Z}\sum_{Free}\left[P\left(Searched \wedge known \wedge Free\right)\right] \qquad (17.7)$$

17.15 Specification

Specification is where the *preliminary knowledge* (see Section 17.13) of a programmer is specified in a Bayesian program.

17.16 Subroutines

Subroutine calls in Bayesian programming consist in specifying a form of one distribution appearing in the *decomposition* of a *Bayesian program* π_1 as a *question* (see Section 17.14) to another *Bayesian program* π_2 (see Chapter 9 for details and examples).

When a question is asked to π_1, during the necessary inferences, each time the form corresponding to the question to π_2 has to be evaluated, it triggers supplementary inferences to answer this question.

This mechanism allows the conception of hierarchical models where a Bayesian program can be conceived using other Bayesian programs, eventually written by others, as its elementary components.

17.17 Variable

Variables in *Bayesian programming* are defined formally as a set of mutually exclusive and exhaustive logical propositions. More intuitively this corresponds to the concept of discrete variables. Continuous variables when necessary are discretized (see the discussion on this matter in Section 16.9).

These variables have absolutely no intrinsic character of randomness. They are not "random variables" formally defined as functions from the set of events into $\mathbb{R}$ (or $\mathbb{R}^n$ for a random vector). The knowledge about these variables may be probabilistic but it is not the nature of the variable itself. Indeed, in Bayesian programming either a variable has a known value which appears on the right part of a conditioning symbol, or it is known by a probability distribution on its possible values.

When writing a *Bayesian program* the choice of the relevant variables that should appear in this program is the most difficult part. When the appropriate set of variables has been selected a large part of the modeling work has been done. However, this remark is not specific to probabilistic modeling but is true for any kind of modeling work.

Bibliography

D. Adams. *The Hitchhiker's Guide to the Galaxy.* Del Rey, 1995.

S. Aji and R. McEliece. The Generalized Distributive Law. *IEEE Trans. Information Theory*, 46(2), 2000.

D. Alais and D. Burr. The ventriloquist effect results from near-optimal bimodal integration. *Current Biology*, 14:257–262, February 2004.

M. Aly and A. P. Yonelinas. Bridging consciousness and cognition in memory and perception: Evidence for both state and strength processes. *PLoS ONE*, 7(1), 2012.

M. Amavizca. *Reconstruction 3D du bassin humain a partir d'images medicales multimodales incompletes. Application a l'assistance de la chirurgie de la prothese totale de la hanche (PTH).* PhD thesis, Inst. Nat. Polytechnique de Grenoble, 2005.

M. Amavizca. 3D Human Hip Volume Reconstruction with Incomplete Multimodal Medical Images. In P. Bessière, C. Laugier, and R. Siegwart, editors, *Probabilistic Reasoning and Decision Making in Sensory-Motor Systems.* Springer, 2008.

T. J. Anastasio, P. E. Patton, and K. Belkacem-Boussaid. Using Bayes' rule to model multisensory enhancement in the superior colliculus. *Neural Computation*, 12(5):1165–87, 2000.

S. Arnborg and G. Sjödin. Bayes Rules in Finite Models. In *Proceedings of European Conference on Artificial Intelligence*, pages 571–575, 2000.

S. Arnborg, D. G. Corneil, and A. Proskurowski. Complexity of finding embeddings in a k-tree. *SIAM J. Algebraic Discrete Methods*, 8:277–284, April 1987. ISSN 0196-5212.

S. Arulampalam, S. Maskell, N. Gordon, and T. Clapp. A Tutorial on Particle Filter for Online Nonlinear/Non-Gaussian Bayesian Tracking. *IEEE Transactions on Signal Processing*, 50(2), 2002.

G. Bachelard. *La formation de l'esprit scientifique : Contribution a une psychanalyse de la connaissance objective.* Librairie philosophique J. Vrin, 1938.

M. S. Banks. Neuroscience: what you see and hear is what you get. *Current Biology*, 14(6):236–238, 2004.

A. Barker, D. Brown, and W. Martin. Bayesian Estimation and the Kalman Filter. Technical Report IPC-TR-94-002, University of Virginia, 1994.

P. W. Battaglia, R. A. Jacobs, and R. N. Aslin. Bayesian integration of visual and auditory signals for spatial localization. *Journal of the Optical Society of America A*, 20(7):1391–1397, 2003.

Y. Bengio and P. Frasconi. An Input/Output HMM Architecture. In G. Tesauro, D. Touretzky, and T. Leen, editors, *Advances in Neural Information Processing Systems 7*, pages 427–434. MIT Press, Cambridge, MA, 1995.

J. Bernouilli. *Ars Conjectandi.* Thurneysen Brothers, 1713.

P. Bessière. Procédé de détermination de la valeur à donner à différents paramètres d'un système, 2002. European Patent N=EP1525520.

P. Bessiere, E. Dedieu, and E. Mazer. Representing Robot/Environment Interactions Using Probabilities: the"Beam in the Bin" Experiment. In *PerAc'94 (From Perception to Action)*, Lausanne, Suisse, 1994.

P. Bessiere, E. Dedieu, and O. Lebeltel. Wings Were Not Designed to Let Animals Fly. In *Third European Conference on Artificial Evolution (Megève, France)*, volume 1363 of *Lecture Notes in Computer Science*, pages 237–250. Springer-Verlag, 1997.

P. Bessière, J.-M. Ahuactzin, O. Aycard, D. Bellot, F. Colas, C. Coué, J. Diard, R. Garcia, C. Koike, O. Lebeltel, R. LeHy, O. Malrait, E. Mazer, K. Mekhnacha, C. Pradalier, and A. Spalanzani. Survey: Probabilistic Methodology and Techniques for Artefact Conception and Development. Technical Report RR-4730, INRIA Rhône-Alpes, Montbonnot, France, 2003.

P. Bessière, C. Laugier, and R. Siegwart. *Probabilistic Reasoning and Decision Making in Sensory-Motor Systems.* Springer, 2008.

L.-J. Boë, L. Ménard, J. Serkhane, P. Birkholz, B. J. Kröger, P. Badin, G. Captier, M. Canault, and N. Kielwasser. La croissance de l'instrument vocal: Contrôle, modélisation, potentialités acoustiques et conséquences perceptives. *Revue française de linguistique appliquée*, XIII(2):59–80, 2008.

E. Borel. *Le hasard.* Felix Alcan, Paris, France, 1914.

C. Boutilier, T. Dean, and S. Hanks. Decision theoretic planning: structural assumptions and computational leverage. *Journal of Artificial Intelligence Research (JAIR)*, 11:1–94, 1999.

G. E. P. Box and N. R. Draper. *Empirical Model-Building and Response Surfaces.* Wiley, 1987.

G. E. P. Box and M. E. Muller. A note on the generation of random normal deviates. *Annals Math. Stat.*, 29:610–611, 1958.

P. J. Brockwell and R. A. Davis. *Introduction to Time Series and Forecasting (Second Edition).* Springer-Verlag, 2000.

W. Burgard, D. Fox, D. Hennig, and T. Schmidt. Estimating the Absolute Position of a Mobile Robot Using Position Probability Grids. In *Proceedings of the Thirteenth National Conference on Artificial Intelligence and the Eighth Innovative Applications of Artificial Intelligence Conference*, pages 896–901, Menlo Park, August 1996. AAAI Press / MIT Press.

F. Colas. *Perception des objets en mouvementComposition bayésienne du flux optique et du mouvement de l'observateur.* These, Institut National Polytechnique de Grenoble — INPG, Jan. 2006.

F. Colas, P. Bessière, J. Droulez, and M. Wexler. Bayesian Modelling of Perception of Structure from Motion. In P. Bessière, C. Laugier, and R. Siegwart, editors, *Probabilistic Reasoning and Decision Making in Sensory-Motor Systems*, pages 301–328. Springer, 2008a.

F. Colas, J. Droulez, M. Wexler, and P. Bessière. A unified probabilistic model of the perception of three-dimensional structure from optic flow. *Biological Cybernetics*, pages 132–154, 2008b.

F. Colas, F. Flacher, T. Tanner, P. Bessière, and B. Girard. Bayesian models of eye movement selection with retinotopic maps. *Biological Cybernetics*, 100(3):203–14, Mar. 2009. BACS FP6-IST-027140.

F. Colas, J. Diard, and P. Bessière. Common Bayesian models for common cognitive issues. *Acta Biotheoretica*, 58(2-3):191–216, 2010.

G. F. Cooper. The computational complexity of probabilistic inference using Bayesian belief networks. *Artif. Intell.*, 42(2-3):393–405, 1990.

A. Corana, C. Martini, and S. Ridella. Minimizing multimodal functions of continuous variables with the simulated annealing algorithm. *ACM Transactions on Mathematical Software*, 13:262–280, 1987.

C. Coué. *Modèle bayésien pour l'analyse multimodale d'environnements dynamiques et encombrés : Application à l'assistance à la conduite en milieu urbain.* These, Institut National Polytechnique de Grenoble — INPG, 2003.

C. Coué, C. Pradalier, C. Laugier, T. Fraichard, and P. Bessière. Bayesian occupancy filtering for multitarget tracking: an automotive application. *International Journal of Robotics Research*, 25(1):19–30, Jan. 2006.

R. T. Cox. Probability, frequency, and reasonable expectation. *American Journal of Physics*, 14:1–13, 1946.

R. T. Cox. *The Algebra of Probable Inference.* John Hopkins Univ. Press: Baltimore, 1961.

R. T. Cox. *Of Inference and Inquiry — An Essay in Inductive Logic.* MIT Press, Cambridge, MA, 1979.

P. Dagum and M. Luby. Approximating probabilistic inference in Bayesian belief networks is NP-hard. *Artif. Intell.*, 60(1):141–153, 1993.

P. Dangauthier, A. Spalanzani, and P. Bessière. Statistical methods and genetic algorithms for prior knowledge selection. In *Actes du congrès francophone de Reconnaissance des Formes et Intelligence Artificielle*, Toulouse (FR), France, Jan. 2004.

P. Dangauthier, P. Bessière, and A. Spalanzani. Auto-Supervised Learning in the Bayesian Programming Framework. In *Proc. of the IEEE International Conference on Robotics and Automation (ICRA), Barcelona (Spain)*, pages 1–6, 2005.

P. Dangauthier, R. Herbrich, T. Minka, and T. Graepel. TrueSkill Through Time: Revisiting the History of Chess. In M. Press, editor, *Advances in Neural Information Processing Systems*, Vancouver, Canada, 2007.

P.-C. Dangauthier. *Fondations, méthode et applications de l'apprentissage bayésien.* These, Institut National Polytechnique de Grenoble — INPG, Dec. 2007.

A. Darwiche. *The Complexity of Probabilistic Inference.* Cambridge University Press, 2009.

A. Darwiche and G. Provan. Query DAGs: a practical paradigm for implementing belief-network inference. *J. Artif. Int. Res.*, 6(1):147–176, May 1997. ISSN 1076-9757.

T. Dean and K. Kanazawa. Probabilistic Temporal Reasoning. In *AAAI*, pages 524–529, 1988.

T. Dean and K. Kanazawa. A model for reasoning about persistence and causation. *Computational Intelligence*, 5(3):142–150, 1989.

R. Dechter. Bucket Elimination: A Unifying Framework for Reasoning. *Artificial Intelligence*, pages 41–85, 1999.

R. Dechter and I. Rish. A Scheme for Approximating Probabilistic Inference. In *In Proceedings of Uncertainty in Artificial Intelligence (UAI97)*, pages 132–141, 1997.

E. Dedieu. *La représentation contingente.* PhD thesis, Inst. Nat. Polytechnique de Grenoble, September 1995.

E. Dedieu and E. Mazer. An Approach to Sensorimotor Relevance. In F. Varela and P. Bourgine, editors, *Proc. of the 1st European Conference on Artificial Life (ECAL91), Paris (France).* MIT Press / Bradford Books, 1992.

F. Dellaert, D. Fox, W. Burgard, and S. Thrun. Monte Carlo Localization for Mobile Robots. In *Proc. of the IEEE Int. Conf. on Robotics and Automation*, Detroit, MI, May 1999.

A. P. Dempster, N. M. Laird, and D. B. Rubin. Maximum likelihood from incomplete data via the EM algorithm. *Journal of the Royal Statistical Society. Series B*, 39(1):1–38, 1977.

R. Descartes. *Discours de la méthode.* Imprimerie Ian Maire, 1637.

J. Diard. *La carte bayséienne — Un modèle probabiliste hiérarchique pour la navigation en robotique mobile.* Thèse de doctorat, Institut National Polytechnique de Grenoble, Grenoble, France, Janvier 2003.

J. Diard and P. Bessière. Bayesian Maps: Probabilistic and Hierarchical Models for Mobile Robot Navigation. In P. Bessière, C. Laugier, and R. Siegwart, editors, *Probabilistic Reasoning and Decision Making in Sensory-Motor Systems*, volume 46 of *Springer Tracts in Advanced Robotics*, pages 153–176. Springer-Verlag, 2008.

J. Diard, P. Bessière, and E. Mazer. Hierarchies of Probabilistic Models of Navigation: the Bayesian Map and the Abstraction Operator. In *Proceedings of the IEEE International Conference on Robotics and Automation (ICRA04)*, pages 3837–3842, New Orleans, LA, USA, 2004.

J. Diard, E. Gilet, E. Simonin, and P. Bessière. Incremental learning of Bayesian sensorimotor models: from low-level behaviours to large-scale structure of the environment. *Connection Science*, 22(4):291–312, Dec. 2010.

K. Drewing and M. Ernst. Integration of force and position cues for shape perception through active touch. *Brain Research*, 1078:92–100, 2006.

D. Dubois and H. Prade. Possibility theory, probability theory and multiple-valued logics: a clarification. *Ann. Math. Artif. Intell.*, 32(1-4):35–66, 2001.

A. Einstein. Physics and reality. *Journal of the Franklin Institute*, 1936.

M. O. Ernst and M. S. Banks. Humans integrate visual and haptic information in a statistically optimal fashion. *Nature*, 415(6870):429–33, 2002.

J. Ferreira. *Bayesian cognitive models for 3D structure and motion multimodal perception.* PhD thesis, Universidade de Coimbra, Faculdade de Ciencaias E Tecnologia, July 2011.

J. Ferreira, J. Lobo, P. Bessière, M. Castelo-Branco, and J. Dias. A Bayesian Framework for Active Artificial Perception. *IEEE Transactions on Systems, IEEE Transactions on Systems, Man, and Cybernetics, Part B*, 99:1–13, 2012.

J. Fourier. Eloge historique de M. le Marquis de Laplace. *Académie Royale des Sciences*, 1829.

D. Fox, W. Burgard, F. Dellaert, and S. Thrun. Monte Carlo Localization: Efficient Position Estimation for Mobile Robots. In *Proc. of the Sixteenth National Conference on Artificial Intelligence (AAAI'99)*, 1999.

B. J. Frey. *Graphical Models for Machine Learning and Digital Communication.* MIT Press, 1998.

R. S. Garcia-Ramirez. *Programmation bayesienne des bras manipulateurs.* PhD thesis, Inst. Nat. Polytechnique de Grenoble, May 2003.

W. S. Geisler and D. Kersten. Illusions, perception and Bayes. *Nature Neuroscience*, 5(6):598–604, 2002.

S. Geman and D. Geman. Stochastic Relaxation, Gibbs Distributions and the Bayesian Restoration of Images. *IEEE Transactions on Pattern Analysis and Machine Intelligence*, 6:721–741, 1984.

S. Gepshtein and M. S. Banks. Viewing geometry determines how vision and haptics combine in size perception. *Current Biology*, 13(6):483–488, 2003.

J. Geweke. Monte Carlo Simulation and Numerical Integration. In H. Amman, D. Kendrick, and J. Rust, editors, *Handbook of Computational Economics*, volume 13, pages 731–800. Elsevier North-Holland, Amsterdam, 1996.

Z. Ghahramani. *An Introduction to Hidden Markov Models and Bayesian Networks*, pages 9–42. World Scientific Publishing Co., Inc., River Edge, NJ, USA, 2002.

Z. Ghahramani, D. M. Wolpert, and M. I. Jordan. Computational Models of Sensorimotor Integration. In P. G. Morasso and V. Sanguineti, editors, *Self-Organization, Computational Maps and Motor Control*, pages 117–47. Elsevier, 1997.

E. Gilet. *Modélisation Bayésienne d'une boucle perception-action : application à la lecture et à l'écriture.* These, Université Joseph-Fourier — Grenoble I, Oct. 2009.

E. Gilet, J. Diard, and P. Bessière. Bayesian action–perception computational model: interaction of production and recognition of cursive letters. *PLoS ONE*, 6(6):e20387, 2011.

N. D. Goodman, V. K. Mansinghka, D. M. Roy, K. Bonawitz, and J. B. Tenenbaum. Church: A language for generative models. In *UAI*, pages 220–229, 2008.

A. Gopnik and L. Schulz. Mechanisms of theory formation in young children. *Trends in Cognitive Sciences*, 8(8):371–377, 2004.

S. J. Gould. The Streak of Streaks. *The New York Review of Books*, 35(13), 1988.

A. Greenspan. *The Age of Turbulence*. Penguin Press, 2007.

A. Haith, C. Jackson, C. Miall, and S. Vijayakumar. Unifying the Sensory and Motor Components of Sensorimotor Adaptation. In *Advances in Neural Information Processing Systems (NIPS 2008)*, 2008.

J. Y. Halpern. A counterexample to theorems of Cox and Fine. *Journal of Artificial Intelligence Research*, 10:67–85, 1999a. ISSN 1076-9757.

J. Y. Halpern. Cox's theorem revisited. *Journal of Artificial Intelligence Research*, 11:429–435, 1999b. ISSN 1076-9757.

A. C. Harvey. *Forecasting, Structural Time Series Models and the Kalman Filter*. Cambridge University Press, Cambridge, 1992.

M. Hauskrecht, N. Meuleau, L. Boutilier, L. Kaelbling, and T. Dean. Hierarchical Solution of Markov Decision Processes Using Macro-Actions. In *Proceedings of the 14th Conference on Uncertainty in Artificial Intelligence*, pages 220–229, 1998.

J. M. Hillis, S. J. Watt, M. S. Landy, and M. S. Banks. Slant from texture and disparity cues: optimal cue combination. *Journal of Vision*, 4:967–992, 2004.

J. H. Holland. *Adaptation in Natural and Artificial Systems*. University of Michigan Press, Ann Arbor, MI, 1975.

T. S. Jaakkola and M. I. Jordan. Variational probabilistic inference and the QMR-DT network. *J. Artif. Intellig. Res. (JAIR)*, 10:291–322, 1999.

R. A. Jacobs. Optimal integration of texture and motion cues to depth. *Vision Research*, 39:3621–9, 1999.

E. T. Jaynes. On the rationale of maximum-entropy methods. *Proc. IEEE*, 70(9):939–952, 1982.

E. T. Jaynes. *Probability Theory: the Logic of Science.* Cambridge University Press, 2003.

F. Jensen. *An Introduction to Bayesian Networks.* UCL Press, 1996.

F. V. Jensen, S. L. Lauritzen, and K. G. Olesen. Bayesian updating in recursive graphical models by local computations. *Comput. Stat. Quarterly*, 4:269–282, 1990.

M. I. Jordan. *Learning in Graphical Models.* MIT Press, 1999. Edited Volume.

M. I. Jordan and R. A. Jacobs. Hierarchical mixtures of experts and the EM algorithm. *Neural Computation*, 6(2):181–214, 1994.

M. I. Jordan and Y. Weiss. Graphical Models: Probabilistic Inference. In M. A. Arbib, editor, *Handbook of Neural Networks and Brain Theory.* MIT Press, Cambridge, MA, 2002.

R. Jürgens and W. Becker. Perception of angular displacement without landmarks: evidence for Bayesian fusion of vestibular, optokinetic, podokinesthetic, and cognitive information. *Experimental Brain Research*, 174:528–543, 2006.

L. P. Kaelbling, M. Littman, and A. Cassandra. Planning and acting in partially observable stochastic domain. *Artificial Intelligence*, 101(1-2):99–134, 1998.

R. E. Kalman. A New Approach to Linear Filtering and Prediction Problems. *Transactions of the ASME–Journal of Basic Engineering*, 82(Series D):35–45, 1960.

A. Keller. The Fast Calculation of Form Factors Using Low Discrepancy Point Sequence. In *Proc. of the 12th Spring Conf. on Computer Graphics*, pages 195–204, Bratislava, 1996.

C. Kemp and J. Tenenbaum. The discovery of structural form. *Proc. Natl. Acad. Sci. USA*, 105(31):10687–10692, 2008.

D. Kersten, P. Mamassian, and A. Yuille. Object perception as Bayesian inference. *Annual Review of Psychology*, 55:271–304, 2004.

T. Kiemel, K. Oie, and J. Jeka. Multisensory fusion and the stochastic structure of postural sway. *Biological Cybernetics*, 87:262–277, 2002.

D. C. Knill and W. Richards. *Perception as Bayesian Inference.* MIT Press, Cambridge, MA, 1996.

C. Koike. *Bayesian Approach to Action Selection and Attention Focusing. Application in Autonomous Robot Programming.* Thèse de doctorat, Inst. Nat. Polytechnique de Grenoble, Grenoble (FR), November 2005.

C. Koike, P. Bessière, and E. Mazer. Bayesian Approach to Action Selection and Attention Focusing. In P. Bessière, C. Laugier, and R. Siegwart, editors, *Probabilistic Reasoning and Decision Making in Sensory-Motor Systems*, volume 46 of *STAR*. Springer Verlag, 2008.

D. Koller and A. Pfeffer. Object-oriented Bayesian networks. In *Proceedings of the Thirteenth Conference on Uncertainty in Artifical Intelligence*, pages 302–313. Morgan Kaufmann publishers, 1997.

K. P. Körding and D. M. Wolpert. Bayesian integration in sensorimotor learning. *Nature*, 427:244–7, 2004.

K. P. Körding, U. Beierholm, W. J. Ma, S. Quartz, J. B. Tenenbaum, and L. Shams. Causal inference in multisensory perception. *PLoS one*, 2(9): e943, 2007.

B. J. Kuipers. The spatial semantic hierarchy. *Artificial Intelligence*, 119 (1–2):191–233, 2000.

M. S. Landy, L. T. Maloney, E. B. Johnston, and M. Young. Measurement and modeling of depth cue combination: in defense of weak fusion. *Vision Research*, 35:389–412, 1995.

T. Lane and L. P. Kaelbling. Toward Hierarchical Decomposition for Planning in Uncertain Environments. In *Proceedings of the 2001 IJCAI Workshop on Planning under Uncertainty and Incomplete Information*, pages 1–7, 2001.

P. S. Laplace. *Essai philosophique sur les probabilités*. Gauthier-Villars, 1814.

K. B. Laskey and S. M. Mahoney. Network Fragments: Representing Knowledge for Constructing Probabilistic Models. In *Proceedings of the thirteenth conference on uncertainty in artifical intelligence*, pages 334–341. Morgan Kaufmann publishers, 1997.

J. Laurens. *Modélisation bayésienne des interactions visuo-vestibulaires*. PhD thesis, Université Pierre et Marie Curie, November 2006.

J. Laurens and J. Droulez. Bayesian processing of vestibular information. *Biological Cybernetics*, 96:389–404, 2007.

J. Laurens and J. Droulez. Bayesian Modeling of Visuo-Vestibular Interactions. In P. Bessière, C. Laugier, and R. Siegwart, editors, *Probabilistic Reasoning and Decision Making in Sensory-Motor Systems*, volume 46 of *STAR*. Springer Verlag, 2008.

S. Lauritzen and D. J. Spiegelhalter. Local computations with probabilities on graphical structures and their application to expert systems (with discussion). *Journal of the Royal Statistical Society series B*, 50:157–224, 1988.

S. L. Lauritzen. *Graphical Models.* Oxford University Press, Oxford, UK, 1996.

R. Le Hy. *Programmation et apprentissage bayésien de comportements pour personnages synthétiques — application aux personnages de jeux vidéos.* These, Institut National Polytechnique de Grenoble — INPG, Apr 2007.

R. Le Hy and P. Bessière. Playing to Train Your Video Game Avatar. In *Probabilistic Reasoning and Decision Making in Sensory-Motor Systems*, pages 263–276. Springer, 2008.

R. Le Hy, A. Arrigoni, P. Bessière, and O. Lebeltel. Teaching Bayesian behaviours to video game characters. *Robotics and Autonomous Systems*, 47: 177–185, 2004.

O. Lebeltel. *Programmation Bayésienne des Robots.* Thèse de doctorat, Institut National Polytechnique de Grenoble, Grenoble, France, Septembre 1999.

O. Lebeltel, J. Diard, P. Bessière, and E. Mazer. A Bayesian Framework for Robotic Programming. In A. Mohammad-Djafari, editor, *Twentieth International Workshop on Bayesian Inference and Maximum Entropy Methods in Science and Engineering (Maxent 2000)*, pages 625–637, Melville, New York, USA, 2000. American Institute of Physics Conference Proceedings.

O. Lebeltel, P. Bessière, J. Diard, and E. Mazer. Bayesian robot programming. *Advanced Robotics*, 16(1):49–79, 2004.

J. Leonard, H. Durrant-Whyte, and I. Cox. Dynamic map-building for an autonomous mobile robot. *The Intl. J. of Robotics Research*, 11(4):286–298, 1992.

P. Leray. *Réseaux bayésiens : apprentissage et modélisation de systèmes complexes.* Habilitation a diriger les recherches, Université de Rouen, Rouen, France, November 2006.

Z. Li and B. D'Ambrosio. Efficient inference in Bayes networks as a combinatorial optimization problem. *International Journal of Approximate Reasoning*, 11(1):55–81, 1994. ISSN 0888-613X.

D. V. Lindley. Bayesian Analysis in Regression Problem. In D. L. Meyer and R. O. Collier, editors, *Bayesian Statistics.* Peacok, 1970.

H.-A. Loeliger. An introduction to factor graphs. *Signal Processing Magazine, IEEE*, 21(1):28–41, Jan. 2004. ISSN 1053-5888.

D. Lunn, C. Jackson, N. Best, A. Thomas, and D. Spiegehalter. *The BUGS Book: A Practical Introduction to Bayesian Analysis.* Chapman and Hall, 2012.

D. G. C. MacKay. Introduction to Monte Carlo Methods. In *Proc. of an Erice summer school, ed. M. Jordan*, 1996.

D. J. MacKay. *Information Theory, Inference, and Learning Algorithms.* Cambridge University Press, 2003.

S. Maeda. Compensatory Articulation During Speech: Evidence from the Analysis and Synthesis of Vocal-Tract Shapes Using an Articulatory Model. In W. J. Hardcastle and A. Marchal, editors, *Speech Production and Speech Modelling*, pages 131–149. Dordrecht: Kluwer, 1990.

E. Mazer, J. M. Ahuactzin, and P. Bessière. The Ariadne's clew algorithm. *J. Artif. Intellig. Res. (JAIR)*, 9:295–316, 1998.

A. Mccallum, K. Schultz, and S. Singh. Factorie: Probabilistic programming via imperatively defined factor graphs. In *Advances in Neural Information Processing Systems 22*, pages 1249–1257, 2009.

G. McLachlan and D. Peel. *Finite Mixture Models.* Wiley, 2000.

K. Mekhnacha. *Méthodes probabilistes baysiennes pour la prise en compte des incertitudes géométriques: application à la CAO-robotique.* Thèse de doctorat, Institut National Polytechnique de Grenoble (INPG), Grenoble (FR), juillet 1999.

K. Mekhnacha and P. Bessière. BCAD: A Bayesian CAD System for Geometric Problems Specification and Resolution. In *Probabilistic Reasoning and Decision Making in Sensory-Motor Systems*, pages 205–231. Springer, 2008.

K. Mekhnacha, E. Mazer, and P. Bessière. A Robotic CAD System Using a Bayesian Framework. In *Proc. of the IEEE/RSJ Int. Conf. on Intelligent Robots and Systems (IROS 2000)*, volume 3 of *Best Paper Award*, pages 1597–1604, France, 2000.

K. Mekhnacha, E. Mazer, and P. Bessière. The design and implementation of a Bayesian CAD modeler for robotic applications. *Advanced Robotics*, 15 (1):45–69, 2001.

K. Mekhnacha, J.-M. Ahuactzin, P. Bessière, E. Mazer, and L. Smail. Exact and approximate inference in ProBT. *Revue d'Intelligence Artificielle*, 21/3: 295–332, 2007.

N. Metropolis, A. W. Rosenblusth, M. N. Rosenblusth, A. Teller, and E. Teller. Equation of state by fast computing machines. *Journal of Chemical Physics*, 21:1087–1092, 1953.

B. Milch, B. Marthi, and S. Russell. Blog: Relational Modeling with Unknown Objects. In *ICML 2004 Workshop on Statistical Relational Learning and Its Connections*, pages 67–73, 2004.

A. Mohammad-Djafari, J.-F. Bercher, and P. Bessière. *Bayesian Inference and maximum Entropy Methods in Science and Engineering.* American Institute of Physics, 2010.

E. Morin. *La méthode.* Seuil, 1981.

C. Moulin-Frier. *Rôle des relations perception-action dans la communication parlée et l'émergence des systèmes phonologiques : étude, modélisation computationnelle et simulations.* These, Université de Grenoble, June 2011.

C. Moulin-Frier, J.-L. Schwartz, J. Diard, and P. Bessière. Emergence of Articulatory-Acoustic Systems from Deictic Interaction Games in a "Vocalize to Localize" Framework. In *Primate Communication, and Human Language Vocalisation, Gestures, Imitation and Deixis in Humans and Non-Humans*, pages 193–220. John Benjamins, 2011.

C. Moulin-Frier, R. Laurent, P. Bessière, J.-L. Schwartz, and J. Diard. Adverse conditions improve distinguishability of auditory, motor and percepto-motor theories of speech perception: an exploratory Bayesian modeling study. *Language and Cognitive Processes*, 27(7-8 Special Issue: Speech Recognition in Adverse Conditions):1240–1263, 2012.

K. Murphy. *Dynamic Bayesian networks: representation, inference and learning.* PhD thesis, UC Berkley, Computer Science Division, July 2002.

R. M. Neal. Probabilistic inference using Markov Chain Monte Carlo methods. Research Report CRG-TR-93-1, Dept. of Computer Science, University of Toronto, 1993.

R. M. Neal, M. J. Beal, and S. T. Roweis. Inferring State Sequences for Non-Linear Systems with Embedded Hidden Markov Models. In S. Thrun, editor, *Advances in Neural Information Processing Systems 16.* MIT Press, Cambridge, MA, 2003.

A. Nefian and M. Hayes. Face Recognition Using an Embedded HMM. In *Proceedings of the IEEE Conference on Audio and Video-based Biometric Person Authentication*, pages 19–24, 1999.

J. Park. MAP Complexity Results and Approximation Methods. In *Proc. of the 17th Conference on Uncertainty in Artificial Intelligence (UAI)*, pages 388–396, 2002.

J. Pearl. *Probabilistic Reasoning in Intelligent Systems: Networks of Plausible Inference.* Morgan Kaufmann, 1988.

X. Perrin. *Semi-Autonomous navigation of an assistive robot using low throughput interfaces.* PhD thesis, ETH Zurich, 2009.

X. Perrin, R. Chavarriaga, F. Colas, R. Siegwart, and J. Millan. Brain-coupled interaction for semi-autonomous navigation of an assistive robot. *Robotics and Autonomous Systems*, 58:1246–1255, 2010.

A. Pfeffer. IBAL: A Probabilistic Rational Programming Language. In *In Proc. 17th IJCAI*, pages 733–740. Morgan Kaufmann Publishers, 2001.

A. Pfeffer. Creating and Manipulating Probabilistic Programs with Figaro. In *UAI Workshop on Statistical Relational Artificial Intelligence (StarAI)*, 2012.

J. Pineau and S. Thrun. High-Level Robot Behaviour Control with POMDPs. In *AAAI Workshop on Cognitive Robotics*, 2002.

Z. Pizlo. Perception viewed as an inverse problem. *Vision Research*, 41(24): 3141–61, 2001.

T. Poggio. Vision by man and machine. *Scientific American*, 250:106–116, 1984.

H. Poincaré. *La science et l'hypothèse.* Flammarion, Paris, 1902.

H. Poincaré. *Calcul des probabilités.* Gauthier-Villars, 1912.

D. Poole. Probabilistic Horn abduction and Bayesian networks. *Artificial Intelligence*, 64:81–129, 1993.

D. Poole. The independent choice logic for modelling multiple agents under uncertainty. *Artificial Intelligence*, 94:7–56, 1997.

D. Poole. Probabilistic Programming Languages: Independent Choices and Deterministic Systems. In R. Dechter, H. Geffner, and J. Y. Halpern, editors, *Heuristics, Probability and Causality: A Tribute to Judea Pearl*, pages 253–269. College Publications, 2010.

C. Pradalier. *Navigation intentionnelle d'un robot mobile.* These, Institut National Polytechnique de Grenoble — INPG, Sept. 2004.

C. Pradalier and P. Bessière. The CyCab: Bayesian Navigation on Sensory–Motor Trajectories. In *Probabilistic Reasoning and Decision Making in Sensory-Motor Systems*, pages 51–75. Springer, 2008.

C. Pradalier, F. Colas, and P. Bessière. Expressing Bayesian Fusion as a Product of Distributions: Applications in Robotics. In *Proc. IEEE Int. Conf. on Intelligent Robots and Systems*, 2003.

C. Pradalier, J. Hermosillo, C. Koike, C. Braillon, P. Bessière, and C. Laugier. An Autonomous Car-Like Robot Navigating Safely Among Pedestrians. In *Proc. of the IEEE Int. Conf. on Robotics and Automation*, New Orleans, LA (US), France, 2004.

C. Pradalier, J. Hermosillo, C. Koike, C. Braillon, P. Bessière, and C. Laugier. The CyCab: a car-like robot navigating autonomously and safely among pedestrians. *Robotics and Autonomous Systems*, 50(1):51–68, 2005.

L. R. Rabiner. A Tutorial on Hidden Markov Models and Selected Applications in Speech Recognition. *Proc. of the IEEE*, 77(2):257–286, February 1989.

L. R. Rabiner and B.-H. Juang. *Fundamentals of Speech Recognition*, chapter Theory and implementation of Hidden Markov Models, pages 321–389. Prentice Hall, Englewood Cliffs, NJ, 1993.

L. D. Raedt, A. Kimmig, and H. Toivonen. ProbLog: A Probabilistic Prolog and its Application in Link Discovery. In *In Proceedings of 20th International Joint Conference on Artificial Intelligence*, pages 2468–2473. AAAI Press, 2007.

G. Ramel. *Analyse du contexte à l'aide de méthodes probabilistes pour l'interaction hommes-robots.* PhD thesis, Ecole Polytechnique Fédérale de Lausanne (EPFL), April 2006.

G. Ramel and R. Siegwart. Probabilistic Contextual Situation Analysis. In *Probabilistic Reasoning and Decision Making in Sensory-Motor Systems.* Spinger, 2008.

J. Rett. *Robot-human interface using Laban movement analysis inside a Bayesian framework.* PhD thesis, University of Coimbra, June 2008.

J. Rett, J. Dias, and J.-M. Ahuactzin. Bayesian reasoning for Laban movement analysis used in human-machine interaction. *Int. J. Reasoning-based Intelligent Systems*, 2(1):13–35, 2010.

C. Robert. An entropy concentration theorem: applications in artificial intelligence and descriptive statistics. *Journal of Applied Probabilities*, 37: 303–313, 1990.

J. A. Robinson. A machine-oriented logic based on the resolution principle. *Journal of the ACM*, 12(1):23–41, Jan. 1965.

J. A. Robinson. *Logic: Form and Function.* North-Holland, 1979.

J. A. Robinson and E. E. Silbert. LOGLISP: An Alternative to PROLOG. *Machine Intelligence, Edinburgh Univ. Press*, 1(10):399–419, 1982a. Edinburgh Univ. Press, Edinburgh.

J. A. Robinson and E. E. Silbert. LOGLISP: Motivation, Design and Implementation. In K. L. Clark and S.-A. Tärnlund, editors, *Logic Programming*, pages 299–313. Academic Press, London, 1982b.

R. Y. Rubinstein. *Simulation and the Monte Carlo Method.* John Wiley and Sons, 1981.

T. Sato and Y. Kameya. Parameter learning of logic programs for symbolic-statistical modeling. *Journal of Artificial Intelligence Research*, page 454, 2001.

Y. Sato, T. Toyoizumi, and K. Aihara. Bayesian inference explains perception of unity and ventriloquism aftereffect: Identification of common sources of audiovisual stimuli. *Neural Computation*, 19(12):3335–3355, 2007.

J.-L. Schwartz, J. Serkhane, P. Bessière, and L.-J. Boë. La robotique de la parole, ou comment modéliser la communication par gestes orofaciaux. *Primatologie*, 6:329–352, 2004.

J. Serkhane, J.-L. Schwartz, L.-J. Boë, B. Davis, P. Bessière, and E. Mazer. Etude comparative de vocalisations de bébés humains et de bébés robots. In *XXIVème Journées d'Etude sur la Parole (JEP), LORIA et ATLIF, Nancy (France)*, 2002.

J. Serkhane, J.-L. Schwartz, and P. Bessière. Simulating Vocal Imitation in Infants, using a Growth Articulatory Model and Speech Robotics. In *International Congress of Phonetic Sciences (ICPhS), Barcelona, Spain*, page x, 2003.

J. Serkhane, J.-L. Schwartz, and P. Bessière. Building a talking baby robot: A contribution to the study of speech acquisition and evolution. *Interaction Studies*, 6(2):253–286, 2005.

J. Serkhane, J.-L. Schwartz, L.-J. Boë, B. Davis, and C. Matyear. Infants' vocalizations analyzed with an articulatory model: A preliminary report. *Journal of Phonetics*, 35(3):321–340, Mar. 2007.

J. Serkhane, J.-L. Schwartz, and P. Bessière. Building a Talking Baby Robot: A Contribution to the Study of Speech Acquisition and Evolution. In P. Bessière, editor, *Probabilistic Reasoning and Decision Making in Sensory-Motor Systems*, pages 329–357. Springer, 2008.

J. E. Serkhane. *Un bébé androïde vocalisant: Etude et modélisation des mécanismes d'exploration vocale et d'imitation orofaciale dans le développement de la parole.* PhD thesis, Inst. Nat. Polytechnique de Grenoble, November 2005.

R. D. Shachter, B. D'Ambrosio, and B. A. Del Favero. Symbolic Probabilistic Inference in Belief Networks. In *Proceedings of the Eighth National Conference on Artificial Intelligence*, AAAI'90, pages 126–131. AAAI Press, 1990.

A. F. Smith and G. O. Roberts. Bayesian computation via the Gibbs sampler and related Monte Carlo methods. *Journal of the Royal Statistical Society B*, 55:3–23, 1993.

A. Stocker and E. Simoncelli. A Bayesian Model of Conditioned Perception. In J. Platt, D. Koller, Y. Singer, and S. Roweis, editors, *Advances in Neural Information Processing Systems*, pages 1409–1416. MIT Press, Cambridge, MA, 2008.

G. Synnaeve. *Bayesian programming and learning for multi-player video games: application to RTS AI.* PhD thesis, Institut National Polytechnique de Grenoble — INPG, October 2012.

G. Synnaeve and P. Bessière. Bayesian Modeling of a Human MMORPG Player. In *30th International Workshop on Bayesian Inference and Maximum Entropy*, Chamonix, France, July 2010.

G. Synnaeve and P. Bessière. A Bayesian Model for Opening Prediction in RTS Games with Application to StarCraft. In *Proceedings of 2011 IEEE CIG*, Seoul, Corée, République De, Sept. 2011a.

G. Synnaeve and P. Bessière. A Bayesian Model for RTS Units Control Applied to StarCraft. In *Proceedings of IEEE CIG 2011*, page 000, Seoul, Corée, République De, Sept. 2011b.

G. Synnaeve and P. Bessière. A Bayesian Model for Plan Recognition in RTS Games Applied to StarCraft. In AAAI, editor, *Proceedings of the Seventh Artificial Intelligence and Interactive Digital Entertainment Conference (AIIDE 2011)*, Proceedings of AIIDE, pages 79–84, Palo Alto, États-Unis, Oct. 2011c.

A. Tapus. *Topological SLAM — Simultaneous localization and mapping with fingerprints of places.* PhD thesis, Ecole Polytechnique Fédérale de Lausanne (EPFL), October 2005.

A. Tapus. Topological SLAM Using Fingerprints of Places. In P. Bessière, C. Laugier, and R. Siegwart, editors, *Probabilistic Reasoning and Decision Making in Sensory-Motor Systems.* Springer, 2008.

C. Tay, K. Mekhnacha, C. Chen, M. Yguel, and C. Laugier. An efficient formulation of the Bayesian occupation filter for target tracking in dynamic environments. *International Journal of Autonomous Vehicles*, 2007.

M. Tay, K. Mekhnacha, M. Yguel, C. Coué, C. Pradalier, C. Laugier, T. Fraichard, and P. Bessière. The Bayesian Occupation Filter. In *Probabilistic Reasoning and Decision Making in Sensory-Motor Systems*, pages 77–98. Springer, 2008.

J. B. Tenenbaum, C. Kemp, T. L. Griffiths, and N. D. Goodman. How to grow a mind: statistics, structure, and abstraction. *Science*, 331(6022):1279–1285, 2011.

S. Thrun. Towards Programming Tools for Robots that Integrate Probabilistic Computation and Learning. In *ICRA*, pages 306–312. IEEE, 2000a.

S. Thrun. Probabilistic algorithms in robotics. *AI Magazine*, 21(4):93–109, 2000b.

S. Thrun, W. Burgard, and D. Fox. *Probabilistic Robotics.* MIT Press, 2005.

J. W. Tukey. The future of data analysis. *Annals of Mathemtical Statistics*, 33(1):1–67, 1962.

A. Vallentin. *Einstein: A Biography.* Weidenfeld and Nicolson, 1954.

H. van der Kooij, R. Jacobs, B. Koopman, and H. Grootenboer. A multisensory integration model of human stance control. *Biological Cybernetics*, 80: 299–308, 1999.

S. Vasudevan. *Spatial cognition for mobile robots: A hierarchical probabilistic concept-oriented representation of space.* PhD thesis, ETH Zurich, 2008.

S. Vasudevan and R. Siegwart. Bayesian space conceptualization and place classification for semantic maps. *Robotics and Autonomous Systems*, 56: 522–537, 2008.

A. Viterbi. Error bounds for convolutional codes and an asymptotically optimum decoding algorithm. *IEEE Trans. Information Theory*, 13(2):260–269, 1967.

Voltaire. *Dictionnaire philosophique ou la raison par l'alphabet.* Flammarion, Paris, 1993–1764.

Voltaire. *Philosophical Dictionary.* Penguin Books, 2005.

Y. Weiss, E. P. Simoncelli, and E. H. Adelson. Motion illusions as optimal percepts. *Nature Neuroscience*, 5(6):598–604, 2002.

D. Wu and C. Butz. On the Complexity of Probabilistic Inference in Singly Connected Bayesian Networks. In *Proceedings of the 10th International Conference on Rough Sets, Fuzzy Sets, Data Mining, and Granular Computing*, RSFDGrC'05, pages 581–590, Berlin, Heidelberg, 2005. Springer-Verlag.

F. Xu and V. Garcia. Intuitive statistics by 8-month-old infants. *Proc. Natl. Acad. Sci. USA*, 105(13):5012–5015, 2008.

C. Yuan and M. J. Druzdzel. Importance sampling algorithms for Bayesian networks: principles and performance. *Mathematical and Computer Modeling*, 43:1189–1207, 2006.

A. L. Yuille and H. H. Bülthoff. *Bayesian Decision Theory and Psychophysics*, pages 123–161. MIT Press, Cambridge, MA, 1996.

L. A. Zadeh. Fuzzy sets. *Information and Control*, 8(3):338–353, 1965.

L. A. Zadeh. Fuzzy Logic and Its Application to Approximate Reasoning. In *IFIP Congress*, pages 591–594, 1974.

L. A. Zadeh. Fuzzy logic and approximate reasoning. *Synthese*, 30(3-4):407–428, 1975.

N. L. Zhang and D. Poole. Exploiting causal independence in Bayesian network inference. *Journal of Artificial Intelligence Research*, 5:301–328, 1996.

L. H. Zupan, D. M. Merfeld, and C. Darlot. Using sensory weighting to model the influence of canal, otolith and visual cues on spatial orientation and eye movements. *Biological Cybernetics*, 86(3):209–230, 2002.

Index

Boldface page numbers indicate words appearing in section titles.